Phylogenetic Networks

Concepts, Algorithms and Applications

The evolutionary history of species is traditionally represented using a rooted phylogenetic tree. However, when reticulate events such as hybridization, horizontal gene transfer or recombination are believed to be involved, phylogenetic networks that can accommodate non-treelike evolution have an important role to play.

This book provides the first interdisciplinary overview of phylogenetic networks. Beginning with a concise introduction to both phylogenetic trees and phylogenetic networks, the fundamental concepts and results are then presented for both rooted and unrooted phylogenetic networks. Current approaches and algorithms available for computing phylogenetic networks from different types of datasets are then discussed, accompanied by examples of their application to real biological datasets. The book also summarizes the algorithms used for drawing phylogenetic networks, along with the existing software for their computation and evaluation.

All datasets, examples and other additional information and links are available from the book's companion website at: www.phylogenetic-networks.org.

DANIEL H. HUSON is Professor of Algorithms in Bioinformatics at Tübingen University. He has authored numerous papers in bioinformatics, biology and mathematics, and is the main author of the widely used computer programs Dendroscope, MEGAN and SplitsTree.

REGULA RUPP received her Ph.D. in Mathematics from Bern University in 2006. Between 2007 and 2009 she held a postdoctoral research position at Tübingen University, working with Daniel H. Huson in developing robust methods for computing phylogenetic networks from real biological data.

CELINE SCORNAVACCA is a postdoctoral researcher working on algorithms for phylogenetic networks with Daniel H. Huson at Tübingen University. She received her Ph.D. in Computer Science from Montpellier University in 2009.

T0350445

Phylogenetic Networks

Concepts, Algorithms and Applications

Daniel H. Huson

Eberhard-Karls-Universität Tübingen, Germany

Regula Rupp

Eberhard-Karls-Universität Tübingen, Germany

Celine Scornavacca

Eberhard-Karls-Universität Tübingen, Germany

CAMBRIDGE
UNIVERSITY PRESS

CAMBRIDGE
UNIVERSITY PRESS

University Printing House, Cambridge CB2 8BS, United Kingdom

Cambridge University Press is part of the University of Cambridge.

It furthers the University's mission by disseminating knowledge in the pursuit of
education, learning and research at the highest international levels of excellence.

www.cambridge.org
Information on this title: www.cambridge.org/9780521755962

© D. H. Huson, R. Rupp and C. Scornavacca 2010

First published 2010
Reprinted 2011

A catalogue record for this publication is available from the British Library

Library of Congress Cataloguing in Publication data
Huson, Daniel H.
Phylogenetic networks : concepts, algorithms and applications / Daniel H. Huson, Regula Rupp,
Celine Scornavacca.
    p.   cm.
Includes bibliographical references and index.
ISBN 978-0-521-75596-2 (hardback)
1. Cladistic analysis – Data processing.   2. Cladistic analysis – Mathematics.   3. Phylogeny.
I. Rupp, Regula.   II. Scornavacca, Celine.   III. Title.
QH83.H87   2011
578.01′2 – dc22   2010037669

ISBN 978-0-521-75596-2 Hardback

Contents

Preface

The evolutionary history of a set of species is usually described by a rooted phylogenetic tree. The concept of a rooted tree is very simple and has proved to be extremely useful in many application domains. However, *the truth is rarely pure and never simple.*[1]

By definition, phylogenetic trees are well suited to represent evolutionary histories in which the main events are speciations (at the internal nodes of the tree) and descent with modification (along the edges of the tree). But such trees are less suited to model mechanisms of *reticulate evolution* [219], such as horizontal gene transfer, hybridization, recombination or reassortment. Moreover, mechanisms such as incomplete lineage sorting, or complicated patterns of gene duplication and loss, can lead to incompatibilities that cannot be represented on a tree. Although the analysis of individual genes or short stretches of genomic sequence often gives strong support to a phylogenetic tree, different genes or sequence segments usually support different trees.

While it is generally undisputed that bifurcating speciation events and descent with modifications are major forces of evolution, there is also a growing belief that reticulate events play an important role in the shaping of evolutionary histories, too [55, 61, 111, 173].

Horizontal gene transfer (HGT), the direct transfer of genes from one organism to another, is known to occur very frequently in the prokaryotic world, the main mechanisms being transformation, conjugation and transduction [13, 28, 189, 231]. Because of horizontal gene transfer, even between quite distantly related species, phylogenetic trees on the same taxa based on different genes may be very incongruent and the questions arise of how to define the concept of a species tree for a set of prokaryotic taxa, and then how to infer it? One answer may be to represent the evolutionary history of a set of prokaryotes by an appropriate rooted phylogenetic *network* that encompasses the different gene histories [54, 153, 157].

---

[1] Oscar Wilde, The Importance of Being Earnest, 1895.

In a more general setting, horizontal gene transfers are considered together with gene duplication and loss events [56, 197]. Here the goal is to reconcile incongruent gene trees with a given species tree under a model of *duplication, loss and transfer* (DLT) [100].

Speciation by hybridization is a widespread phenomenon in plants [201, 202], but also occurs in some other types of organisms [153, 165]. In allopolyploidization, two individuals from distinct species hybridize and merge their sets of chromosomes. In rare cases this produces a new fertile species that is reproductively isolated from the parent species. In diploid hybridization, two parents from different species each supply a gamete and produce a diploid hybrid. In very rare cases the hybrid may become reproductively isolated from the parents and then evolve as a distinct species. Phylogenies involving hybrid species are more informative when they explicitly include postulated hybridization events.

Recombination and gene conversion produce new combinations of genetic material through pairing and shuffling of very similar DNA sequences [190]. It is usually considered a mechanism that belongs within the realm of population genetics, which deals with the statistical analysis of the inheritance and prevalence of genes in populations [118]. In this context, the evolution of sequences under the *coalescent-with-recombination* model gives rise to an *ancestral recombination graph* (ARG) [89, 108], which is used for statistical inference and is beyond the scope of this book. However, as sequencing technologies advance and more projects aim at the full (re)sequencing of many individuals, strains and species [242], the fields of phylogenetic analysis and population genetics are drawing closer together. Being able to explicitly represent recombination in a network is of value to both fields [98, 128, 169, 194].

When interspecific recombination occurs, it may result in different histories for different segments of an individual gene and thus impact the performance of phylogenetic tree reconstruction methods [198, 207, 211]. In such cases, a network reconstruction method may be more suitable.

Reassortment is akin to recombination in that it involves the swapping of genetic material between individual organisms. Many viruses, such as influenza A, have segmented genomes. The evolution of such viruses involves mutations, of course. Moreover, when the viruses co-infect a host cell, then segments of their genomes can be swapped in a process called reassortment [47]. Hence, a phylogenetic tree will not always suffice to correctly represent the evolutionary history of a population of such viruses in a host, and sometimes a network representation will be more appropriate [21].

In an essay entitled "Mathematics is biology's next microscope, only better; biology is mathematics' next physics, only better" [53], the author poses "five biological challenges that could stimulate, and benefit from, major innovations in mathematics." The third challenge is: "Replace the tree of life with a network or tapestry to represent lateral transfers of heritable features such as genes, genomes, and prions."

While there is a great need for practical and reliable computational methods for inferring rooted phylogenetic networks to *explicitly* describe evolutionary scenarios involving reticulate events, generally speaking, such methods do not yet exist, or have not yet matured enough to become standard tools.

In contrast, there exist a number of established computational methods for inferring *unrooted* phylogenetic networks, which are used to *abstractly* describe reticulate evolution by providing a visualization of incompatible evolutionary pathways. Among the most widely used are methods for computing split networks [9], median networks [11] quasi-median networks [10], and other types of haplotype networks [52]. Such methods are not only used in phylogenetic analysis, but also in phylogeography and population genetics, as well.

The phylogenetic analysis of molecular sequences using phylogenetic trees is an established field and there are a number of books that describe the different approaches in detail, such as [77, 163, 217]. Population genetics is a similarly mature discipline that has been treated in a number of books, including [102, 108]. Both phylogeny and population genetics remain very active areas of research that are developing further.

Although there has been a number of book chapters published on the subject of phylogenetic networks (for example [123, 181, 182, 184, 214]), to the best of our knowledge, this is the first book that is solely dedicated to the topic.

The overall aim of our book is to give an introduction to the field of phylogenetic networks. As bioinformaticans, we sit between the mathematicians who develop theories and concepts for modeling and calculating phylogenetic networks, and the biologists who are focused on understanding the evolution of the organisms that they are interested in, using the concepts, algorithms and tools that we help to provide for them. Hence, while the content of this book is complementary to *Phylogenetics* [217] by Semple and Steel and *Inferring Phylogenies* [77] by Felsenstein, we have aimed at making the exposition in our book less mathematical than the former, while being more formal and algorithmic than the latter.

The book has three parts. In Part I, Introduction, we first describe some basic concepts, mainly from elementary graph theory. We then give introductions to sequence alignment and to phylogenetic analysis using trees. Our hope is that this will make the book a self-contained introduction both to phylogenetic trees and networks, to a degree. In the last chapter of Part I we give an introduction to phylogenetic networks, providing a high-level overview of the area. The details are then given in the remaining two parts of the book.

In Part II, Theory, we develop the theoretical underpinning first for splits and unrooted networks, and then for clusters and rooted networks. Here we attempt to develop a unified treatment of a number of different aspects of both types of networks.

In Part III, Algorithms and Applications, we systematically describe many of the existing algorithms for computing phylogenetic networks. The chapters here are organized by type of input that the algorithms work on. For many of the algorithms we briefly summarize their applications that have been reported in the literature. The last chapter gives an overview of some of the software that is available for computing phylogenetic networks from biological data.

We would like to acknowledge the advice and support that we have received from a number of colleagues. First, and foremost, we would like to thank Andreas Dress in Shanghai, who introduced D. H. to the topic in the early nineties, who has been a source of great inspiration and is the "father" of a whole generation of bio-mathematicians and bio-informaticians in Germany and abroad. Second, D. H. would like to thank Tandy Warnow in Austin, Texas, for two very valuable post-doc years working with her on phylogenetics. Third, we would like to thank Mike Steel, Pete Lockhart and David Bryant in New Zealand, with whom D. H. has worked closely on phylogenetic networks over a number of years.

We are also very grateful to the following colleagues for helpful discussions and for comments on different parts of the manuscript: Elizabeth S. Allman, Thomas Bonfert, Magnus Bordewich, David Bryant, Sydney Cameron, Tobias Dezulian, Johannes Fischer, Olivier Gascuel, Stefan Grünewald, Dan Gusfield, Jotun Hein, Mike Hendy, Leo van Iersel, Tobias Klöpper, Pete Lockhart, Bill Martin, David Morrison, Luay Nakhleh, Kay Nieselt, David Penny, Christian Rausch, John A. Rhodes, Stephan C. Schuster, Charles Semple, Yun Song, Mike Steel, Ali Tofigh, Gabriel Valiente, Detlef Weigel and Jim Whitfield.

We thank the Newton Institute of Cambridge University for hosting the *Phylogenetics* research program in 2007, which was jointly organized by Vincent Moulton, Mike Steel and D. H. The program gave us an excellent opportunity to draft and discuss a first outline of this book. We used early versions of our manuscript as the basis of two courses on *Phylogenetic Networks* at Tübingen University in 2008 and 2009 and would like to thank the participating students for their useful feedback. We are grateful to the Deutsche Forschungsgemeinschaft for financial support of our research on phylogenetic networks.

Finally, D. H. would like to thank his wife, Elke Grieswelle, and his two sons, Marlon and Moritz, for their love and support. R. R. would to like thank all her friends for their encouragement and especially her husband, Bernhard Nemec, for his love and patience. C. S. would like to thank her friends and family for their encouragement, support and affection.

Part I

Introduction

In the first part of this book we give an introduction to basic concepts from graph theory and systematics (Chapter 1). We briefly discuss the problem of aligning molecular sequences (Chapter 2) and give a more detailed introduction to the computation of phylogenetic trees from aligned sequences and distances (Chapter 3). Finally, we give a brief introduction to the computation of phylogenetic networks, which also serves as an overview for the material presented in the second and third parts of the book (Chapter 4).

Chapters 2 and 3 are provided for the sake of completeness and reference. They can be skipped by readers who have a basic knowledge of phylogenetic analysis.

Basics

In this chapter we introduce some basic concepts and results from mathematics and biological taxonomy.

1.1 Overview

The focus of this chapter is on the introduction of *graphs* (see Figure 1.1). Graphs come in two flavors, *undirected* and *directed*. One type of undirected graph that plays an important role in this book is *unrooted trees*. In the context of directed graphs, we discuss the concept of a *directed, acyclic graph (DAG)* and then introduce *rooted trees* as an important example. We introduce a number of different kinds of *traversals* of trees or DAGs that are often used in algorithms. This chapter concludes by introducing the concepts of *taxa, clusters, clades* and *splits*.

1.2 Undirected and directed graphs

A graph is a convenient and popular way of representing relationships between objects: in a graph, objects are represented by *nodes* and relationships between objects are represented by *edges*. There are two types of graphs, *undirected* and *directed*, and we make use of both types throughout the book. First, we define undirected graphs:

Definition 1.2.1 (Undirected graph) *An* undirected graph $G = (V, E)$ *consists of a finite set of* nodes V *and a finite set of* edges E, *where each edge* $e \in E$ *is of the form* $\{v, w\}$, *with* $v, w \in V$ *(see Figure 1.2a).*

For an edge $e = \{v, w\}$ we call v and w the *endpoints* of e and we say that e *connects* v and w. We also say that both v and w are *incident* to e. We say that two edges are *adjacent*, if they share an endpoint, and two nodes are adjacent if they are connected by an edge. The *degree* of a node v is the number of edges that are incident to v.

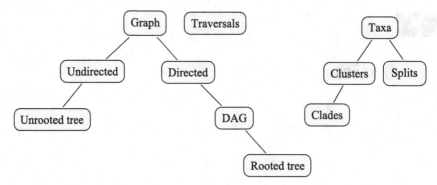

Figure 1.1   Overview of the main concepts introduced in this chapter.

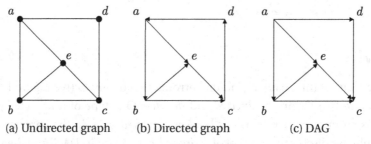

(a) Undirected graph   (b) Directed graph   (c) DAG

Figure 1.2   (a) An undirected graph with five nodes labeled $a, \ldots, e$. Here, each node is shown as a small disk •, however, we usually refrain from emphasizing nodes in this way. Edges are represented by lines connecting the incident nodes. (b) A directed graph with five nodes $a, \ldots, e$. An arrow from node v to node w represents the directed edge (v, w). (c) A DAG (directed acyclic graph) in which all edges are directed away from the node a.

Unless otherwise stated, we usually assume that each pair of nodes is connected by at most one edge (thus excluding so-called *multi-graphs*) and that every edge is incident to two different nodes (thus excluding so-called *self-loops*). Graphs with these two additional properties are called *simple*.

So, in an undirected graph, an edge consists of an unordered pair of two different nodes. In a *directed graph*, an edge consists of an *ordered* pair of nodes and thus has a *direction*:

Definition 1.2.2 (Directed graph)   *A directed graph (or digraph for short) $G = (V, E)$ consists of a finite set of nodes V and a finite set of edges E, where each edge $e \in E$ is of the form (v, w), with $v, w \in V$ (see Figure 1.2b).*

In a directed graph we say that the edge $e = (v, w)$ is *directed* from v to w, and we call v the *source* and w the *target* of e. We say that e is an *out-edge* of v and an *in-edge* of w. The *indegree* of a node u is the number of in-edges of u and the *outdegree* of u is the number of out-edges of u. The *degree* of u is the sum of its indegree and outdegree.

Let $G = (V, E)$ be a directed or undirected graph. A *subgraph* $G' = (V', E')$ of a graph G is a graph whose node and edge sets are subsets of those of G, that is, $V' \subseteq V$ and $E' \subseteq E$, such that the edges in E' only contain nodes in V'. Let U be a subset of nodes of G. The subgraph $G|_U = (U, E|_U)$ *induced* by U has node set U and *induced* edge set $E|_U$ consisting of all edges in G whose endpoints both lie in U.

A *path* (or *undirected path*) P of length $\text{len}(P) = k \geq 1$ is a sequence of nodes and edges

$$P = (v_0, e_1, v_1, e_2, v_2, \ldots, e_k, v_k) \tag{1.1}$$

such that the nodes v_{i-1} and v_i are the endpoints of the edge e_i, for $1 \leq i \leq k$, and no edge occurs more than once in P. Such a path is said to *connect* its two *endpoints* v_0 and v_k.

In some settings, we may understand a path P to consist only of the edges that it contains, in other settings it may be taken to contain only its nodes.

A *cycle* (or *undirected cycle*) is a path in which the first node equals the last node $v_0 = v_k$, and for which this is the only node that occurs more than once. An undirected graph is called *acyclic* if it contains no cycles. In a directed graph we can additionally define directed paths and cycles: a *directed path* is a path P in which every edge e_i is directed from v_{i-1} to v_i, for $i = 1, \ldots, k$. Similarly, a *directed cycle* is a cycle with this property. Note that in a directed graph it may happen that two nodes v and w are connected by a path, but not by a directed path. For example, in Figure 1.2c, nodes b and d are connected by a path, but not by a directed path. A directed graph is called *acyclic* if it contains no directed cycles. Directed acyclic graphs play an important role for phylogenetic networks and thus are discussed later in more detail.

Let G be a directed graph. When necessary, we can also think of G as an undirected graph, simply by ignoring the direction of its edges. A special case arises when there are two nodes v and w such that both (v, w) and (w, v) are edges of G. In this case, the two directed edges are replaced by a single undirected one $\{v, w\}$. Similarly, if G is an undirected graph, then we can convert G into a directed graph by choosing a direction for each edge in G.

When modifying a graph $G = (V, E)$, to *delete* an edge e means simply to remove e from E. However, to *delete* a node v, one must remove v from V and also delete all edges that are incident to v. If v is a node of degree two in an undirected graph, then v is called *suppressible* and to *suppress* v means to first connect the two nodes adjacent to v by a new edge and then to delete v. Similarily, in a directed graph, a node v can be suppressed when it has both indegree one and outdegree

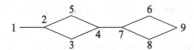

Figure 1.3   This graph G has three cut nodes, namely 2, 4 and 7. The graph has four biconnected components, containing the node sets $\{1, 2\}$, $\{2, 3, 4, 5\}$, $\{4, 7\}$ and $\{6, 7, 8, 9\}$, respectively. The graph is bipartite, as every edge connects an even-numbered node to an odd-numbered one.

one, by first connecting the source node of the in-edge to the target node of the out-edge by a new edge and then deleting v.

To *contract* an edge $e = (v, w)$ means to delete e and then to merge the two nodes v and w, in other words, to replace the two nodes v and w by a single node u and to change all edges incident to v and w so that they become incident to u, instead. A graph $G = (V, E)$ is called a *refinement* of another graph $G' = (V', E')$ if we can contract edges of G in such a way that the resulting graph equals G'.

A graph $G = (V, E)$ is called *bipartite* if and only if its set of nodes can be partitioned into two subsets V_1 and V_2, with $V = V_1 \cup V_2$ and $V_1 \cap V_2 = \emptyset$, such that for every edge $e \in E$ one of the endpoints lies in V_1 and the other endpoint lies in V_2. Equivalently, a graph is bipartite if one can color the set of nodes using two colors such that the two endpoints of any edge have different colors.

Any two nodes v and w of G are said to be *connected*, if there exists an undirected path that starts at v and ends at w. A graph $G = (V, E)$ is called *connected* if every pair of nodes $v, w \in V$ is connected by some undirected path. A *connected component* of a graph G is a maximal connected subgraph. For example, the graph displayed in Figure 1.4 has precisely three connected components. A *cut node* or *cut edge* is a node or edge, respectively, whose removal disconnects the graph.

In the literature a graph G is sometimes also called *weakly connected*, if any two nodes are joined by an undirected path, and *strongly connected*, if any two nodes are joined by a directed path in both directions. Throughout this book we only make use of the former definition and refer to such graphs as connected.

A *biconnected component* is a maximal subgraph that is induced by a set of edges and does not contain a cut node. In consequence, a graph G is called a biconnected component, if either it consists of only one node, or it consists of two nodes joined by a single edge, or it has more than two nodes and any two nodes v and w are connected by at least *two* different paths that are node-disjoint (except at v and w). For an example of this concept, and some of the others just defined, see Figure 1.3.

Exercise 1.2.3 (Number of cut nodes and components) *Determine the number of cut nodes and biconnected components in Figure 1.4. Is the graph bipartite?*

Figure 1.4   Graph with ten nodes labeled 1–10 and nine edges.

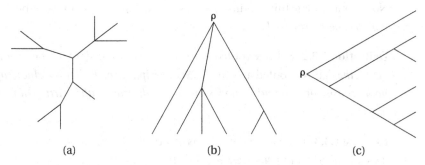

Figure 1.5   (a) An unrooted tree with 13 nodes and 12 edges. Exactly 8 nodes are leaves and 5 are internal nodes. (b) A rooted tree with 9 nodes and 8 edges, the root here being the topmost node ρ. (c) A rooted tree with 11 nodes and 10 edges, the root here being the left-most node ρ. This tree is bifurcating, as all internal nodes have outdegree 2.

Throughout this book, we always refer to the constituents of a graph as *nodes* and *edges*, but remark that in the literature nodes are also called *vertices* (usually when they are contained in an undirected graph) and edges are also called *arcs* (usually when they are directed) or branches (usually when they are contained in a *phylogenetic tree*).

1.3 Trees

A tree is a special kind of graph:

Definition 1.3.1 (Tree)   *A tree is a connected graph with no (undirected) cycles.*

In this book, we distinguish between *unrooted* and *rooted* trees. By an *unrooted tree* we mean any graph satisfying Definition 1.3.1, see Figure 1.5a. In a *rooted tree T*, one node ρ is declared the *root*, see Figure 1.5b. In the case of such a *rooted tree* $T = (V, E, \rho)$, we usually think of the graph as being a directed graph in which all edges are directed away from the root.

In an unrooted tree, all nodes of degree 1 are called *leaves* and all other nodes are called *internal nodes*. In a rooted tree, all nodes of outdegree 0 are called *leaves*, all nodes of outdegree $\geq$ 1 are called *internal nodes* and there is only one node with indegree 0, namely the root ρ.

When drawing rooted trees, the directions of the edges are not indicated explicitly, e.g. by drawing arrows, but rather implicitly by special placement of the root, usually at the top, left or bottom of the drawing, with the tree extending to the bottom, right or top, respectively, see Figure 1.5b–c.

An unrooted tree is called *bifurcating* if every internal node has degree 3. A rooted tree is called *bifurcating* if every internal node v has outdegree 2, see Figure 1.5c. Nodes that violate this condition are called *multifurcating* nodes. A bifurcating tree is also called a *resolved tree*.

Definition 1.3.2 (Subtree rooted at v)   *Let T be a rooted tree and v a node in T. We define the* subtree rooted at v *to be the subgraph T_v of T that is induced by the set of nodes consisting of v and all nodes that lie on directed paths starting at v. The root of T_v is v.*

Exercise 1.3.3 (Equivalent definitions of a tree)   *Prove that the following statements are equivalent for any undirected graph $T = (V, E)$:*

  (i)   *T is a tree.*
 (ii)   *Any two nodes v, $w \in V$ of T are connected by exactly one path P.*
(iii)   *T is connected and it is not possible to delete any edge without producing a graph that is not connected.*
(iv)   *T is connected and the number of nodes exceeds the number of edges by one, that is $|V| = |E| + 1$.*
 (v)   *T is acyclic and the number of nodes exceeds the number of edges by one, that is $|V| = |E| + 1$.*
(vi)   *T is acyclic and it is not possible to add an edge without creating a cycle.*

1.4 Rooted DAGs

Just as trees provide the mathematical basis for phylogenetic trees in Chapter 3, a generalization of rooted trees called rooted *DAGs* forms the basis for rooted phylogenetic networks presented in Chapter 6. A DAG is a special kind of directed graph:

Definition 1.4.1 (Rooted DAG)   *A directed acyclic graph (DAG) is a directed graph that is free of directed cycles. A DAG is called* rooted *if it contains precisely one node of indegree 0, called the* root *(see Figure 1.2c).*

If G is a DAG, then for any two nodes v and w we call v an *ancestor* of w, and w a *descendant* of v, if there exists a directed path from v to w. We call v a *parent* of w, and w a *child* of v if there exists an edge $e = (v, w)$ from v to w. If w is a descendant of v, then we also say that w is *lower* than v. It follows from the definition of a DAG that a node v cannot be simultaneously both a descendant and also an ancestor of any other node w. Similarly, we say that an edge $e = (v, w)$ is *lower* than an edge $f = (v', w')$, if v is lower than w'. We say that two nodes v and w are *incomparable*, if neither node is lower than the other. Similarly, two edges e and f are called *incomparable* if neither is lower than the other.

Lemma 1.4.2 (All nodes are descendants of the root)   *In a rooted DAG all nodes are descendants of the root ρ. In particular, a rooted DAG is always (weakly) connected.*

Proof   Consider some node v. If v is not the root, then there exists a parent v_1 of v. Again, either v_1 is the root, or v_1 has a parent v_2. Repeating this argument we obtain a chain of nodes $v, v_1, v_2, \ldots, v_r$. Because G is acyclic, any two nodes in this chain are distinct. As the set of nodes of G is finite, the chain must end at some node with indegree 0, which is ρ. Following the chain of nodes and edges used from ρ to v provides a directed path.   □

A useful characterization of a rooted tree is the following:

Lemma 1.4.3 (DAG with indegrees one is rooted tree)   *A DAG is a rooted tree if and only if it contains precisely one node of indegree 0, which is the root, and all other nodes have indegree one.*

Exercise 1.4.4 (Proof)   *Prove Lemma 1.4.3.*

From this characterization and Lemma 1.4.2, it follows for any DAG G with root ρ that, if we delete all but one in-edge for every node v of indegree ≥ 2 in G, then the remaining graph is a tree with root ρ.

In analogy to the definition of the *subtree rooted at* a node v, we define:

Definition 1.4.5 (Sub-DAG rooted at v)   *Let G be a DAG and v a node in G. We define the* sub-DAG *rooted at v as the subgraph G_v of G that is induced by the set of nodes consisting of v and all descendants of v. The root of G_v is v.*

1.5 Traversals of trees and DAGs

The term *tree traversal* refers to the process of systematically examining, that is visiting and processing, every node in a tree exactly once. Such traversals are classified by the order in which the nodes are examined. In the following definition we assume that the trees are rooted and bifurcating. However, all the algorithms can be easily adapted to the case of non-bifurcating trees.

Definition 1.5.1 (Tree traversals)   *Let T be a bifurcating rooted tree. We distinguish the following types of tree traversals:*

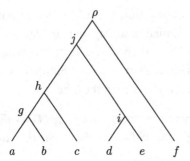

Figure 1.6   A bifurcating tree rooted at node ρ, with other nodes labeled $a, \ldots, j$.

Traversal	Order in which nodes are examined
preorder	$\rho, j, h, g, a, b, c, i, d, e, f$
postorder	$a, b, g, c, h, d, e, i, j, f, \rho$
inorder	$a, g, b, h, c, j, d, i, e, \rho, f$
breadth-first	$\rho, j, f, h, i, g, c, d, e, a, b$

Figure 1.7   Different traversals of the tree depicted in Figure 1.6.

- Preorder *traversal: Examine the root of T. Then traverse its left subtree in preorder, then traverse its right subtree in preorder.*
- Postorder *traversal: First traverse the left subtree of the root in postorder, then traverse the right subtree in postorder, and finally examine the root.*
- Inorder *traversal: First traverse the left subtree of the root in inorder, then examine the root, and then traverse its right subtree in inorder.*
- Breadth-first *traversal: Put the root in a queue. Repeat the following until the queue is empty: Pop a node off the beginning of the queue and examine it. Then, if the node has any children, push them onto the end of the queue.*

The different traversals give rise to different orders in which the nodes are examined. For example, consider the tree shown in Figure 1.6. The order in which nodes are examined in each of the traversals is listed in Figure 1.7.

Preorder, postorder and breadth-first traversals apply more generally to rooted DAGs:

Definition 1.5.2 (DAG traversals)   *Let G be a rooted DAG. We distinguish between the following types of DAG traversals:*

- Preorder *traversal: Examine the root ρ. While ρ has any child v that has not yet been examined, traverse the sub-DAG rooted at v in preorder.*
- Postorder *traversal: While the root ρ has any child v that has not yet been examined, traverse the sub-DAG rooted at v in postorder. Examine ρ.*

■ Breadth-first *traversal: Put the root in a queue and mark it. Repeat the following until the queue is empty: Pop a node off the beginning of the queue and examine it. Then, if the node has any children that have not yet been marked, mark them and push them onto the end of the queue.*

Exercise 1.5.3 (DAG for given traversals) *Construct a rooted DAG with nodes labeled $a, b, \ldots, i$ such that the three traversals visit the nodes in the following order:*

■ *preorder: $i, a, d, g, h, e, b, f, c$*
■ *postorder: $g, h, d, e, a, f, b, c, i$*
■ *breadth-first: $i, a, b, c, d, e, f, g, h$*

Construct a second rooted DAG that gives rise to the same traversals.

A preorder traversal is also called a *top-down algorithm* and a postorder traversal is also called a *bottom-up algorithm*.

1.6 Taxa, clusters, clades and splits

Throughout this book, we use $\mathcal{X} = \{x_1, \ldots, x_n\}$ to denote a set of *taxa*, in which each *taxon* x_i represents some species, group or individual organism whose evolutionary history is of interest to us. For example, $\mathcal{X} = \{x_1, x_2, \ldots\}$ might denote a set of mammals, with x_1 representing gorillas, x_2 representing seals, etc.

We use expressions such as "*C* is a cluster *on $\mathcal{X}$*", "*T* is a phylogenetic tree *on $\mathcal{X}$*", "*D* is a distance matrix *on $\mathcal{X}$*" or "*A* is a multiple sequence alignment *on $\mathcal{X}$*", to indicate that the corresponding set of taxa is $\mathcal{X}$.

Let $\mathcal{X}$ be a set of taxa. A *cluster* is any subset of $\mathcal{X}$, excluding both the empty set $\emptyset$ and the full set $\mathcal{X}$.

A *phylogeny* describes the evolutionary history of a set of taxa and the ultimate goal of any phylogenetic analysis is to reconstruct some part of the *tree of life*, the evolutionary history of all life on Earth.

In this context two species are called *related*, if they share a recent common ancestor. We say that a species *a* is *more closely related* to a species *b* than to a species *c*, if the last common ancestor of *a* and *b* is more recent that the last common ancestor of *a* and *c*. For example, humans are more closely related to mice (last common ancestor about 100 million years ago) than they are to fruit flies (last common ancestor about 600 million years ago).

A group of organisms is called a *clade* or a *monophyletic group*, if it contains all (and only) the descendants of a common ancestor and the ancestor itself. In a *monophyletic cluster C* (of extant species) on $\mathcal{X}$, all species are more closely related to each other than to any taxon outside of *C*. In other words, all species in *C* are descended from a common ancestor, which is not an ancestor of any other taxon in $\mathcal{X}$. For example, the set of all marsupials forms a monophyletic group within the class of all mammals.

A cluster C on $\mathcal{X}$ that does not contain all species descended from the last common ancestor of all its members is called a *paraphyletic group*. For example, the group of all reptiles is a paraphyletic group because it does not contain the group of birds or the mammals, both of which descended from ancestors embedded well within the phylogenetic tree of reptiles.

Similarly, a cluster C on $\mathcal{X}$ is called a *polyphyletic group*, if it consists of species that are descended from multiple last common ancestors, which each have other descendants that are not contained in C. For example, the term worms encompasses both nematodes and earth worms (Lumbricidae), which are evolutionarily so distant from each other that they belong to different phyla, the Nematoda and the Annelida.

The main role of phylogenetic analysis is to compute a set of clusters on $\mathcal{X}$ such that each cluster is monophyletic, and not a paraphyletic or polyphyletic group, and then to represent the clusters as a rooted phylogenetic tree or network, as described in detail in this book.

Let $\mathcal{X}$ be a set of taxa. A *split* is any bipartitioning of $\mathcal{X}$ into two non-empty subsets A and B of $\mathcal{X}$, such that $A \cap B = \emptyset$ and $A \cup B = \mathcal{X}$. As we shall see, splits arise naturally in the context of unrooted phylogenetic trees or networks. The concepts of clusters and splits are closely related: while clusters are employed to represent clades in rooted histories, splits are used to represent clades and their complements in unrooted settings.

2

Sequence alignment

This chapter is provided for the sake of completeness and reference. It can be skipped by readers who have a basic knowledge of phylogenetic analysis.

The comparison of biomolecular sequences is one of the most fundamental operations in computational biology. Two DNA or protein sequences that have a high similarity are usually presumed to be *homologous*, that is, to have evolved from a common ancestral sequence. Both high similarity and also homology of protein sequences often imply that the structure of the proteins is similar, which in turn usually implies that the proteins have a similar function.

Phylogenetic trees and networks are generally computed from aligned DNA or protein sequences and so the first step in an evolutionary study is often to build such an alignment.

2.1 Overview

Figure 2.1 shows the relationships between some of the main concepts introduced in this chapter. The focus of this chapter is on how to compute an alignment of molecular sequences. We discuss how to produce a *pairwise alignment* of two sequences. Then we focus on how to a compute *multiple sequence alignment* of a set of sequences using a *progressive alignment* approach.

2.2 Pairwise sequence alignment

The basic idea of alignment is to write two sequences one above the other so as to maximize the number of similar or identical bases or amino acids that occur underneath each other. For example, consider the two sequences $a =$ Y E S T E R D A Y and $b =$ E A S T E R S. Here is one possible alignment:

$$a^* = \text{Y} \quad \text{E} \quad - \quad \text{S} \quad \text{T} \quad \text{E} \quad \text{R} \quad \text{D} \quad \text{A} \quad \text{Y}$$
$$b^* = - \quad \text{E} \quad \text{A} \quad \text{S} \quad \text{T} \quad \text{E} \quad \text{R} \quad \text{S} \quad - \quad -$$

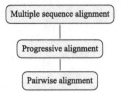

Figure 2.1   Overview of the main concepts introduced in this chapter.

and here is a second one:

$$a^* = \text{Y} \quad \text{E} \quad - \quad \text{S} \quad \text{T} \quad \text{E} \quad \text{R} \quad \text{D} \quad \text{A} \quad \text{Y}$$
$$b^* = - \quad \text{E} \quad \text{A} \quad \text{S} \quad \text{T} \quad \text{E} \quad \text{R} \quad - \quad \text{S} \quad -$$

Each of these two alignments was obtained by inserting *gaps* (shown as dashes) into the two sequences, a and b, so as to produce two new sequences of the same length, a^* and b^*. It is important to note that the order of symbols in either sequence is not changed. In consequence, removal of all gaps from an aligned sequence c^* reproduces the original sequence c.

Definition 2.2.1 (Pairwise alignment)   *Suppose we are given two sequences a and b over some alphabet Σ, for example $\Sigma = \{A, C, G, T\}$ for DNA sequences. A pairwise (global) alignment of a and b is obtained by inserting gaps ("–") into either or both sequences so that the resulting sequences a^* and b^* are of equal length and thus can be written underneath (or "opposite") each other in such a way that each symbol in the one string is opposite to exactly one symbol in the other string.*

In a pairwise alignment of two sequences we usually require that no two dashes appear together in a column. However, if the pairwise alignment is obtained as the restriction of a *multiple sequence alignment* (as defined in the next section) to two sequences, then columns containing only gaps may occur.

There are two questions that we now want to consider. First, how do we score an alignment? Second, how to compute the optimal score for two sequences and an alignment that attains that optimal score?

Given a pairwise alignment, we can either score the distance between the two sequences, or their similarity. A popular distance measure is the *edit distance* that counts the number of *insertion, deletion* and *replacement* operations required to convert the one sequence into the other. In an alignment of sequences a and b, the number of insertions is given by the number of symbols in b that are aligned to a gap in a, the number of deletions is given by the number symbols in a that are aligned to a gap in b, and the number of replacements is given by the number of symbols in a that are aligned to a different symbol in b. For example, the edit distance for the two alignments shown at the beginning of this section is 5 in both cases.

Although DNA alignments are sometimes scored using the edit distance, alignments of molecular sequences, especially protein sequences, are usually scored using a similarity score based on a *substitution matrix* that provides a score for each aligned pair of residues (or nucleotides):

Definition 2.2.2 (Substitution matrix) *A substitution matrix S over an alphabet $\Sigma = \{y_1, y_2, \ldots, y_t\}$ of size t is a $t \times t$ matrix in which each entry $s(y, z)$ assigns a score to the substitution of the symbol y by the symbol z in an alignment.*

For DNA sequences, we have $\Sigma = \{A, C, G, T\}$ and $t = 4$, and the simplest possible substitution matrix treats all nucleotides identically and assigns a score of $+1$ to a pair of identical nucleotides and a score of -1 to the substitution of one nucleotide by a different one.

There are two types of nucleotides in DNA, namely *pyrimidines*, which include cytosine and thymine, and *purines*, which include adenine and guanine. For biochemical reasons, *transitions*, which are substitutions between two pyrimidines or between two purines, occur more frequently than *transversions*, which are substitutions between a purine and a pyrimidine. A more sophisticated substitution matrix for DNA will assign different scores to these different types of substitutions.

A popular family of substitution matrices for amino acid sequences are the *BLOSUM* (an acronym for Blocks Substitution Matrix) matrices. These were empirically derived from *blocks*, which are gap-free alignments of highly conserved regions of proteins that are provided by the BLOCKS database [110].

To construct such a substitution matrix from a database of aligned sequences, one first counts the frequency f_s of each symbol s in the database and the frequency $f_{s,t}$ of each pair of aligned symbols (s, t) in the database. The frequency f_s provides an estimation of the probability p_s that the symbol s occurs in a sequence. The frequency $f_{s,t}$ provides an estimation of the probability $p_{s,t}$ that the two symbols s and t may substitute each other in two related sequences. Now, consider a non-gapped alignment of the form:

$$a = a_1 a_2 \ldots a_n$$
$$b = b_1 b_2 \ldots b_n. \tag{2.1}$$

Under the *null hypothesis* that the two sequences are unrelated, the probability of this alignment arising is described by a *random model R*: each symbol s occurs independently with some probability p_s, and the probability of obtaining this alignment is given by the product:

$$P(a, b \mid R) = \prod_i p_{a_i} \prod_j p_{b_j}. \tag{2.2}$$

The alternative hypothesis is the *match* model M that states that the two sequences are related (homologous). Under this model, aligned pairs of residues (s, t) occur with joint probability $p_{s,t}$. In this case, the probability for the whole alignment is:

$$P(a, b \mid M) = \prod_i p_{a_i b_i}.$$ (2.3)

The ratio of these two expressions gives a measure of the relative likelihood that the sequences are related (model M) as opposed to being unrelated (model R). This ratio is called the *odds ratio*:

$$\frac{P(a, b \mid M)}{P(a, b \mid R)} = \frac{\prod_i p_{a_i b_i}}{\prod_i p_{a_i} \prod_i p_{b_i}} = \prod_i \frac{p_{a_i b_i}}{p_{a_i} p_{b_i}}.$$ (2.4)

To obtain an additive scoring scheme, we take the logarithm to get the *log-odds ratio*:

$$\log\left(\frac{P(a, b \mid M)}{P(a, b \mid R)}\right) = \log\left(\prod_i \frac{p_{a_i b_i}}{p_{a_i} p_{b_i}}\right) = \sum_i \log\left(\frac{p_{a_i b_i}}{p_{a_i} p_{b_i}}\right),$$ (2.5)

and thus the entries of the substitution matrix are given by:

$$s(y, z) = \log\left(\frac{p_{yz}}{p_y p_z}\right),$$ (2.6)

for each pair of symbols y and z.

Different members of the BLOSUM family of matrices were produced by considering alignments of sequences with different levels of identity. The most commonly used matrix is *BLOSUM62*. To create this matrix, sequences are clustered so that any two sequences of $\geq$ 62% sequence identity are placed in the same cluster and all counts are based on comparisons between pairs of sequences contained in different clusters. The BLOSUM62 matrix is scaled so that its values are in *half-bits*, that is, the log-odds score is taken as $s'(y, z) = 2 \times \log_2\left(\frac{p_{yz}}{p_y p_z}\right)$ and then rounded to the nearest integer, see Figure 2.2.

Substitution matrices of the type considered here are based on gap-free alignments and do not provide any insight into how to score the alignment of a symbol against a gap in an alignment. Nevertheless, gaps are considered undesirable and are thus penalized, as we now discuss.

There are two main schemes for scoring gaps in pairwise alignments. Given a run of gaps of length g, the *linear gap penalty* scheme assigns a penalty that depends linearly on the length g of the run, thus:

$$\gamma(g) = -gd,$$ (2.7)

where d is a fixed positive *gap penalty*. The *affine gap penalty* scheme assigns a slightly higher penalty to the first gap in a run of gaps, and then penalizes

	A	R	N	D	C	Q	E	G	H	I	L	K	M	F	P	S	T	W	Y	V
A	4	-1	-2	-2	0	-1	-1	0	-2	-1	-1	-1	-1	-2	-1	1	0	-3	-2	0
R	-1	5	0	-2	-3	1	0	-2	0	-3	-2	2	-1	-3	-2	-1	-1	-3	-2	-3
N	-2	0	6	1	-3	0	0	0	1	-3	-3	0	-2	-3	-2	1	0	-4	-2	-3
D	-2	-2	1	6	-3	0	2	-1	-1	-3	-4	-1	-3	-3	-1	0	-1	-4	-3	-3
C	0	-3	-3	-3	9	-3	-4	-3	-3	-1	-1	-3	-1	-2	-3	-1	-1	-2	-2	-1
Q	-1	1	0	0	-3	5	2	-2	0	-3	-2	1	0	-3	-1	0	-1	-2	-1	-2
E	-1	0	0	2	-4	2	5	-2	0	-3	-3	1	-2	-3	-1	0	-1	-3	-2	-2
G	0	-2	0	-1	-3	-2	-2	6	-2	-4	-4	-2	-3	-3	-2	0	-2	-2	-3	-3
H	-2	0	1	-1	-3	0	0	-2	8	-3	-3	-1	-2	-1	-2	-1	-2	-2	2	-3
I	-1	-3	-3	-3	-1	-3	-3	-4	-3	4	2	-3	1	0	-3	-2	-1	-3	-1	3
L	-1	-2	-3	-4	-1	-2	-3	-4	-3	2	4	-2	2	0	-3	-2	-1	-2	-1	1
K	-1	2	0	-1	-3	1	1	-2	-1	-3	-2	5	-1	-3	-1	0	-1	-3	-2	-2
M	-1	-1	-2	-3	-1	0	-2	-3	-2	1	2	-1	5	0	-2	-1	-1	-1	-1	1
F	-2	-3	-3	-3	-2	-3	-3	-3	-1	0	0	-3	0	6	-4	-2	-2	1	3	-1
P	-1	-2	-2	-1	-3	-1	-1	-2	-2	-3	-3	-1	-2	-4	7	-1	-1	-4	-3	-2
S	1	-1	1	0	-1	0	0	0	-1	-2	-2	0	-1	-2	-1	4	1	-3	-2	-2
T	0	-1	0	-1	-1	-1	-1	-2	-2	-1	-1	-1	-1	-2	-1	1	5	-2	-2	0
W	-3	-3	-4	-4	-2	-2	-3	-2	-2	-3	-2	-3	-1	1	-4	-3	-2	11	2	-3
Y	-2	-2	-2	-3	-2	-1	-2	-3	2	-1	-1	-2	-1	3	-3	-2	-2	2	7	-1
V	0	-3	-3	-3	-1	-2	-2	-3	-3	3	1	-2	1	-1	-2	0	-3	-1	4

Figure 2.2   The BLOSUM62 substitution matrix. Each row and each column corresponds to one of the twenty standard amino acids, which is indicated by its single-letter code. Positive entries indicate substitutions that are favored in an alignment of related sequences, whereas negative values indicate substitutions usually only found in random alignments.

(a)
```
GSAQVKGHGKKVADALTNAVAHVDDMPNALSALSDLHAHKL
GSAQVKGHGKK-----VA--D---A-SALSDLHAHKL
```

(b)
```
GSAQVKGHGKKVADALTNAVAHVDDMPNALSALSDLHAHKL
GSAQVKGHGKKVADA----------SALSDLHAHKL
```

Figure 2.3   (a) An alignment produced using a linear gap score, and (b) an alignment in which all gaps occur in a single run, as produced using an affine gap score.

subsequent gaps linearly:

$$\gamma(g) = -d - (g - 1)e, \tag{2.8}$$

where d is called the *gap open penalty* and e is called the *gap extension penalty*, with $d > e > 0$, see Figure 2.3.

Exercise 2.2.3 (Scoring of alignments) *Determine the score of the two alignments presented at the beginning of Section 2.2, using the BLOSUM62 scoring matrix and a linear gap penalty of $d = 4$.*

There are three main versions of the pairwise alignment problem. Suppose we are given two biomolecular sequences a and b. If they have similar length and are similar along their whole length, then we require an alignment that extends from end-to-end; this is called a *global alignment*, as defined above. If, on the other hand, there is significant similarity only over a portion of the two sequences, perhaps because the two sequences contain a conserved domain, but are otherwise unrelated, then what is required is an alignment that covers only the region of similarity; this is called a *local alignment*. Finally, in the context of sequence assembly, where the goal is to assemble a longer sequence from short overlapping fragments, alignments are required that do not penalize gaps at either end of the alignment; these are known as *overlap alignments*.

Altogether, there are these three main variants of the pairwise alignment problem, each of which can be combined either with the goal of minimizing distance or maximizing similarity – using some chosen substitution matrix – and also there is the choice of whether to use a linear or affine gap penalty.

These and most other variations of the pairwise alignment problem can all be solved efficiently using different variants of the same basic dynamic programming approach. We illustrate this for the case of computing a global alignment to maximize a similarity score using linear gap penalties. This algorithm is known as the *Needleman-Wunsch* algorithm [186].

The basic idea is to iteratively build an optimal alignment from optimal alignments for smaller prefixes of the two sequences. Suppose we are given two sequences $a = a_1 a_2 \ldots a_m$ and $b = b_1 b_2 \ldots b_n$. Also, we require a substitution matrix $s(\cdot, \cdot)$ and a gap penalty $d > 0$. We shall compute a matrix

$$F : \{0, 1, 2, \ldots, m\} \times \{0, 1, 2, \ldots, n\} \to \mathbb{R} \tag{2.9}$$

in which $F(i, j)$ equals the best score of the alignment of the two prefixes $a_1 a_2 \ldots a_i$ and $b_1 b_2 \ldots b_j$.

To this end, we first set $F(0, 0) = 0$ and then set the boundary values $F(i, 0) = -i \times d$ for $i = 0, 1, \ldots, m$ and $F(0, j) = -j \times d$ for $j = 0, 1, \ldots, n$. The value set for $F(i, 0)$ reflects the fact that the first i symbols of sequence a is aligned to i leading gaps in sequence b, for any alignment that is computed via $F(i, 0)$.

Now the rest of the matrix can be filled from top-left to bottom-right by computing $F(i, j)$ from $F(i - 1, j - 1)$, $F(i - 1, j)$ and $F(i, j - 1)$, see Figure 2.4. There are three ways in which an alignment can be extended up to (i, j), see Figure 2.5:

- a_i aligns to b_j, leading to a score of $F(i - 1, j - 1) + s(a_i, b_j)$,
- a_i aligns to a gap, leading to a score of $F(i - 1, j) - d$, or
- b_j aligns to a gap, leading to a score of $F(i, j - 1) - d$.

We obtain $F(i, j)$ as the largest score arising from these three options:

$$F(i, j) = \max \begin{cases} F(i - 1, j - 1) + s(a_i, b_j) \\ F(i - 1, j) - d \\ F(i, j - 1) - d. \end{cases} \tag{2.10}$$

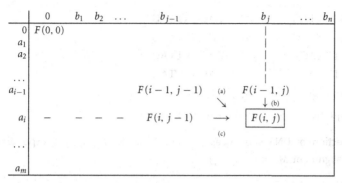

Figure 2.4   Dynamic programming matrix used to compute an optimal global alignment. The value for cell $F(i, j)$ is computed from the values in the predecessor cells, namely (a) from $F(i - 1, j - 1)$, (b) from $F(i - 1, j)$ or (c) from $F(i, j - 1)$.

```
A G A a_i          A A G A a_i          T G a_i - -
A G - b_j          A G - b_j -          T G C C b_j
```

(a) a_i aligns to b_j (b) a_i aligns to a gap   (c) b_j aligns to a gap

Figure 2.5   The three ways in which an alignment of the prefixes $a_1 \ldots a_i$ and $b_1 \ldots b_j$ can be obtained from shorter alignments.

The final value $F(m, n)$ contains the score of the best global alignment between a and b. An alignment that attains this value can be found by tracing back through the matrix, at each cell $F(i, j)$ moving backward to the cell $F(i - 1, i - j)$, $F(i - 1, j)$ or $F(i, j - 1)$ that was used to set the value at $F(i, j)$. In each of these cases, we report a match between symbols a_i and b_j, the alignment of a_i against a gap, or the alignment of a gap against b_j, respectively.

Note that in the case of a *tie*, in which the value of a cell can be obtained in more than one way from a previous cell, different choices lead to different alignments of the same score.

Exercise 2.2.4 (Application of Needleman-Wunsch)   *Apply the Needleman-Wunsch algorithm to the two sequences $a =$ YEASTS and $b =$ EASTERS, using the BLOSUM62 matrix and a gap penalty of $d = 3$.*

A similar approach, called the *Smith-Waterman algorithm*, can be used to compute a local alignment between two sequences [218], and a simple modification leads to an algorithm for overlap alignments. All these algorithms, including further modifications to accommodate affine gap penalties, run in time that is in proportion

a_1	CAGGATTAG	a_1^*	CAGGATTAG
a_2	CAGGTTTAG	a_2^*	CAGGTTTAG
a_3	CATTTTAG	a_3^*	CA-TTTTAG
a_4	ACGTTAA	a_4^*	-ACG-TTAA
a_5	ATGTTAA	a_5^*	-ATG-TTAA

     (a) Sequences   (b) Multiple alignment

Figure 2.6   (a) A collection of DNA sequences $A = (a_1, \ldots, a_5)$ and (b) a corresponding multiple sequence alignment $M = (a_1^*, \ldots, a_5^*)$.

to the product of the lengths of the two sequences (that is, in $O(mn)$) and can be modified further to require only linear space.

2.3 Multiple sequence alignment

In phylogenetics, a set of taxa $\mathcal{X} = \{x_1, \ldots, x_n\}$ is often represented by a collection of molecular sequences, $A = (a_1, \ldots, a_n)$, where the i-th sequence a_i was obtained from taxon x_i and corresponds to some specific gene or locus. For phylogenetic analysis, care is taken to ensure that the sequences are homologous, that is, have evolved from a common ancestor sequence.

The sequences in A differ from each other due to evolutionary events such as insertions, deletions and mutations. To facilitate phylogenetic analysis, these sequences are usually aligned by inserting gaps into each sequence such that the resulting collection of sequences

$$
M = \begin{cases}
a_{11}^* & a_{12}^* & \cdots & a_{1m}^* \\
a_{21}^* & a_{22}^* & \cdots & a_{2m}^* \\
& & \cdots & \\
a_{n1}^* & a_{n2}^* & \cdots & a_{nm}^*
\end{cases}
\tag{2.11}
$$

all have the same length m, forming a *multiple sequence alignment* of length m, see Figure 2.6. We are particularly interested in finding a multiple sequence alignment that achieves an optimal score according to an appropriate scoring scheme.

As discussed in the previous section, a pairwise sequence alignment of amino acid sequences is usually scored with the help of a substitution matrix such as a BLOSUM matrix, which assigns an empirically derived score to each pair of aligned amino acids; positive for favorable substitutions and negative for unfavorable ones. The score of the whole alignment is given by the sum of scores for each pair of aligned characters.

Multiple sequence alignments are usually scored using the *sum of pairs* approach: the score of a multiple sequence alignment is given by the sum of scores of all

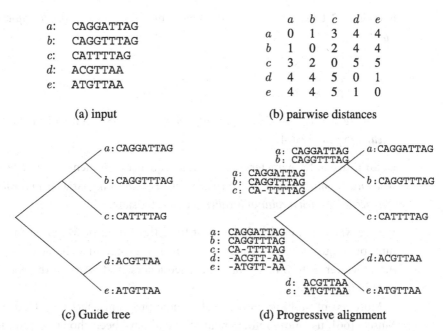

<table>
<tr><td></td><td>a</td><td>b</td><td>c</td><td>d</td><td>e</td></tr>
<tr><td>a</td><td>0</td><td>1</td><td>3</td><td>4</td><td>4</td></tr>
<tr><td>b</td><td>1</td><td>0</td><td>2</td><td>4</td><td>4</td></tr>
<tr><td>c</td><td>3</td><td>2</td><td>0</td><td>5</td><td>5</td></tr>
<tr><td>d</td><td>4</td><td>4</td><td>5</td><td>0</td><td>1</td></tr>
<tr><td>e</td><td>4</td><td>4</td><td>5</td><td>1</td><td>0</td></tr>
</table>

a: CAGGATTAG
b: CAGGTTTAG
c: CATTTTAG
d: ACGTTAA
e: ATGTTAA

(a) input

(b) pairwise distances

(c) Guide tree

(d) Progressive alignment

Figure 2.7   Outline of the progressive alignment approach. For a given set of input sequences (a), a distance matrix is computed (b), and from this a phylogenetic *guide tree* is derived (c). Then, sequences at the leaves of the tree are aligned to produce profiles at the internal nodes of the tree, which in turn are aligned to each other, continuing in this manner until the final alignment of all sequences is obtained at the root of the guide tree (d).

pairwise alignments induced by pairs of sequences in the multiple sequence alignment. The problem of computing a multiple sequence alignment that optimizes the sum of pairs score is known to be computationally hard [240] and thus, in practice, heuristics such as *progressive alignment* are used.

The basic idea of *progressive alignment* is to build a multiple sequence alignment incrementally, or *progressively*, by first aligning pairs of similar sequences, and then aligning sequences to *profiles* (in this context, a *profile* is simply a multiple sequence alignment of a subset of the input sequences) and then finally aligning profiles with profiles to obtain a multiple sequence alignment of the total set of input sequences, see Figure 2.7.

One of the most popular implementations of this approach is ClustalW [235]. In a basic outline, first all $\binom{n}{2}$ pairwise alignments of the n input sequences are computed to produce distances between all pairs of taxa. Second, these distances are provided as input to the neighbor-joining tree reconstruction method, described later in Section 3.14, to build a phylogenetic tree T, which is called the *guide tree*. Third, in a postorder traversal of the guide-tree, all sequences are progressively aligned so as to obtain the final alignment.

Exercise 2.3.1 (Multiple sequence alignment) *Given the multiple sequence alignment*

```
Y   E   -   S   T   E   R   D   A   Y
-   E   A   S   T   E   R   -   S   -
Y   E   A   S   T   -   -   -   -   -
```

determine the score of the alignment, using the BLOSUM62 scoring matrix and a linear gap penalty of 4.

Definition 2.3.2 (Characters in a multiple sequence alignment) *Let M be a multiple sequence alignment on* $\mathcal{X}$. *Each column of M is called a* character *and each symbol that occurs in such a column is called a* character state.

We use M_i to denote the character at the i-th column of M. For example, in the multiple sequence alignment shown in Figure 2.6 the first character M_1 takes on the character state C in the first three sequences, and takes on the *gap* character state in the last two.

More recent multiple sequence alignment programs, such as T-Coffee [188] or Muscle [66], use more refined heuristics and have been shown to produce better alignments than ClustalW.

Chapter notes

Most of the content of this chapter is based on teaching notes from our course on *Algorithms in Bioinformatics* taught at Tübingen University in recent years. Those notes, in turn, are based on the cited papers and on the book "Biological Sequence Analysis" by Durbin and colleagues [64].

3

Phylogenetic trees

This chapter is provided for the sake of completeness and reference. It can be skipped by readers who have a basic knowledge of phylogenetic analysis.

Phylogenetic analysis aims at uncovering the evolutionary relationships between different species or taxa, to obtain an understanding of the evolution of life on Earth. Phylogenetic trees are widely used to address this task and are usually computed from molecular sequences. They also have applications in many other areas. For example, they are used to determine the age and rate of diversification of taxa, to understand the evolutionary history of gene families, in sequence-analysis methods to allow *phylogenetic footprinting*, in epidemiology to trace the origin and transmission of infectious diseases, or to study the co-evolution of hosts and parasites.

The main focus of this book is on phylogenetic networks. However, as phylogenetic trees generalize to phylogenetic networks and also to make the book reasonably self-contained, in this chapter we give a brief introduction to some of the main methods used to infer phylogenetic trees.

3.1 Overview

Figure 3.1 shows the relationships between some of the main concepts introduced in this chapter. The focus of this chapter is on how to compute *unrooted phylogenetic trees*. Usually, the process of phylogenetic inference is begun with a *multiple sequence alignment*. From this, one can pursue either a distance-based analysis, or a sequenced-based one.

In a distance-based analysis of DNA sequences, first the *Hamming distances* between pairs of sequences are computed. These distances are then exposed to a *distance correction* that is based on some appropriate *model of evolution*. The resulting *distance matrix* is then provided to a method such as *UPGMA*, *neighbor-joining* or *FastME*, to produce an unrooted phylogenetic tree. The latter two are both

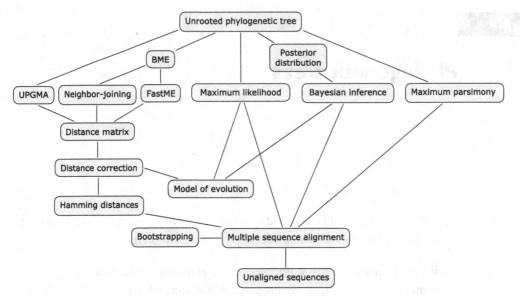

Figure 3.1   Overview of the main concepts introduced in this chapter. In this diagram, data "flows" from bottom to top.

heuristics for computing the *balanced minimum evolution* (BME) tree. Distance methods are often applied to distance matrices obtained in other ways, too.

For a sequence-based analysis, there are three main types of approaches to choose from. A *maximum parsimony* method searches for a phylogenetic tree that explains the given dataset using a minimum number of observable mutations. A *maximum likelihood* approach aims at determining a tree that maximizes the likelihood of generating the given dataset, under a given model of evolution. Bayesian methods attempt to compute the *posterior distribution* of trees, based on the given input data, a specified model of evolution and a presumed prior distribution of phylogenetic trees.

The posterior distribution provided by a Bayesian method provides a means of evaluating the support of computed groups of taxa. For other methods, *bootstrapping* can be used to perform such evaluations.

3.2 Phylogenetic trees

A *phylogenetic tree on* $\mathcal{X}$ is obtained by labeling the leaves of a tree T (a connected graph with no cycles, see Definition 1.3.1) by the taxon set $\mathcal{X}$, as illustrated in Figure 3.2.

Using the terminology introduced in Chapter 1, we refer to the constituents of a phylogenetic tree as *nodes* and *edges*, but remark that edges are also called *branches* in the context of phylogenetic trees.

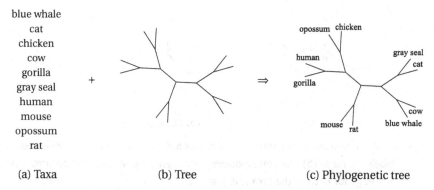

blue whale
cat
chicken
cow
gorilla
gray seal
human
mouse
opossum
rat

+   ⇒

opossum chicken
human   gray seal
gorilla   cat
cow
mouse rat blue whale

(a) Taxa   (b) Tree   (c) Phylogenetic tree

Figure 3.2   A set of taxa $\mathcal{X}$ (a), here a set of mammals and chicken, are used to label the leaves of a tree (b), giving rise to a phylogenetic tree T on $\mathcal{X}$, shown in (c).

We use λ to denote the *taxon labeling* that maps each taxon $x \in \mathcal{X}$ onto some node v of T. If $\lambda(x) = v$, then we say something like v *is labeled by* x, or v *obtains the label* x. We always assume that all leaves of the given tree T obtain some label via λ and we usually assume that internal nodes remain unlabeled and that no leaf obtains more than one label. Moreover, we usually require that all nodes have degree $\neq 2$:

Definition 3.2.1 (Phylogenetic tree)   *Given a set of taxa $\mathcal{X}$, a phylogenetic tree T on $\mathcal{X}$ consists of a tree $T = (V, E)$, in which all nodes have degree $\neq 2$, together with a taxon labeling $\lambda : \mathcal{X} \to V$ that assigns exactly one taxon to every leaf and none to any internal node.*

Note that we usually abuse notation in this way to let T denote both a phylogenetic tree and also the graph-theoretical tree on which it is based.

In mathematical phylogeny, the more general concept of an $\mathcal{X}$-*tree* is sometimes considered, which is obtained by weakening the requirements imposed on the tree, by allowing nodes of degree 2, and by weakening the requirements imposed on the taxon labeling λ, so as to demand only that every node of degree 0 or 2 obtains at least one label, thus allowing (multiple) labels on both leaves and internal nodes [216].

Our focus is on *phylogenetic trees*, rather than $\mathcal{X}$-*trees* (and then later on *phylogenetic networks*, rather than $\mathcal{X}$-*networks*), for ease of exposition. This is not a limitation, as any $\mathcal{X}$-tree can be converted into a corresponding phylogenetic tree simply by introducing a new leaf for every taxon that appears as the label of an internal node, or as the label of a leaf node that is labeled by more than one taxon, as illustrated in Figure 3.3.

As discussed in the previous chapter, trees can be either *unrooted* or *rooted*, and these two definitions carry over to phylogenetic trees (and $\mathcal{X}$-trees). Although we

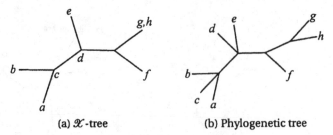

(a) $\mathcal{X}$-tree

(b) Phylogenetic tree

Figure 3.3 (a) An $\mathcal{X}$-tree in which some internal nodes have labels, c and d, and one node has multiple labels, g and h. (b) The corresponding phylogenetic tree obtained by introducing one new leaf edge for each of the taxa c, d, g and h.

usually forbid nodes of degree 2 in phylogenetic trees, in a rooted phylogenetic tree we allow the *root* to be of degree 2:

Definition 3.2.2 (Rooted phylogenetic tree)   *Given a set of taxa $\mathcal{X}$, a rooted phylogenetic tree consists of a rooted tree $T = (V, E, \rho)$ and a taxon labeling $\lambda : \mathcal{X} \rightarrow V$ that assigns exactly one taxon to every leaf and none to any internal node. All nodes, except ρ, must have degree $\neq 2$.*

Figure 3.2c shows an example of an unrooted phylogenetic tree. From a theoretical and algorithmic point of view, unrooted phylogenetic trees are sometimes easier to work with than rooted ones. However, in biology, rooted phylogenetic trees are usually of more interest, as the placement of the root defines a direction of time along the phylogenetic tree, namely away from the root, and because rooted phylogenetic trees explicitly define clades or clusters of putatively related taxa.

An unrooted phylogenetic tree can be *rooted* simply by declaring one of its nodes to be the root, or by inserting a new root node into the interior of one of the edges of the tree. In practice, a popular way to determine where to *root* a phylogenetic tree is by using an *outgroup*. An outgroup is a taxon that is closely related to the main group of taxa under consideration, but lies outside of it. The root is placed on the edge leading to the outgroup. In Figure 3.4, we show a rooted version of the example depicted in Figure 3.2, which is rooted using the outgroup *chicken*.

Exercise 3.2.3 (Rooting an unrooted tree)   *What is the number of rooted phylogenetic trees that can be obtained by rooting the unrooted tree depicted in Figure 3.3b in all possible ways?*

In many cases, the edges of a phylogenetic tree are each assigned a *weight*, or *length*, which often indicates the number of mutations that have happened along an edge or is correlated to evolutionary time in some other way:

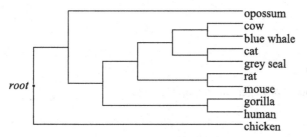

Figure 3.4   A rooted phylogenetic tree on mammals; the position of the root was determined by the outgroup *chicken*.

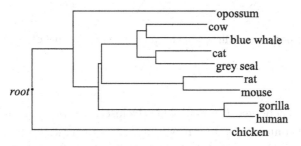

Figure 3.5   A rooted phylogenetic tree on mammals in which the edge lengths represent the number of differences observed in the mtDNA sequence of the species. In this type of diagram, called a phylogram, only the horizontal portion of an edge represents length (see Chapter 13 for further details).

Definition 3.2.4 (Edge weights) *A phylogenetic tree T is called an edge-weighted tree if we are given a map ω that assigns a non-negative weight or length $\omega(e)$ to every edge e of the tree.*

Note that assigning a length of 0 to an edge is equivalent to contracting the edge. In drawings, the weights of edges in a phylogenetic tree are often indicated by drawing the edges to scale, rather than writing the lengths explicitly next to the edges, see Figure 3.5.

3.3 The number of phylogenetic trees

How fast does the number of possible different phylogenetic trees T grow as a function of n, the size of the given taxon set $\mathcal{X}$?

Let us first determine the number of nodes and edges in a phylogenetic tree T on n taxa. To make the calculations easier, we shall assume that T is bifurcating, noting that a non-bifurcating tree has fewer nodes and edges.

Any bifurcating, unrooted phylogenetic tree on $n = 4$ taxa has 6 nodes and 5 edges, see Figure 3.6. Suppose $n > 4$. Then T can be clearly obtained from a

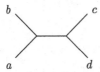

Figure 3.6   A bifurcating unrooted phylogenetic tree on four taxa.

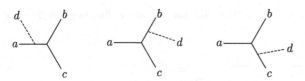

Figure 3.7   All possible ways of adding a fourth taxon d to an unrooted phylogenetic tree on three taxa $\{a, b, c\}$.

bifurcating phylogenetic tree T' on $n - 1$ taxa by inserting a new node v into some edge e of T' and connecting v to a new leaf w. This increases both the number of nodes and the number of edges in T' by 2. Hence, by induction, T has $2n - 2$ nodes and $2n - 3$ edges.

Now we can address the problem of counting the number of bifurcating phylogenetic trees on n taxa. If $n = 3$, then there is clearly one such tree. By adding a new leaf to any of the edges in this tree we obtain all three possible bifurcating phylogenetic trees on 4 taxa, see Figure 3.7. Similarly, adding a new leaf to any edge of any of these three trees we obtain all 3×5 bifurcating phylogenetic trees on 5 taxa, etc.

In general, we can obtain every bifurcating phylogenetic tree on n taxa by adjoining a new leaf and edge to some bifurcating phylogenetic tree T' on the first $n - 1$ taxa, in exactly one way. Since any such phylogenetic tree T' has $2(n - 1) - 3 = 2n - 5$ edges, it follows that there are

$$U(n) = (2n - 5)!! = 3 \cdot 5 \cdot 7 \cdot \cdots \cdot (2n - 5) \tag{3.1}$$

different bifurcating phylogenetic trees on n taxa.

The function $U(n)$ grows very rapidly with n. For example, $U(10) \approx 2$ million and $U(20) \approx 2.2 \times 10^{20}$. This makes exhaustive strategies for inferring phylogenetic trees infeasible for datasets involving more than 10–12 taxa, say.

Exercise 3.3.1 (Counting trees)   *What is the number of different bifurcating rooted phylogenetic trees on n taxa? Are there more or fewer trees if multifurcations are allowed? Are there more or fewer trees if we do not distinguish the labels?*

3.4 Models of DNA evolution

How do sequences evolve? As we shall see, a model of evolution is required by tree reconstruction methods that aim at determining the best tree for explaining a given set of sequences. To give a basic understanding of such models, in this section we discuss the *Jukes-Cantor model* of evolution [144], which describes a simple scenario in which sequences evolve via mutations and speciation, see also [230].

The main idea is to start with a random DNA sequence placed at the root of a rooted, bifurcating phylogenetic tree. This sequence is then *evolved* along the tree: mutations are applied to the sequences along the edges of the tree, whereas speciation events are implicitly given by the tree as the bifurcations at internal nodes. There are no insertion or deletion events and thus all sequences produced under this model have the same length.

Definition 3.4.1 (Jukes-Cantor model)   *Let T_0 be a rooted phylogenetic tree, called the* model tree. *The* Jukes-Cantor *model of evolution makes the following assumptions:*

(i)   *The possible states for each site are* A, C, G *and* T.
(ii)   *The sequence length is an input parameter and the state of each site of the initial sequence at the root is drawn independently and uniformly at random from the set of possible states.*
(iii)   *The sites evolve identically and independently (that is, all sites evolve following the same rules, but independently of each other) along the edges of the tree T_0 from the root ρ at a fixed rate u.*
(iv)   *With each edge $f \in E$ we associate a duration $\tau(f)$ and the expected number of mutations per site along f is given by $u\tau(f)$. The probabilities of change to each of the three remaining states are equal.*

How does a site "evolve along" an edge f under the Jukes-Cantor model? Consider a site i that has nucleotide $P \in \{A, C, G, T\}$ at the source of edge f, how does one determine the base $Q \in \{A, C, G, T\}$ at the target node of f?

Under the Jukes-Cantor model, the evolutionary event $C_3 =$ *the nucleotide at site i changes to one of the other three bases* occurs at a fixed rate u. Hence, the event $C_4 =$ *nucleotide changes to any base* (including "changing to the same base") can be taken to be $\frac{4}{3}u = 4\frac{u}{3}$. The probability that event C_4 *does not* occur within time t follows a Poisson distribution with parameter $k = 0$ and is given by $e^{-\frac{4}{3}ut}$. So, the probability that C_4 *does* occur in time t equals $(1 - e^{-\frac{4}{3}ut})$. When the event C_4 occurs, the probability of ending in any particular base is $\frac{1}{4}$.

This implies that the probability that an event of type C_4 occurs and changes from a given base $P \in \{A, C, G, T\}$ to some specific base Q within time t is:

$$\text{Prob}(Q \mid P, t) = \frac{1}{4}\left(1 - e^{-\frac{4}{3}ut}\right). \tag{3.2}$$

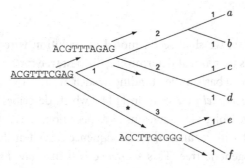

Figure 3.8　Under the Jukes-Cantor model of evolution, a random DNA sequence is generated at the root of the tree (underlined) and is then evolved along a model tree (in the direction of the arrows), for example giving rise to the two other sequences shown here. Each edge e of the tree is labeled by its duration $\tau(e)$.

The probability that $P = Q$ after time t is given by the sum of $\text{Prob}(Q = P \mid P, t)$ and the probability that C_4 does not occur:

$$\text{Prob}(Q = P \mid P, t) = \frac{1}{4}\left(1 - e^{-\frac{4}{3}ut}\right) + e^{-\frac{4}{3}ut} = \frac{1}{4}\left(1 + 3e^{-\frac{4}{3}ut}\right). \quad (3.3)$$

Thus, we obtain a *probability-of-change formula* for the probability of an *observable* change occurring at any given site in time t:

$$\text{Prob}(\text{change} \mid t) = 1 - \text{Prob}(Q = P \mid P, t)$$

$$= 1 - \frac{1}{4}\left(1 + 3e^{-\frac{4}{3}ut}\right)$$

$$= \frac{3}{4}\left(1 - e^{-\frac{4}{3}ut}\right). \quad (3.4)$$

So, when a sequence evolves along an edge f, the probability that the two sequences at the opposite ends of f differ at position i depends on the mutation rate u and the duration $t = \tau(f)$, and is given by the probability-of-change formula.

In Figure 3.8 we show a simple example in which the sequence at the root of the tree (underlined), evolves along the tree in the direction of the arrows. For example, assuming a mutation rate of $u = 0.1$, the probability of change along the edge marked "$*$", which has duration 3, is given by

$$\frac{3}{4}\left(1 - e^{-\frac{4}{3}\times 0.1 \times 3}\right) = \frac{3}{4}(1 - e^{-0.4}) \approx 0.247. \quad (3.5)$$

Exercise 3.4.2 (Probability of change, mutation rate)　*With the tree of Figure 3.8 and a mutation rate of $u = 0.1$, what is the probability that the first character of the sequence at leaf a is the same as the first character of the sequence at the root?*

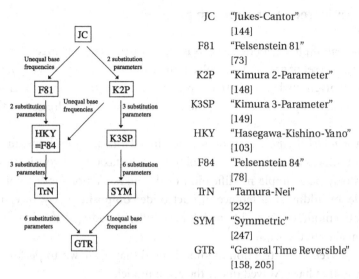

JC	"Jukes-Cantor" [144]
F81	"Felsenstein 81" [73]
K2P	"Kimura 2-Parameter" [148]
K3SP	"Kimura 3-Parameter" [149]
HKY	"Hasegawa-Kishino-Yano" [103]
F84	"Felsenstein 84" [78]
TrN	"Tamura-Nei" [232]
SYM	"Symmetric" [247]
GTR	"General Time Reversible" [158, 205]

Figure 3.9   A classification of the most important DNA models. Starting with the simple Jukes-Cantor model, more general models can be obtained by allowing unequal base frequencies or more than one substitution parameter. The most general model of this type is the GTR model that allows unequal base frequencies and prescribes a different substitution parameter for each of the six pairs of different bases.

In summary, the Jukes-Cantor model of DNA evolution assumes that all four bases (A, C, G and T) occur with equal frequencies ($= 0.25$) and that changes from one base to another occur at the same rate between all bases. There are many ways to relax these conditions to obtain more general models. For example, if we let the bases occur with different and arbitrary frequencies (although they have to sum to 1), and allow two different rates of change, one for transitions (that is, changes between A and G or between C and T) and a second one for transversions (all other changes), then we obtain the so-called *Hasegawa-Kishino-Yano model*. Both the Jukes-Cantor model and the Hasegawa-Kishino-Yano model are special cases of the *general time reversible model*, see Figure 3.9.

Another important way to obtain more general models of DNA evolution is to allow different substitution rates at different positions in the sequence. For example, this is often done by defining a discrete collection of rate classes using a *Gamma distribution*, which is popular because it has one main parameter, α, that determines the shape of the distribution [246]. When $\alpha < 1$, then the distribution is exponentially shaped and asymptotic to both the vertical and horizontal axes. When $\alpha = 1$, then the result is an exponential distribution. For $\alpha > 1$, the Gamma distribution assumes a unimodal, but skewed shape.

3.5 The phylogenetic tree reconstruction problem

A main goal of phylogenetic analysis is to reconstruct the phylogenetic tree along which a given set of species has evolved. In nearly all cases, there is no way to verify whether a given phylogenetic tree represents the *true tree* along which sequences or species actually evolved. Thus, the problem is usually addressed indirectly, as follows.

First, a *model of sequence evolution* is developed for which one hopes that it covers the most important aspects of the evolution of the taxa under consideration. Such models may be a simple modification of the Jukes-Cantor model, obtained, for example, by adding a difference in nucleotide composition, say, or can be quite substantial modifications, obtained, for example, by adding dependencies between different sites in the sequence.

Then, a tree reconstruction method is used that is known to perform well for sequences that have evolved under the given model.

In practice, there is a trade-off that needs to be taken into account: a more simple model of evolution is usually computationally less expensive and requires less data to get a robust result, whereas a more elaborate model of evolution is usually computationally more expensive and requires more data. However, a more detailed model should provide more reliable results. On the other hand, the variance of the solution increases with more parameters.

An important property of any such model of evolution is that it be *identifiable*, meaning that, in the idealized situation that the sequences under consideration have infinite length, it must, in principle, allow one to determine precisely which model tree a given set of sequences was generated on. When identifiability fails, this is because the same set of sequences can be generated with the same probability on two different model trees. Simple models, such as the Jukes-Cantor model, are usually identifiable. As models become more realistic or complex, care must be taken not to lose identifiability. In current research, mathematicians are interested in designing models that are as general as possible, while still being identifiable.

How well any phylogenetic tree can be reconstructed from sequences depends on the length of the sequences provided. For identifiable models, an important concept is *statistical consistency*: a tree reconstruction method is called *statistically consistent* with respect to a given (identifiable) model of evolution, if it is guaranteed to correctly reconstruct the phylogenetic tree from sequences that evolved along that phylogenetic tree under the model, given long enough sequences.

There exist many different approaches to the tree reconstruction problem [77]. These are usually classified into two main groups:

- *Sequence-based methods* usually search for a phylogenetic tree T that optimally explains a given multiple sequence alignment M.
- *Distance-based methods* usually construct a phylogenetic tree T from a given distance matrix D.

We discuss some of the most important methods in the following sections.

3.6 Sequence-based methods

The first main group of tree reconstruction methods are called *sequence-based methods*. In this type of approach, the input consists of a multiple sequence alignment M on $\mathcal{X}$ and a phylogenetic tree T is determined for M, usually by performing a search in tree space to find an optimal phylogenetic tree or trees.

In the following, we discuss three main approaches: *maximum parsimony* [67], *maximum likelihood* [68], and *Bayesian inference* [119].

3.7 Maximum parsimony

The *maximum parsimony method* is one of the most widely used sequence-based tree reconstruction methods. In science, the principle of *maximum parsimony* can be stated as a preference for the least complex explanation for an observation. In phylogenetic analysis, the *maximum parsimony problem* is to find a phylogenetic tree that explains a given set of aligned sequences using a minimum number of *evolutionary events*.

In the simplest form of this approach, the evolutionary events to be minimized are nucleotide mutations. In this context, the *difference* between two aligned sequences $x = (x_1, \ldots, x_m)$ and $y = (y_1, \ldots, y_m)$ is given by the number of positions at which they differ,

$$\text{diff}(x, y) = |\{i \mid x_i \neq y_i\}|, \tag{3.6}$$

which is also known as the (unnormalized) *Hamming distance*.

Suppose we are given a multiple alignment of sequences $M = (a_1, a_2, \ldots, a_n)$ of length m and a corresponding phylogenetic tree T on $\mathcal{X}$. If we assign a hypothetical ancestral sequence of length m to every internal node in T, then we can obtain a score for T together with this assignment, by summing over all differences $\text{diff}(x, y)$ between any two sequences x and y that are assigned to the two ends of some edge in T. The minimum value obtainable in this way is called the *parsimony score* $PS(T, M)$ of T and M:

Definition 3.7.1 (Parsimony score of a tree) *Suppose we are given a multiple sequence alignment M of length m and a corresponding phylogenetic tree T on $\mathcal{X}$.*

The parsimony score *of T and M is defined as:*

$$PS(T, M) = \min_{\alpha} \sum_{\{x,y\}} \text{diff}(x, y), \tag{3.7}$$

where the minimum is taken over all possible assignments α of hypothetical ancestor sequences of length m to the internal nodes of T and the sum is taken over all pairs of sequences x, y that are assigned to opposite ends of some edge f of T.

3.7.1 The small parsimony problem

The task of computing $PS(T, M)$ is known as the *small parsimony problem*. This problem can be solved efficiently by the *Fitch algorithm* [79] if the phylogenetic trees considered are bifurcating and mismatches between different character states are all weighted equally. For trees with multifurcations, a generalization of this algorithm, called the *Fitch-Hartigan* algorithm [101] can be used. In even more general settings, for example, when changes between different states are weighted differently, Sankoff's algorithm can be applied [210] in a modification of (3.7).

To give a flavor of such algorithms, we describe the Fitch algorithm. The Fitch algorithm requires that the given phylogenetic tree has a root of outdegree two. However, as the placement of the root does not influence the achievable score, this requirement can be met by simply inserting a root into an arbitrary edge of the input tree, if necessary.

Assume that we are given a multiple sequence alignment M and a bifurcating (rooted) phylogenetic tree T on $\mathcal{X}$. We score each character (that is, column of the alignment) separately and then obtain the parsimony score $PS(T, M)$ by summing over all characters.

Consider a fixed character c. The algorithm proceeds in two passes. In a *bottom-up pass*, each node v of the tree is labeled by a set of character states $S(v)$ in a post-order traversal of the tree, as follows: First, each leaf is assigned the character-state associated with the given taxon represented by the leaf. Now, consider some internal node v with children w_1 and w_2. We set $S(v) = S(w_1) \cap S(w_2)$, if the intersection of $S(w_1)$ and $S(w_2)$ is non-empty, and otherwise we set $S(v) = S(w_1) \cup S(w_2)$. The parsimony score for the tree T and the character c equals the number of times that the union operation is performed in the bottom-up pass.

To produce a labeling α of the internal nodes of T by character-states that attains the optimal score computed by the bottom-up pass, a *top-down pass* is performed that starts at the root and then proceeds in a preorder traversal toward the leaves of the tree: First, an arbitrary character state $\alpha(\rho) \in S(\rho)$ is chosen for the root node ρ. Then, for every child w of a node v we set its label $\alpha(w)$ to $\alpha(v)$, if $\alpha(v) \in S(w)$, and otherwise we choose an arbitrary character state from $S(w)$.

The Fitch algorithm runs in time proportional to the length of the alignment times the number of taxa. The bottom-up pass computes the optimal score $PS(T, M)$. The top-down pass provides at least one labeling α that attains that score, but cannot be used to generate all possible optimal labellings.

Exercise 3.7.2 (Fitch algorithm) *Construct an example for which the Fitch algorithm cannot be used to generate all possible optimal labellings.*

3.7.2 The large parsimony problem

Now that we know how to determine the parsimony score on a given phylogenetic tree, let us look at the problem of computing the following value:

Definition 3.7.3 (Parsimony score of an alignment) *Given a multiple sequence alignment $M = (a_1, \ldots, a_n)$ on $\mathcal{X}$, its parsimony score is defined as*

$$PS(M) = \min\{PS(T, M) \mid T \text{ is a phylogenetic tree on } \mathcal{X}\}. \tag{3.8}$$

The task of computing $PS(M)$ is known as the *large parsimony problem*. Potentially, we need to consider all $(2n - 5)!!$ possible unrooted phylogenetic trees. Unfortunately, it is most probably not possible to find a solution much faster, as the maximum parsimony problem is known to be NP-hard.

3.7.3 Branch-and-bound approach

Recall how we obtained an expression for the number $U(n)$ of unrooted phylogenetic tree topologies on a set of n taxa $\mathcal{X} = \{x_1, x_2, \ldots, x_n\}$: Starting with the unique phylogenetic tree with 3 leaves, there are three ways of adding an extra edge with a new leaf to obtain a phylogenetic tree on 4 taxa. This new phylogenetic tree has $(2n - 3) = 5$ edges. Hence, there are 5 ways to obtain a new phylogenetic tree with 5 leaves, and so forth.

More precisely, we can obtain any tree T_{i+1} on $\{x_1, \ldots, x_i, x_{i+1}\}$ by adding an extra edge with a leaf labeled x_{i+1} to some (unique) tree T_i on $\{x_1, \ldots, x_i\}$. Hence, we can construct the set of *all* possible trees on n taxa by adding one leaf at a time in all possible ways, thus systematically generating a complete *enumeration tree* of the trees.

Let $M = (a_1, \ldots, a_n)$ be a multiple sequence alignment on $\mathcal{X}$. A simple, but crucial observation in the setting of maximum parsimony is that if we add a new taxon x_{i+1} to a phylogenetic tree T_i to obtain a new phylogenetic tree T_{i+1} then the parsimony score of T_{i+1} can only be equal to, or worse than, the parsimony score of T_i.

This gives us a way to bound the search through the complete enumeration tree [109]: If the *local* parsimony score of the current incomplete tree T' is worse than

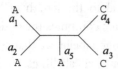

Figure 3.10 The parsimony score of this tree is 1 for the indicated character states.

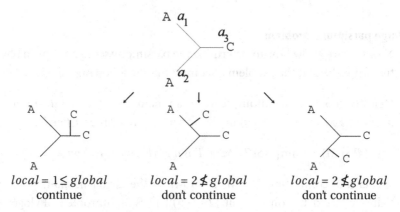

$local = 1 \leq global$ $local = 2 \nleq global$ $local = 2 \nleq global$
continue   don't continue   don't continue

Figure 3.11 In branch-and-bound, we first consider all three ways of adding a fourth taxon a_4 to a phylogenetic tree T on $\{a_1, a_2, a_3\}$. The local parsimony score of the resulting subtrees is compared with the best global parsimony score found so far (equal to 1, in this example) to decide which subtrees merit further inspection.

the best *global* score for any complete tree seen so far, then we do not need to generate the enumeration subtree below T'.

In practice, using this *branch-and-bound* strategy, one can obtain exact solutions for datasets of 20 or slightly more sequences, depending on the sequence length and complexity of the data. In general, a good starting strategy is to first compute an initial phylogenetic tree T_0 for the data, e.g. using the neighbor-joining method (described in Section 3.14), and then to initialize the *global* bound to the parsimony score of T_0.

As an example, suppose we have a multiple sequence alignment $M = (a_1, a_2, \ldots, a_5)$ of length one, and the characters are $a_1 = A$, $a_2 = A$, $a_3 = C$, $a_4 = C$, and $a_5 = A$. Assume that the initial phylogenetic tree on M is as shown in Figure 3.10, which gives a global upper bound of 1 for position i. The first step in generating the enumeration tree is shown in Figure 3.11. Only the first of the three trees fulfills the bound criterion and so we do not generate any further trees from the other two trees. The second step in generating the enumeration tree is shown in Figure 3.12. The first three trees are optimal. Note that the bound criterion in the first step reduced the number of full trees to be considered by two thirds.

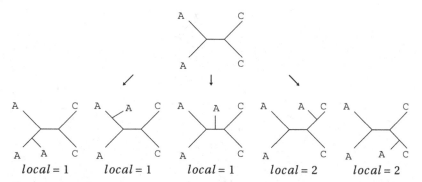

Figure 3.12 The five different phylogenetic trees obtained by adding a fifth taxon with character state A.

3.7.4 Stepwise addition and star-decomposition heuristics

We now discuss a simple greedy heuristic for approximating the maximum parsimony tree called the *stepwise-addition heuristic* [73]. As in the branch-and-bound method, we build a phylogenetic tree T by adding one leaf after the other, however, in each step we choose an optimal position for adding in the new leaf-edge.

More precisely, given a multiple sequence alignment $M = (a_1, a_2, \ldots, a_n)$ on $\mathcal{X} = \{x_1, \ldots, x_n\}$, we start with a phylogenetic tree T_2 that has two leaves labeled x_1 and x_2. Now suppose that at step i we have constructed a phylogenetic tree T_i for the first i taxa $(x_1, x_2, \ldots, x_i)$. Then, for each edge e in T_i we consider a new phylogenetic tree T_i^e that is obtained by inserting a new node v in e and joining it via a new edge f to a new leaf w with label x_{i+1}. We choose the tree with the best parsimony score, that is we set $T_{i+1} = \arg \min_e \{PS(T_i^e, M)\}$.

An alternative approach is the *star-decomposition heuristic*. In this strategy, we start with the *star tree* on all n taxa that has one internal node and n leaf nodes. At each step, the optimality criterion is evaluated for every possible tree obtained by pulling out a pair of edges incident to the central node. The best tree found at one step is then used as the basis of the next step, as indicated in Figure 3.13.

Whereas the branch-and-bound approach is guaranteed to find an optimal solution, the stepwise-addition and star-decomposition heuristics are not and the result that they produce depends on the order in which the sequences are processed or ties are resolved.

3.8 Branch-swapping methods

Let $M = (a_1, \ldots, a_n)$ be a multiple sequence alignment on $\mathcal{X}$. An important and widely-used heuristic approach to solving the large parsimony problem on M, and also other optimization problems on phylogenetic trees, is to explore the space of bifurcating phylogenetic trees on $\mathcal{X}$. This is done by moving from one phylogenetic

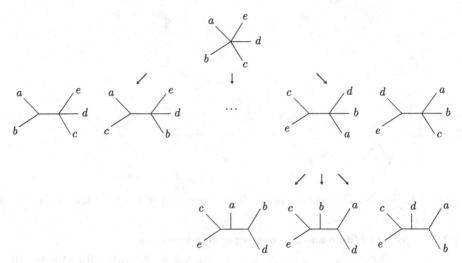

Figure 3.13 In the star-decomposition heuristic, one repeatedly *pairs off* two edges that are incident to the central node that provide the best improvement of the parsimony score, until a resolved phylogenetic tree is obtained.

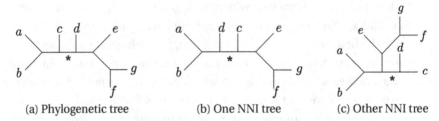

| (a) Phylogenetic tree | (b) One NNI tree | (c) Other NNI tree |

Figure 3.14 Nearest neighbor interchange (NNI). (a) An unrooted phylogenetic tree T. The four subtrees attached to the two ends of the edge marked "*" can be interchanged in four different ways, leading to two distinct phylogenetic trees, shown in (b) and (c).

tree T to another T' by applying a *branch-swapping* operation that rearranges a part of T.

More precisely, the idea is to generate a number of neighboring phylogenetic trees in this way and then to move to the one that has the best parsimony score. This is repeated a number of times until no further improvement can be achieved.

The simplest branch-swapping operation is the *nearest neighbor interchange (NNI)* [178, 203], in which two of four subtrees that are attached to a common edge are swapped in one of two possible ways, as illustrated in Figure 3.14.

A more general branch-swapping method is called *subtree prune and regraft (SPR)*, in which a subtree is pruned from the given phylogenetic tree and re-grafted at a different location of the phylogenetic tree, as shown in Figure 3.15 [191].

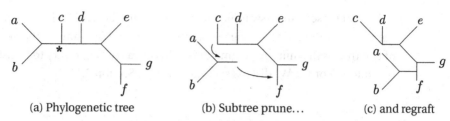

(a) Phylogenetic tree (b) Subtree prune... (c) and regraft

Figure 3.15 Subtree prune and regraft (SPR). (a) An unrooted phylogenetic tree T. The subtree consisting of three edges and the two leaves labeled a and b is pruned at the location marked "$*$" to produce the two subtrees shown in (b), and then reattached into the edge leading to the leaf labeled f (c).

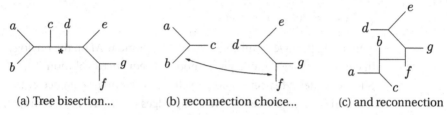

(a) Tree bisection... (b) reconnection choice... (c) and reconnection

Figure 3.16 Tree bisection and reconnection (TBR). The phylogenetic tree T shown in (a) is bisected into two subtrees by deleting the edge marked "$*$". As illustrated in (b), one possible choice is to reattach the two subtrees by a new edge joining the middles of the two leaf edges associated with taxa b and f. The resulting new phylogenetic tree is shown in (c).

The most general branch-swapping method considered here is called *tree bisection and reconnection (TBR)*. To apply this method to a phylogenetic tree, first bisect the tree into two subtrees by completely removing an edge from the tree. Then, to reattach the tree, join the two subtrees by a new edge that leads from somewhere in the one part to somewhere in the other, see Figure 3.16.

For a bifurcating phylogenetic tree T on $\mathcal{X}$, let NNI(T), SPR(T) and TBR(T) denote the set of all phylogenetic trees that can be obtained by applying exactly one NNI, SPR or TBR operation to T, respectively. It is not too difficult to see that the following inclusions hold:

$$\text{NNI}(T) \subseteq \text{SPR}(T) \subseteq \text{TBR}(T). \tag{3.9}$$

Let n denote the number of taxa. It can be shown that the cardinality of NNI(T) and SPR(T) depend only on n. While the size of NNI(T) grows linearly with n, the size of SPR(T) grows quadratically. The size of TBR(T) depends not only on n, but also on the actual topology of the phylogenetic trees considered.

Exercise 3.8.1 (Space of trees is connected) *Show that the space of all bifurcating phylogenetic trees (without edge lengths) on n taxa is connected in the sense that any one such tree T_1 can be transformed into a second one, T_2, by a finite series of NNI (and thus also SPR or TBR) operations.*

Each of these branch-swapping methods can also be used to define a distance on the set of all bifurcating phylogenetic trees on n taxa. For example, the *SPR distance* of two trees is the minimum number of SPR operations necessary to transform one tree into the other. We discuss such distances in Section 3.16.

3.9 Maximum likelihood estimation

Let $M = (a_1, \ldots, a_n)$ be a multiple sequence alignment of length m on $\mathcal{X}$. The basic idea of the *maximum likelihood estimation (MLE)* method is to determine a phylogenetic tree T, together with edge lengths ω, that maximizes the *likelihood*

$$L(T) = P(M \mid T, \omega, \mathcal{M}) \tag{3.10}$$

of generating the given multiple sequence alignment M on the phylogenetic tree T with edge lengths ω under a given model of sequence evolution $\mathcal{M}$.

Such a model $\mathcal{M}$ specifies how to choose the initial sequence at the root of the tree T and how sequences evolve along edges of the tree. For example, the Jukes-Cantor model described in Section 3.4 specifies that at any position i of the root sequence, all four nucleotides appear with the same probability of $\frac{1}{4}$. The mutation rate is u. Moreover, the length $\omega(e)$ of any edge e is interpreted as a duration and the probability that an observable change occurs along an edge of length t is given by the probability-of-change formula: $P(\text{change} \mid t) = \frac{3}{4}\left(1 - e^{-\frac{4}{3}ut}\right)$.

In practice, the models employed are more general than the simple Jukes-Cantor model, usually allowing unequal nucleotide frequencies and different transition probabilities between different nucleotides. The models are generally *time reversible* which implies that the optimal score attainable is independent of the location of the root.

A main advantage of maximum likelihood estimation is that it provides a systematic framework for explicitly incorporating assumptions and knowledge about the process that generated the given data. Although evolutionary models are only a rough estimate of real-world biological evolution, in practice, maximum likelihood methods are believed to be quite robust to violations of the assumptions made in the models.

As with maximum parsimony, the main draw-back of this approach is that finding an optimal phylogenetic tree is NP-hard [50]. In fact, the situation may be in some sense worse: for a given phylogenetic tree T, the complexity of the problem of defining edge lengths ω on T so as to maximize the likelihood $P(M \mid T, \omega, \mathcal{M})$ is currently unknown.

Assume we are given the multiple sequence alignment indicated in Figure 3.17 as input. There are three possible unrooted phylogenetic trees on a set of four taxa. To simplify the calculation of the maximum likelihood score for such a phylogenetic

	1	2		i				m
a_1	A	C	...	G	C	G	...	A
a_2	A	C	...	G	C	C	...	G
a_3	A	C	...	T	A	C	...	G
a_4	A	C	...	T	G	G	...	A

Figure 3.17 An example of a multiple sequence alignment M of length m on four sequences.

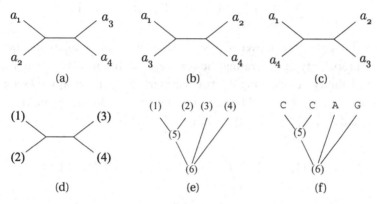

Figure 3.18 There are three different resolved phylogenetic trees on four taxa, shown in (a), (b) and (c). In the context of maximum likelihood estimation, any such phylogenetic tree (d) is rooted using one of its internal nodes (e) and each individual character is considered in turn (f).

tree T we root it by choosing one of its internal nodes to be the root ρ, see Figure 3.18. We discuss how to compute the maximum likelihood for the first tree. The other trees are processed similarly.

Conceptually, we obtain the likelihood $L(i, T)$ that a specific rooted phylogenetic tree T generated M_i, the i-th character of M, by summing over *all possible* labellings α of the internal nodes of T by character states. For a bifurcating unrooted phylogenetic tree on n taxa, there are $n - 2$ internal nodes, and so there are $k^{(n-2)}$ possible labellings to be considered, where k is the number of possible character states. For our example, we need to consider 16 different possibilities:

$$
L(i, T) = P\left(M_i \middle| \begin{matrix} C & C & A & G \\ & A & & \\ & & A & \end{matrix}\right) + P\left(M_i \middle| \begin{matrix} C & C & A & G \\ & C & & \\ & & A & \end{matrix}\right)
$$

$$
+ \quad \cdots \quad + P\left(M_i \middle| \begin{matrix} C & C & A & G \\ & G & & \\ & & C & \end{matrix}\right)
$$

$$
+ \quad \cdots \quad + P\left(M_i \middle| \begin{matrix} C & C & A & G \\ & T & & \\ & & T & \end{matrix}\right)
$$

In general, for a given rooted phylogenetic tree T and model $\mathcal{M}$, the likelihood of T for a single character M_i is given by

$$
\begin{aligned}
L(i, T) &= P(M_i | T, \omega, \mathcal{M}) \\
&= \sum_{\alpha} P(M_i, \alpha | T, \omega, \mathcal{M}) \\
&= \sum_{\alpha} \left(P(\alpha(\rho)) \prod_{e=\{v,w\}} P_e(\alpha(v), \alpha(w)) \right),
\end{aligned}
\tag{3.11}
$$

where the summation is over all possible labellings α of the internal nodes, $P(\alpha(\rho))$ is the probability of generating the state $\alpha(\rho)$ at the root and $P_e(\alpha(v), \alpha(w))$ is the probability of observing the two states $\alpha(v)$ and $\alpha(w)$ at the two ends of an edge $e = \{v, w\}$. The likelihood that T generated the whole multiple sequence alignment M is obtained as the product of the likelihoods for all positions of M:

$$
L(T) = L(1, T) \cdot L(2, T) \cdots \cdots L(m, T) = \prod_{i=1}^{m} L(i, T).
\tag{3.12}
$$

To obtain a value for $L(i, T)$ we need to be able to efficiently calculate

$$
P(M_i, \alpha \mid T, \omega, \mathcal{M}) = P(\alpha(\rho)) \prod_{e=\{v,w\}} P_e(\alpha(v), \alpha(w))
\tag{3.13}
$$

for any given rooted phylogenetic T with edge lengths ω. We discuss how to do this for the case of the Jukes-Cantor model with mutation rate u. Under this model, the probability of observing any particular nucleotide at the root node is $\frac{1}{4}$. The probability of observing two *different* states $\alpha(v)$ and $\alpha(w)$ at opposite ends of an edge e of length $t = \omega(e)$ is given by

$$
P\left(\alpha(v) \neq \alpha(w) \mid t\right) = \frac{3}{4}\left(1 - e^{-\frac{4}{3}ut}\right),
\tag{3.14}
$$

whereas the probability of observing the *same* state at opposite nodes of e equals

$$
1 - P\left(\alpha(v) \neq \alpha(w) \mid t\right) = \frac{1}{4}\left(1 + 3e^{-\frac{4}{3}ut}\right),
\tag{3.15}
$$

see Section 3.4. For example, the value of

$$
P\left(M_i \left| \begin{array}{c} \text{C} \quad \text{C} \quad \text{A} \quad \text{G} \\ \text{A} \\ \text{A} \end{array} \right. \right)
$$

is given by

$$P(A)\, P_{e_1}(A, C)\, P_{e_2}(A, C)\, P_{e_3}(A, A)\, P_{e_4}(A, A)\, P_{e_5}(A, G)$$

$$= \frac{3^3}{4^6} \left(1 - e^{-\frac{4}{3}ut_1}\right)\left(1 - e^{-\frac{4}{3}ut_2}\right)\left(1 + 3e^{-\frac{4}{3}ut_3}\right)\left(1 + 3e^{-\frac{4}{3}ut_4}\right)\left(1 - e^{-\frac{4}{3}ut_5}\right),$$

$$(3.16)$$

where $e_1, \ldots, e_5$ are the five edges in T and $t_i = \omega(e_i)$ for $i = 1, \ldots, 5$.

One can efficiently compute the likelihood for a given bifurcating, rooted phylogenetic tree T with edge lengths ω using *Felsenstein's algorithm* [73]. In this approach, for each node v in T and every possible character state a, in a post-order traversal of T, one recursively computes the conditional likelihood $L_v(a)$ for the subtree T_v rooted at v, under the condition that the state of v is a, that is, we have $\alpha(v) = a$.

More precisely, for any internal node v with children w_1 and w_2, the main recursion is

$$L_v(a) = \left(\sum_b P_{\{v, w_1\}}(a, b) L_{w_1}(b)\right) \left(\sum_b P_{\{v, w_2\}}(a, b) L_{w_2}(b)\right), \quad (3.17)$$

where summation is over all possible states b and, as above, $P_{\{v, w\}}(a, b)$ is the probability of observing states a and b at the two ends of the edge $\{v, w\}$ under the given model $\mathcal{M}$.

To initialize the computation, for each leaf v and every symbol a, we set $L_v(a) = 1$, if the node v is labeled by the character state a, and $L_v(a) = 0$, otherwise.

The computation of $L(i, T)$ is actually more complicated than we have described so far, because we also have to consider all possible ways of defining edge lengths ω on the tree T. The complexity of determining an optimal choice of ω for a fixed tree is currently unknown. One possible heuristic is to optimize the length of each edge in turn, while keeping all other edge lengths fixed.

In summary, for a given multiple sequence alignment M on $\mathcal{X}$, in maximum likelihood estimation, a result is obtained by searching through all possible phylogenetic trees T and, for each such tree T, computing the maximum likelihood score. The method returns the tree attaining the best score.

3.10 Bootstrap analysis

Both maximum parsimony and maximum likelihood methods aim at producing an optimal phylogenetic tree (or a group of optimal trees). This can be viewed as a single point estimate of the true phylogeny (as opposed to a distribution of trees of similarly high quality) and the question arises how to evaluate the robustness of the different parts of the computed tree. A standard statistical technique for addressing this type of question is called *bootstrapping* [74].

An important concept for the analysis of unrooted phylogenetic trees is that of a *split*, which we discuss in more detail in Chapter 5. Recall that a *split* on $\mathcal{X}$ is a bipartitioning of $\mathcal{X}$ into two non-empty sets A and B, with $A \cap B = \emptyset$ and $A \cup B = \mathcal{X}$. Every edge e of an unrooted phylogenetic tree T gives rise to a different split of $\mathcal{X}$ obtained as the two sets of taxa that label the two different parts of the tree that are separated by e, as illustrated in Figure 5.2. We use $\mathcal{S}(T)$ to denote the set of all splits that can be obtained from an unrooted phylogenetic tree T in this way.

In *nonparametric bootstrapping*, the general approach is to sample replicate datasets from the original input data, to apply the computational method under consideration to each of the replicate datasets and then to count how many times features of the original result are found in the results obtained for the replicates to estimate their support.

More concretely, let T be a phylogenetic tree that we have computed from a multiple sequence alignment M of length m on $\mathcal{X}$, using some phylogenetic tree reconstruction method H. A *bootstrap replicate* is a multiple sequence alignment M' that consists of m columns that have been randomly sampled from the original input M, with replacement. To perform a bootstrapping analysis, a set of 1000, say, bootstrap replicates $M^1, \ldots, M^{1000}$ are produced and then the method H is applied to each replicate in turn, producing a collection of phylogenetic trees $B = (T^1, \ldots, T^{1000})$.

This set of *bootstrap trees* is analyzed to evaluate the statistical robustness of the original output T. The *bootstrap support* of a split S in $\mathcal{S}(T)$ is defined as the percentage of bootstrap trees that contain S.

Bootstrapping provides a measure for assessing how well different portions of a phylogenetic tree are supported. A split that has a low bootstrap support is sensitive to the exact combination of characters in the original input dataset and is deemed statistically unreliable. In practice, a bootstrap support of at least 70%, say, is required for a split to be considered trustworthy.

In the context of maximum likelihood estimation one can also use *parametric bootstrapping*, which operates as follows. Let $\mathcal{M}$ be the assumed model of evolution. For a given multiple alignment M on $\mathcal{X}$, let T be the tree obtained by the maximum likelihood method. Generate bootstrap replicates $M^1, \ldots, M^{1000}$, say, by evolving sequences down the tree T according to the evolutionary model $\mathcal{M}$. Then compute a set of bootstrap trees $B = (T^1, \ldots, T^{1000})$ by applying the maximum likelihood method to each of the bootstrap replicates, and then use these trees to determine the bootstrap support of each split in $\mathcal{S}(T)$, as above.

Bootstrapping multiplies the computational requirements of tree reconstruction by the number of replicates used, which is usually either 100 or 1000.

3.11 Bayesian methods

Bayesian inference of phylogenetic trees is a method whose goal is to estimate the distribution of a quantity called the *posterior probability* of a phylogenetic tree [119]. Generally speaking, the *posterior probability* of a result is the conditional probability of the result being observed, computed *after* seeing a given input dataset. The *prior probability* of a result is the marginal probability of the result being observed, set *before* the input dataset is examined. In other words, the *prior probability distribution* of phylogenetic trees on $\mathcal{X}$ reflects our belief of the distribution of all possible phylogenetic trees on $\mathcal{X}$. This distribution is determined in advance, without knowledge of the actual sequence data to be examined.

Let M be a multiple sequence alignment and T a phylogenetic tree with edge lengths ω, on $\mathcal{X}$. Also, we assume that some fixed model of evolution $\mathcal{M}$ is given. The posterior probability of T can be obtained from the prior probability of T with the help of the likelihood $P(M \mid T) = P(M \mid T, \omega, \mathcal{M})$ using Bayes' Theorem:

Theorem 3.11.1 (Bayes' Theorem)

$$P(T \mid M) = \frac{P(M \mid T) \times P(T)}{P(M)}. \tag{3.18}$$

The posterior probability $P(T \mid M)$ can be interpreted as the probability that the tree T is correct, given the data. In the simplest case, all trees on $\mathcal{X}$ are initially considered to be equally probable and then the prior distribution is called a *flat prior*. In the case of a flat prior, the posterior probability $P(T \mid M)$ is proportional to the likelihood $P(M \mid T)$ (computed under the model $\mathcal{M}$) and the maximum likelihood tree also has maximum posterior probability.

The maximum posterior probability over all trees is usually very low, if we think of tree space as a discrete sample space. In Bayesian tree inference the goal is not to determine a single phylogentic tree of maximum posterior probability, but rather to compute a sample of phylogenetic trees according to the posterior probability distribution of trees. Such a sample of trees is then processed further, for example, to generate a single consensus tree (as defined in Section 3.17) and then to label the splits of such a tree by their posterior probabilities.

The expression for the posterior probability given by Theorem 3.11.1 looks quite simple and we can efficiently compute $P(M \mid T)$ using Felsensteins algorithm, as described in Section 3.9. However, it is usually impossible to solve the complete expression analytically, as computation of the normalizing factor $P(M)$ usually involves summing over all possible phylogenetic tree topologies T' and integrating over all possible edge lengths ω and model parameters μ (of the given model $\mathcal{M}$):

$$P(T \mid M) = \frac{P(M \mid T)P(T)}{P(M)} = \frac{P(M \mid T)P(T)}{\sum_{T'} P(M \mid T')P(T')}, \tag{3.19}$$

where

$$P(M \mid T') = \int_{\omega} \int_{\mu} P(M \mid T', \omega, \mu) P(\omega, \mu) \mathrm{d}\mu \mathrm{d}\omega. \tag{3.20}$$

To avoid this problem, Bayesian inference employs the *Markov chain Monte Carlo (MCMC)* approach, which is based on the idea of sampling from the posterior probability distribution using a suitably constructed chain of results. This is a general approach that is used in a wide range of applications.

In the context of phylogenetic tree reconstruction, the MCMC approach constructs a chain of phylogenetic trees [88, 104, 175]. At each step, a modification of the current phylogenetic tree is proposed and then a probabilistic decision is made whether to *accept* the newly proposed phylogenetic tree or whether to keep the current one.

Let T_i be the phylogenetic tree at position i of the MCMC chain and let T' be a proposed modification of T_i. Using what is called the *Metropolis-Hastings algorithm*, the decision whether to accept T' or to keep T_i is based on the ratio of the posterior probabilities of the two trees, which is given by

$$R = \frac{P(T' \mid M)}{P(T_i \mid M)} = \frac{P(M \mid T')P(T')P(M)}{P(M \mid T_i)P(T_i)P(M)} = \frac{P(M \mid T')P(T')}{P(M \mid T_i)P(T_i)}, \tag{3.21}$$

assuming that the probability of proposing T', given T_i, is the same as the probability of proposing T_i, given T'.

Note that the normalizing factor $P(M)$, which is so difficult to compute, cancels out and this expression is much easier to calculate than the actual posterior probability of either T_i or T'. Based on R, we *accept* the proposed phylogenetic tree T' with probability $\min(1, R)$. In other words, we always accept T', if it has a better posterior probability than T_i, and otherwise we draw a random number p uniformly from the interval $[0, 1]$ and accept the tree T' if $p < R$, and *reject* T', otherwise. In the case of acceptance we set $T_{i+1} = T'$.

The aim is to produce an MCMC chain $\mathcal{Z}$ of bifurcating phylogenetic trees that has the property that the stationary distribution of the trees (including the parameters ω and μ) in the chain approximates the posterior probability distribution of the trees (again, including ω and μ).

For this approach to work well in practice, a number of issues must be taken into account. First, the mechanism for modifying the current phylogenetic tree must be designed with care. On the one hand, the suggested changes should be quite small so as to avoid a drastic drop in posterior probability. On the other hand, the modifications should be large enough so as to ensure that $\mathcal{Z}$ is able to explore a sufficient portion of parameter space in a reasonable amount of time.

Second, as the MCMC chain $\mathcal{Z}$ starts at a random location in parameter space, usually of very low posterior probability, the chain must be allowed to *burn in*, that is, it must be given time to reach *stationarity*, a state in which the chain is actually sampling from the posterior probability. In practice, the first 10 000 elements of $\mathcal{Z}$, say, are excluded from further analysis.

Third, trees that are close to each other in the MCMC chain $\mathcal{Z}$ are highly correlated by construction, as they differ only by small modifications. To avoid this problem of autocorrelation, the final set of output trees is obtained by sparsely sampling $\mathcal{Z}$, retaining only every 1000-th tree, say.

What types of modifications are used in practice? Branch-swapping methods such as NNI (nearest neighbor interchange), SPR (subtree prune and regraft) and TBR (tree bisection and reattachment) are used in the context of maximum parsimony and maximum likelihood methods to search through the space of phylogenetic tree topologies. For the purposes of constructing an MCMC chain we need to couple such an approach with a method for proposing modifications of edge lengths and parameters of the evolutionary model.

3.11.1 The local algorithm

We now describe one approach that is used in practice called the *local algorithm* [160], which is a modification of the NNI method. The algorithm starts by choosing an internal edge (v, w) of the current phylogenetic tree T_i at random. As we assume that T_i is bifurcating, both of the nodes v and w are each connected to two further nodes, which we randomly label a and b for v, and c and d for w, respectively. Let x, y and z denote the distance from a to v, w and d, respectively, as obtained by adding the edge lengths ω of the edges from a to v, v to w, and w to d, appropriately. See Figure 3.19 for details.

The local algorithm changes the length of the path from a to d and modifies the positions of v and w in that path. In more detail, first define the new distance from a to d as $z' = z \times e^{p-0.5}$, where p is a random number uniformly chosen from the interval $(0, 1)$. The lengths of edges on the path from a to d will be modified so that the distance from a to d will be z'.

Second, randomly decide with equal odds whether node v or node w is to be moved by a random amount. If we decide to move node v, then define $x' = q \times z'$ and $y' = y \times z'/z$. Otherwise, define $x' = x \times z'/z$ and $y' = q \times z'$. In both cases, q is a random number uniformly chosen from the interval $(0, 1)$.

Third, to reposition the nodes v and w according to the values just chosen, remove the two nodes v and w from the path connecting a to d and then reinsert them into the path at distances x' and y' from a, respectively.

If $x' < y'$, then the local algorithm does not result in a change of the current tree topology. In this case, only the lengths of the edges from a to v, v to w,

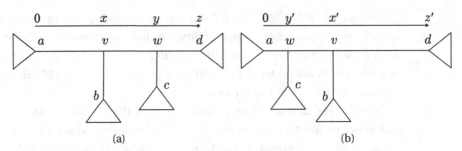

Figure 3.19 A phylogenetic tree (a) before and (b) after a modification by the local algorithm. The distance z from node a to node d is modified by a randomly chosen factor to a new value of z'. In this example, the node w is moved by a random amount along the path from a to d, whereas the relative position of node v on the path from a to d is not changed.

and w to d change. In more detail, we set $\omega(a, v) = x'$, $\omega(v, w) = y' - x'$ and $\omega(w, d) = z' - y'$.

If $x' > y'$, then the modifications result is a new topology in which a and w, w and v, and v and d, respectively, are connected by an edge. In this case, the edge lengths are $\omega(a, w) = y'$, $\omega(w, v) = x' - y'$ and $\omega(v, d) = z' - x'$.

Simple parameters of the evolutionary model, such as the mutation rate μ in the Jukes-Cantor model, are modified by adding a random number uniformly chosen from an appropriate interval centered at 0. For a set of parameters that is constrained to sum to some specific value, such as 1 in the case of probabilities, the values are randomly modified according to a Dirichlet distribution.

A proposal for modifying the tree and a proposal for new values for all parameters of the specified evolutionary model are generated independently and simultaneously, and then accepted or rejected together in a single application of the ratio-based decision step of the Metropolis-Hasting algorithm.

3.11.2 Metropolis-coupled Markov chain Monte Carlo

It is important that the MCMC chain $\mathcal{Z}$ is given the opportunity to visit all relevant parts of parameter space. In consequence, the larger the number of taxa is, and the more parameters the evolutionary model $\mathcal{M}$ has, the longer the chain must be run.

A popular approach for exploring more of parameter space in less time is to use a variant of MCMC called *Metropolis-coupled Markov chain Monte Carlo*, or (MC^3), for short. In this approach k different MCMC chains $\mathcal{Z}_1, \ldots, \mathcal{Z}_k$ are run in parallel. The first chain $\mathcal{Z}_1$ is called the *cold* chain and this is the one from which inferences are made. The other $k - 1$ chains are called *heated* chains that are set up to sample from the original posterior probability distribution raised to a power of

β with $0 < \beta < 1$. Heating a chain has the effect of "flattening out" the probability landscape and the hope is that a heated chain is able to move more freely through parameter space.

For example, in a strategy called *incremental heating*, the i-th chain $\mathcal{Z}_i$ aims at sampling from the distribution $P(T \mid M)^{\beta(i)}$, with

$$\beta(i) = \frac{1}{1 + \lambda(i - 1)}, \tag{3.22}$$

where $\lambda > 0$ is a user-specified *heating* parameter.

After all k chains have moved one step, a swap of states is proposed between two randomly chosen chains $\mathcal{Z}_i$ and $\mathcal{Z}_j$, using a Metropolis-like algorithm. Let T_i denote the current state of chain $\mathcal{Z}_i$, for all $i = 1, \ldots, k$. (Here, by current state we mean the total state consisting of the current tree T, the current length parameters ω and the current model parameters μ.) A swap between $\mathcal{Z}_i$ and $\mathcal{Z}_j$ is accepted with probability

$$\alpha = \frac{P(T_j \mid M)^{\beta(i)} P(T_i \mid M)^{\beta(j)}}{P(T_i \mid M)^{\beta(i)} P(T_j \mid M)^{\beta(j)}}. \tag{3.23}$$

The main idea is that the heated chains act as "scouts" that explore the parameter space. Swapping of chains can help the cold chain to move to a new part of parameter space that it might otherwise have difficulty reaching, perhaps because of a deep valley in the landscape of the original distribution. At the end of the run, the heated chains are discarded and the cold chain is then processed as described above.

3.11.3 Outlook

A major challenge is how to determine whether an MCMC chain $\mathcal{Z}$ has been run long enough so that the chain has *converged*, that is, it provides a good approximation of the posterior distribution of trees. A simple approach is to monitor the likelihood of the phylogenetic trees being added to $\mathcal{Z}$ and to assume that convergence has occurred once there is no further increase in likelihood. A more reliable approach may be to use multiple MCMC chains in parallel and to keep them running until they appear to be sampling from the same distribution.

One of the main attractions of Bayesian inference is that it produces a distribution of phylogenetic trees rather than a point estimate, in many cases using less time than a maximum-likelihood analysis followed by a bootstrap analysis. Areas of ongoing research are how to ensure convergence and also how to choose appropriate prior distributions.

3.12 Distance-based methods

The second main group of tree reconstruction methods are called *distance-based methods*. In this type of approach, first a distance matrix D is computed and then a phylogenetic tree T is constructed from the distance matrix D.

Definition 3.12.1 (Distance function)   *A distance function on a set $\mathcal{X}$ is a function $d : \mathcal{X} \times \mathcal{X} \to \mathbb{R}^{\geq 0}$ that has the following three properties:*

(i)   Identity and separation: *For any two elements $x, y \in \mathcal{X}$ we have $d(x, y) = 0$ if and only if $x = y$.*
(ii)   Symmetry: *For any two elements $x, y \in \mathcal{X}$ we have $d(x, y) = d(y, x)$.*
(iii)   Triangle inequality: *For any three elements $x, y, z \in \mathcal{X}$ we have $d(x, z) \leq d(x, y) + d(y, z)$.*

In practice, empirically derived distances do not always satisfy the identity property, for example, when different species have exactly the same sequence for a given gene, and sometimes they do not even fulfill the triangle inequality. To avoid these issues, throughout this book we use the term *distance matrix* to mean any function for which at least the symmetry property holds, and usually also the triangle inequality. The term *metric* is only used for distance functions that have all three required properties.

Distance-based methods are quite popular because they can be used for many different types of data. It is often easy to define a distance measure for a novel type of data, even when it is difficult or impossible to develop an appropriate model of evolution. An added advantage over sequence-based methods is speed. The neighbor-joining algorithm (described below) can be applied to distance matrices on thousands of taxa in reasonable time.

When sequence data is available, trees built using distance-based methods are often considered only a first, fast approximation, and more elaborate sequence-based methods are then used to obtain a more trusted phylogeny.

Let M be a multiple alignment of DNA sequences on $\mathcal{X}$. The simplest approach is to use the *Hamming distance* $H(a, b)$ (also called *observed p-distance*), defined as the proportion of positions at which two aligned sequences a and b differ. To be more precise, this quantity is often referred to as the *normalized Hamming distance*. The *unnormalized Hamming distance* is defined as the raw number of positions at which two aligned sequences differ.

Exercise 3.12.2 (Hamming distance fulfills triangle inequality)   *Prove that the Hamming distance $H(\cdot, \cdot)$ fulfills the triangle inequality.*

Note that the Hamming distance between two sequences *underestimates* their true evolutionary distance (average number of mutations that took place per site over the elapsed time), as back mutations and multiple mutations at the same position are not counted. To rectify this, a correction formula based on some model of evolution is often used.

For example, in the case of the Jukes-Cantor model, by inverting the probability-of-change formula, we obtain the so-called *Jukes-Cantor transformation*:

$$JC(a, b) = -\frac{3}{4} \ln \left(1 - \frac{4}{3} H(a, b) \right). \tag{3.24}$$

So, if we are given a multiple alignment of sequences that we believe evolved according to the Jukes-Cantor model of evolution, to compute a distance matrix that approximates the true evolutionary distances, we first determine the (normalized) Hamming distance between any two sequences a and b and then apply this transformation to get a corrected value.

Exercise 3.12.3 (Jukes-Cantor transform a metric?)   *Check whether $JC(\cdot, \cdot)$ has all three properties of a metric.*

Let T be a phylogenetic tree on $\mathcal{X}$, with edge lengths ω. We define the *tree distance* (or *patristic distance*) $D_T(a, b)$ between any two taxa $a, b \in \mathcal{X}$ as the sum of lengths $\omega(e)$ of all the edges e on the unique path from the leaf labeled a to the leaf labeled b. Let D be a distance matrix on $\mathcal{X}$. The main goal is to construct a phylogenetic tree T for which the tree distances T_D provide a good approximation of the distances in D.

Let T be a phylogenetic tree on $\mathcal{X}$, with edge lengths ω. The *tree length* of T is defined as the sum of lengths over all edges in T.

Given a distance matrix D on $\mathcal{X}$. How to tell whether D is *tree-like*, that is, whether there exists some phylogenetic tree T (possibly with edges of length 0) such that $D = D_T$? Tree-like distances are also called *additive*, because they can be obtained by adding the lengths of edges along paths in a suitable tree.

This global property of a distance matrix can be determined using the following local criterion:

Definition 3.12.4 (Four-point condition)   *A distance matrix D on $\mathcal{X}$ satisfies the four-point condition if for every quadruple of taxa $w, x, y, z \in \mathcal{X}$ the equation*

$$d(w, x) + d(y, z) \leq \max\{d(w, y) + d(x, z), d(w, z) + d(x, y)\} \tag{3.25}$$

holds (that is, the two larger of the three possible sums are equal).

The following result is a fundamental result in the mathematics of phylogeny [37]:

Theorem 3.12.5 (Additive distances and the four-point condition) *Let D be a distance matrix on $\mathcal{X}$. Then D is additive, if and only if D satisfies the four-point condition.*

In biology, the *molecular clock hypothesis* states that the mutation rate is constant over all sites of the sequences and over all edges of the model tree.

Under this assumption, the leaves of the model tree all have the same distance from the root. In this case the tree is called an *ultrametric tree* and any distance matrix obtained from such a tree is called an *ultrametric*. To tell whether a given distance matrix is an ultrametric, we can use the following local criterion:

Lemma 3.12.6 (Three-point condition for ultrametrics) *A distance matrix D on $\mathcal{X}$ is an ultrametric, if and only if the following* three-point condition *holds: for every triple of taxa $x, y, z \in \mathcal{X}$ we either have that all three distances between pairs of these taxa are equal, that is $d(x, y) = d(x, z) = d(y, z)$, or that two of the distances are equal and larger than the third, that is $d(x, y) = d(x, z) > d(y, z)$, say.*

The most popular method for computing a phylogenetic tree from distances is the *neighbor-joining* method, which is a fast algorithm that uses a hierarchical clustering approach. Neighbor-joining can be viewed as a modification of an older method called *UPGMA*. We first describe UPGMA and then discuss neighbor-joining. We then describe a new method, FastME, which is based on the concept of *balanced minimum evolution*.

3.13 UPGMA

UPGMA stands for *unweighted pair group method using arithmetic averages* [220]. Given a distance matrix D on $\mathcal{X}$, UPGMA produces a rooted phylogenetic tree T with edge lengths.

The method operates by clustering the given taxa, at each stage merging two clusters and at the same time creating a new node in the tree. The tree is assembled bottom-up, first clustering pairs of leaves, then pairs of clustered leaves, etc. Each node is given a height and the length of an edge is obtained as the difference of heights of its two end nodes.

Initially, we are given a distance matrix D that specifies a distance $d(x, y)$ between any two taxa, that is, leaves, x and y, and each taxon is placed in a cluster of its own, which is represented by a node at height 0.

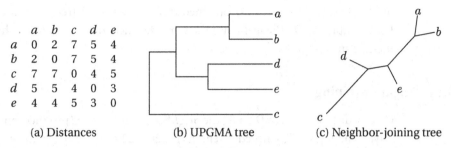

	a	b	c	d	e
a	0	2	7	5	4
b	2	0	7	5	4
c	7	7	0	4	5
d	5	5	4	0	3
e	4	4	5	3	0

(a) Distances (b) UPGMA tree (c) Neighbor-joining tree

Figure 3.20 (a) A distance matrix D, (b) the corresponding UPGMA tree and (c) the corresponding neighbor-joining tree.

We define the distance $d(i, j) = d(C_i, C_j)$ between two disjoint clusters $C_i \subset \mathcal{X}$ and $C_j \subset \mathcal{X}$ as the average distance between pairs of taxa from each cluster:

$$d(i, j) = \frac{1}{|C_i||C_j|} \sum_{x \in C_i, y \in C_j} d(x, y). \tag{3.26}$$

If C_k is the union of two clusters C_i and C_j, and C_l is any other cluster, then

$$d(k, l) = \frac{d(i, l)|C_i| + d(j, l)|C_j|}{|C_i| + |C_j|}. \tag{3.27}$$

This is a useful *update* formula that states how to compute the distance between a new cluster and an existing one in constant time. The cluster C_k is represented by a node at height $\frac{d(i, j)}{2}$ that is connected to the two nodes representing the union of the two clusters C_i and C_j.

The UPGMA algorithm repeatedly merges two closest clusters until only one cluster remains, computing a phylogenetic tree T in $O(n^3)$ steps, with $n = |\mathcal{X}|$. An example is shown in Figure 3.20b.

Exercise 3.13.1 (UPGMA) *Verify that the UPGMA method applied to the distance matrix in Figure 3.20a produces the tree shown in Figure 3.20b.*

By construction, any tree produced by the UPGMA algorithm has the property that all leaves have the same distance to the root. Moreover, we have the following result:

Lemma 3.13.2 (UPGMA on ultra metrics) *Let D be a distance matrix on $\mathcal{X}$. UPGMA computes a tree T with $D_T = D$, if and only if D fulfills the three-point condition, that is, D is an ultrametric.*

The distance matrix D displayed in Figure 3.20 does not fulfill the three-point condition. For example, the distances between c, d and e are 3, 4 and 5, and so all three are not the same, nor are two equal and larger than the third. The lemma implies that the distances D_T obtained from the UPGMA tree T do not equal D. For example, in D_T the leaf c has the same distance to both d and e, which is not the case in D.

3.14 Neighbor-joining

Given a distance matrix D on $\mathcal{X}$, the *neighbor-joining method* produces an unrooted phylogenetic tree T, with edge lengths ω [208, 229]. It is more widely applicable than UPGMA, as it does not assume a molecular clock.

The neighbor-joining algorithm is a modification of the UPGMA algorithm. Both algorithms are agglomerative methods that repeatedly decide which two clusters to join so that their nodes are "neighbors" in the resulting phylogenetic tree. In UPGMA, this decision is based on the current distances and two clusters of smallest distance are chosen. This works correctly, when the distances come from an ultrametric tree, because then closest nodes are indeed also neighbors. In a more general setting, two clusters or nodes may be separated by only a short distance without being true neighbors, see Figure 3.20, and in this case, UPGMA produces an incorrect tree.

To avoid this problem, the neighbor-joining algorithm subtracts the average distance (almost) of each cluster to all other clusters to compensate for the effect of large distances. A pair of clusters C_i, C_j that minimizes the following *neighbor-joining matrix* is chosen to form the next pair of neighbors:

$$N(C_i, C_j) = d(C_i, C_j) - (r_i + r_j), \tag{3.28}$$

where

$$r_i = \frac{1}{|\mathcal{C}| - 2} \sum_{C \in \mathcal{C}} d(C_i, C), \tag{3.29}$$

and $\mathcal{C}$ denotes the set of current clusters. Note that r_i is not precisely the average, as the number of non-zero summands is $|\mathcal{C}| - 1$, not $|\mathcal{C}| - 2$.

There are two further differences between neighbor-joining and UPGMA, namely how the distance matrix is updated after merging two clusters and how the lengths of edges are set.

Let C_i and C_j be two clusters that are about to be merged into a new cluster C_k. If $C_m \in \mathcal{C}$ is some other cluster $\neq C_i, C_j$, then the equations

$$d(C_i, C_m) = d(C_i, C_k) + d(C_k, C_m),$$

$$d(C_j, C_m) = d(C_j, C_k) + d(C_k, C_m) \tag{3.30}$$

and

$$d(C_i, C_j) = d(C_i, C_k) + d(C_j, C_k),$$

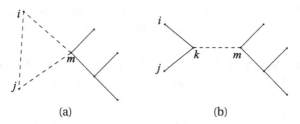

(a)   (b)

Figure 3.21 The dashed lines in (a) represent the distances between the three clusters C_i, C_j and C_m before pairing the nodes i and j. The dashed line in (b) represents the distance between the new cluster $C_k = C_i \cup C_j$ and C_m after pairing the nodes i and j. The neighbor-joining algorithm sets this distance to $d(C_k, C_m) = \frac{1}{2}(d(C_i, C_m) + d(C_j, C_m) - d(C_i, C_j))$.

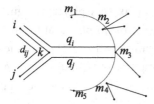

Figure 3.22 The neighbor-joining algorithm sets the length of the edge from i to k equal to $\frac{1}{2}$ of the average distance q_i from C_i to all current clusters C_m with $m \neq i, j$ plus the distance from C_i to C_j, minus the average distance q_j from C_j to all such clusters C_m.

imply

$$d(C_i, C_m) + d(C_j, C_m) = d(C_i, C_k) + d(C_k, C_m) + d(C_j, C_k) + d(C_k, C_m)$$
$$= d(C_i, C_j) + 2d(C_k, C_m), \tag{3.31}$$

and so the distance from C_k to the cluster C_m is given by:

$$d(C_k, C_m) = \frac{1}{2} \left(d(C_i, C_m) + d(C_j, C_m) - d(C_i, C_j) \right), \tag{3.32}$$

see Figure 3.21.

In the final step, when the two remaining clusters are connected, the length of the new edge is set to the distance between the two remaining clusters. In all other cases the neighbor-joining algorithm sets the length of the edge linking the two nodes i and k representing clusters C_i and C_k to $d(C_i, C_k) = \frac{1}{2}(d(C_i, C_j) + r_i - r_j)$. To see why, note that $r_i = \frac{1}{|\mathcal{C}|-2} \sum_{C \in \mathcal{C}} d(C_i, C)$ equals the average distance $q_i = \frac{1}{|\mathcal{C}|-2} \sum_{C_m \in \mathcal{C}, C_m \neq C_i, C_j} d(C_i, C_m)$ from C_i to all other clusters $C_m \neq C_i, C_j$, plus $\frac{d(C_i, C_j)}{|\mathcal{C}|-2}$. Hence, $2d(C_i, C_k) = d(C_i, C_j) + q_i - q_j = d(C_i, C_j) + q_i - q_j + \frac{d(C_i, C_j)}{|\mathcal{X}|-2} - \frac{d(C_i, C_j)}{|\mathcal{X}|-2} = d(C_i, C_j) + r_i - r_j$, which implies $d(C_i, C_k) = \frac{1}{2}(d(C_i, C_j) + r_i - r_j)$, see Figure 3.22.

Exercise 3.14.1 (Neighbor-joining) *Verify that the neighbor-joining tree for the distance matrix in Figure 3.20a is the one depicted in Figure 3.20c.*

Let D be a distance matrix on $\mathcal{X}$, under what conditions does neighbor-joining construct a tree T that correctly represents all distances in D, that is, for which $D_T = D$ holds?

Theorem 3.14.2 (Neighbor-joining on additive distances) *Let D be a distance matrix on $\mathcal{X}$. Neighbor-joining constructs a (unique) unrooted phylogenetic tree T with $D_T = D$ if and only if D is additive.*

The proof of this theorem can be found in the original paper [208].

Please convince yourself that the distance matrix D displayed in Figure 3.20 fulfills the four-point condition. In this case, the lemma implies that the distances D_T represented in the neighbor-joining tree T computed from D equal the distances in D. So, in this example, the neighbor-joining tree shown in 3.20c gives an exact representation of the distance matrix D, while the UPGMA tree shown in 3.20b is incorrect.

Distance matrices considered in practice rarely fulfill the three-point or four-point condition. In such cases neighbor-joining, and, to a lesser extent, also UPGMA, is nevertheless applied and the hope is that the deviations from the required conditions will have only a small influence on which phylogenetic tree is computed by the algorithm.

One question that remained unclear for many years is whether the neighbor-joining algorithm solves, or is an heuristic for, some explicit optimization problem. In other words, does the algorithm aim to optimize some quantity, analogous, say, to the parsimony score that is minimized by the maximum parsimony method? Quite recently, it was shown that neighbor-joining does, in fact, greedily optimize a quantity called the *BME tree length*, which we discuss in the following section.

3.15 Balanced minimum evolution

A new distance-based method called *FastME* has recently been developed within the framework of *balanced minimum evolution* (BME) [57, 58]. As we shall see, both neighbor-joining and FastME are heuristics for finding an optimal *BME tree*. The new approach appears to have three advantages over neighbor-joining. The algorithm is substantially faster, provides more accurate trees and never produces negative edge lengths (which happens quite frequently with neighbor-joining).

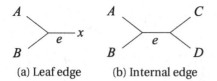

(a) Leaf edge   (b) Internal edge

Figure 3.23 In an unrooted, bifurcating phylogenetic tree, every edge e is either (a) a leaf edge, separating two subtrees containing taxon sets A and B from some taxon x, or (b) an internal edge, separating two subtrees labeled by taxon sets A, B from two other labeled by C and D.

Let D be a distance matrix on $\mathcal{X}$ and assume that we have some unrooted, bifurcating phylogenetic tree T on $\mathcal{X}$. Within the BME setting, every edge e is assigned a *balanced edge length* $\omega(e)$ that is computed from the *balanced average distances* between the taxa that span the edge.

In more detail, to define the *balanced edge length* of an edge e in T there are two cases to consider: either e is a leaf edge or e is an internal edge, as shown in Figure 3.23. In the former case, we define the length of e as:

$$\omega(e) = \frac{1}{2}\left(d^T(A, \{x\}) + d^T(B, \{x\}) - d^T(A, B)\right), \tag{3.33}$$

whereas, in the latter case, we set:

$$\omega(e) = \frac{1}{4}\left(d^T(A, C) + d^T(B, D) + d^T(A, D) + d^T(B, C)\right)$$
$$- \frac{1}{2}\left(d^T(A, B) + d^T(C, D)\right). \tag{3.34}$$

Here, $A, \ldots, D$ denote the subsets of taxa induced by the edge e. Moreover, we use $d^T(A, B)$ to denote the *balanced average distance* from taxon set A to taxon set B in $\mathcal{X}$, which is recursively defined as follows. We set

$$d^T(A, B) = d(a, b), \tag{3.35}$$

when A and B contain only one taxon each, that is, if $A = \{a\}$ and $B = \{b\}$. Otherwise, we set

$$d^T(A, B) = \frac{1}{2}\left(d^T(A, B_1) + d^T(A, B_2)\right), \tag{3.36}$$

if B is made up of two sets B_1 and B_2 that label two different subtrees. We treat the case that A is made up of two such sets A_1 and A_2 analogously.

This averaged distance is called *balanced* because two subsets such as B_1 and B_2 enter the calculation with equal weight, regardless of how many taxa they contain, in contrast, say, to the definition of distances between clusters used in the UPGMA method, see Section 3.13. Note that the length of an edge does not only depend on the distances between taxa that span the edge, but also on the topology of the tree T.

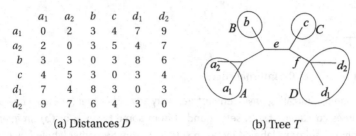

	a_1	a_2	b	c	d_1	d_2
a_1	0	2	3	4	7	9
a_2	2	0	3	5	4	7
b	3	3	0	3	8	6
c	4	5	3	0	3	4
d_1	7	4	8	3	0	3
d_2	9	7	6	4	3	0

(a) Distances D (b) Tree T

Figure 3.24 In (a) we show a distance matrix D on $\mathcal{X} = \{a_1, a_2, b, c, d_1, d_2\}$ and in (b) we show an unrooted phylogenetic tree T on $\mathcal{X}$. The four sets of taxa $A, \ldots, D$ used to compute the balanced edge length for e are indicated by circles. In the text we discuss how to compute the balanced edge length for edge e and in Exercise 3.15.1 the task is to do the same for edge f.

Consider the distance matrix D and phylogenetic tree T shown in Figure 3.24. How is the balanced edge length for the edge e computed? To answer this, we must first calculate all balanced average distances between the indicated taxon sets A, B, C and D, as follows:

$$d^T(A, C) = \frac{d(a_1,c)+d(a_2,c)}{2} = \frac{4+5}{2} = 4.5$$

$$d^T(B, D) = \frac{d(b,d_1)+d(b,d_2)}{2} = \frac{8+6}{2} = 7$$

$$d^T(A, D) = \frac{d^T(\{a_1\}, D)+d^T(\{a_2\}, D)}{2}$$

$$= \frac{d(a_1,d_1)+d(a_1,d_2)+d(a_2,d_1)+d(a_2,d_2)}{4} = 6.75 \qquad (3.37)$$

$$d^T(B, C) = d(b, c) = 3$$

$$d^T(A, B) = \frac{d(a_1,b)+d(a_2,b)}{2} = \frac{3+3}{2} = 3$$

$$d^T(C, D) = \frac{d(c,d_1)+d(c,d_2)}{2} = \frac{3+4}{2} = 3.5$$

Using Equation (3.34) we obtain:

$$\omega(e) = \frac{1}{4}(4.5 + 7 + 6.75 + 3) - \frac{1}{2}(3 + 3.5) = 2.0625. \qquad (3.38)$$

Exercise 3.15.1 (Compute balanced edge length)   *Compute the balanced edge length for the edge f shown in Figure 3.24.*

The following result provides a nice formula for computing the length of a tree with balanced edge lengths [196]:

Lemma 3.15.2 (Tree length formula)   *Let D be a distance matrix on $\mathcal{X}$ and let T be an unrooted, bifurcating phylogenetic tree, together with balanced edge lengths ω computed from D. The tree length of T is given by*

$$l(T) = \sum_{x, y \in \mathcal{X}} 2^{1 - p^T(x,y)} d(x, y), \qquad (3.39)$$

where $p^T(x, y)$ is the number of edges on the path from the leaf labeled x to the leaf labeled y in T.

Note that this formula can be generalized to the case of multifurcating trees, as follows [216]:

$$l(T) = \sum_{x, y \in \mathcal{X}} \prod_{v \in \overset{\circ}{P}_T(x, y)} (d_v - 1)^{-1} d(x, y), \tag{3.40}$$

where $\overset{\circ}{P}_T(x, y)$ denotes the set of all internal nodes on the path from the leaf labeled x to the leaf labeled y and d_v is the degree of v. We can now formulate an optimization goal that provides the basis for a tree-reconstruction method:

Definition 3.15.3 (Balanced minimum evolution tree)   *For a given distance matrix D on $\mathcal{X}$, determine a phylogenetic tree T on $\mathcal{X}$, with balanced edge lengths ω from D, for which the tree length $l(T)$ is minimum over all such trees. We call such a tree T a* balanced minimum evolution tree *(or BME tree) for D.*

Computing the BME tree is a statistically consistent tree reconstruction method [58]. Unfortunately, the problem of finding an optimal balanced minimum evolution tree is believed to be NP-hard. So we need to turn to heuristics.

Interestingly, the neighbor-joining algorithm is in fact a greedy heuristic for computing the BME tree [84]. Recall that in each iteration, the neighbor-joining algorithm considers all possible pairs of taxa and chooses a pair for which the entry in the neighbor-joining matrix is minimum. A careful analysis using the generalization of the tree length formula to multifurcating trees shows that, in each iteration, the pair chosen by the algorithm also maximizes the decrease in the length of the tree computed using balanced edge lengths. Thus we have:

Lemma 3.15.4 (Neighbor-joining is greedy BME)   *The neighbor-joining method selects as neighbors at each step the pair of current taxa that most decreases the tree length, as computed using balanced edge lengths.*

We now turn to the FastME heuristic. The algorithm has two phases. In the first phase, an initial phylogenetic tree is built in a stepwise fashion. In the second phase, the tree is iteratively improved using NNI operations until no further improvement can be attained:

Algorithm 3.15.5 (FastME)   *For a given distance matrix D on $\mathcal{X} = \{x_1, x_2, \ldots, x_n\}$, the FastME algorithm approximates a BME tree for D in two phases.*

In the first phase, the taxa are added to the tree sequentially. Assume we have already inserted the first k taxa $\{x_1, \ldots, x_k\}$ and the resulting tree is T_k. We consider each of *the $2(k-3)$ edges of T_k as possible insertion positions for a new edge joining a new leaf labeled x_{k+1} to T_k and choose the position that gives rise to the smallest new tree length, thus obtaining T_{k+1}.*

The second phase seeks to improve the tree length of T_n by repeatedly performing an NNI swap that provides the best improvement of the tree-length, until no further improvement can be achieved.

For this algorithm to run fast, a crucial observation is that the algorithm does not need to explicitly compute the edge lengths or tree lengths of the trees considered. All that is required is that the change in tree length associated with the insertion of a taxon or with an NNI swap is determined. If all balanced averages have already been computed (which requires $O(n^2)$ steps), then evaluating the tree length change associated with any insertion or NNI move can be done in constant time. This explains why FastME runs significantly faster than the neighbor-joining algorithm in practice.

The balanced minimum evolution approach can be viewed as an improvement over the classical *minimum evolution* approaches. Let D be a distance matrix on $\mathcal{X}$. In the *ordinary least squares* approach, for a given unrooted phylogenetic tree T, edge lengths ω are computed that provide a least squares fit to the distance matrix D. The method then returns such a tree of shortest length. A weakness of this method is that it explicitly assumes that all distances have the same variance, which is usually not true, as the variances of larger distances tend to be larger. To address this problem, in the *weighted least squares approach*, estimates for variances are taken into account [76, 81].

In the BME approach, the variances of all pairwise distances are not assumed to be constant, but rather they are taken to depend on the (topological) distance between taxa. The tree length formula Equation (3.39) implies that large (topological) distances have low weight. BME can be interpreted as a weighted least squares approach and the FastME algorithm runs substantially faster than all previous weighted least square approaches.

3.16 Comparing trees

Given two phylogenetic trees T_1 and T_2 on $\mathcal{X}$, how can their similarity be measured? In the case of unrooted phylogenetic trees, there are two popular measures, namely the *Robinson-Foulds* distance [204] and the *quartet* distance [70].

The Robinson-Foulds distance was originally defined as the number of edge contractions and "decontractions" required to transform one tree into the other. Since each edge corresponds to a split, and we can regard a contraction as removal

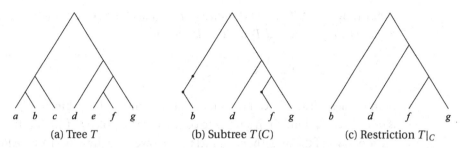

| (a) Tree T | (b) Subtree $T(C)$ | (c) Restriction $T|_C$ |

Figure 3.25 For the rooted phylogenetic tree T on $X = \{a, \ldots, g\}$ displayed in (a), we show (b) the minimal rooted subtree $T(C)$ on $C = \{b, d, f, g\}$ (with nodes of in- and outdegree one shown as black disks) and (c) the restriction $T|_C$ of T to $C = \{b, d, f, g\}$.

of a split and a "decontraction" as addition of a new split, the Robinson-Foulds distance can be defined as follows:

Definition 3.16.1 (Robinson-Foulds distance) *The* Robinson-Foulds distance *between two unrooted phylogenetic trees T_1 and T_2 on X is based on the symmetric difference of the sets of splits represented by the two trees:*

$$d_{RF}(T_1, T_2) = \frac{1}{2}|S(T_1) \triangle S(T_2)|. \tag{3.41}$$

Exercise 3.16.2 (Robinson-Foulds distances) *Compute the Robinson-Foulds distance between each pair of trees shown in Figure 3.26a–d.*

Before introducing the quartet distance, we need to discuss how to restrict a phylogenetic tree onto a subset of its taxa:

Definition 3.16.3 (Restriction of a phylogenetic tree) *Let T be a phylogenetic tree on X and let $C \subset X$ be a subset of taxa. We use $T(C)$ to denote the minimum connected subgraph of T that contains all leaves that are labeled by elements of C. The* restriction of T to C *is defined as the phylogenetic tree $T|_C$ that is obtained from $T(C)$ be suppressing all suppressible nodes.*

An example is shown in Figure 3.25. If T is a rooted phylogenetic tree on X, then both $T(C)$ and $T|_C$ will be rooted, as well. The restricted phylogenetic tree $T|_C$ is sometimes also called the *induced phylogenetic tree*, or the tree induced by C, although it is not the induced graph as defined in Chapter 1.

Let T be an unrooted phylogenetic tree on X. For any set of four distinct taxa $\{a, b, c, d\}$ in X, the tree $T|_{\{a,b,c,d\}}$ induced by the four leaves of T that are labeled by a, b, c and d is called a *quartet tree*. We use $Q(T)$ to denote the set of all such quartet trees obtainable from T. The quartet distance between two unrooted phylogenetic trees is defined as follows:

Definition 3.16.4 (Quartet distance) *The* quartet distance *between two unrooted phylogenetic trees T_1 and T_2 on $\mathcal{X}$ is defined as*

$$d_{quartet}(T_1, T_2) = \frac{1}{2}|\mathcal{Q}(T_1) \triangle \mathcal{Q}(T_2)|. \tag{3.42}$$

Both of these distances can be computed efficiently. The Robinson-Foulds metric is often used in simulation studies to compare the topology of an inferred tree with the topology of the underlying model tree. However, a draw-back of the Robinson-Foulds distance is that the placement of individual taxa has a large influence on the distance; one rogue taxon is enough to produce a large distance between two otherwise very similar trees.

Another approach to comparing two trees is to determine the minimum number of branch-swapping operations required to transform one tree into the other. This gives rise to the so-called *NNI, SPR* and *TBR distances*. All three of these distances are NP-hard to compute and are thus rarely used in practice. However, the computation of these distances is currently an active area of research and this work may lead to better tools for their application.

For a rooted phylogenetic tree T on $\mathcal{X}$, let $\mathcal{C}(T)$ denote the set of all clusters represented by T. In analogy to the Robinson-Foulds distance we define:

Definition 3.16.5 (Cluster distance) *The* cluster distance *between two rooted phylogenetic trees T_1 and T_2 on $\mathcal{X}$ is defined as*

$$d_{cluster}(T_1, T_2) = \frac{1}{2}|\mathcal{C}(T_1) \triangle \mathcal{C}(T_2)|. \tag{3.43}$$

Let T be a rooted phylogenetic tree on $\mathcal{X}$. For any set of three distinct taxa $\{a, b, c\}$ in $\mathcal{X}$, let $T|_{\{a,b,c\}}$ denote the rooted tree induced by the three leaves of T that are labeled by a, b and c. We use $\mathcal{R}(T)$ to denote the set of all such *rooted triple* trees obtainable from T. The rooted triple distance between two rooted phylogenetic trees is defined as follows:

Definition 3.16.6 (Rooted triple distance) *The* rooted triple distance *between two rooted phylogenetic trees T_1 and T_2 on $\mathcal{X}$ is defined as*

$$d_{triple}(T_1, T_2) = \frac{1}{2}|\mathcal{R}(T_1) \triangle \mathcal{R}(T_2)|. \tag{3.44}$$

To compare rooted phylogenetic trees, a rooted version of the SPR branch-swapping method, called *rSPR*, can be used. The *rSPR distance* between two rooted phylogenetic trees T_1 and T_2 on $\mathcal{X}$ is given by the minimum number of rooted SPR operations required to transform one tree into the other. An algorithm for computing the NP-hard rSPR distance is presented in Section 11.5.2.

Phylogenetic trees are often compared as unrooted trees, even when they are rooted, because the placement of the root in the trees is usually uncertain and can have a large effect on the rooted distance measures.

3.17 Consensus trees

In most practical phylogenetic studies, a whole collection of different phylogenetic trees is computed for the set of taxa under investigation. These might be different phylogenetic *gene trees* that were obtained from a number of different gene sequences in a multi-gene study. Or, they might be different phylogenetic trees obtained from the same multiple sequence alignment using different inference methods, such as maximum-parsimony, maximum-likelihood and so forth. Yet another possibility is that the collection of phylogenetic trees was obtained using a method that produces multiple trees, such as Bayesian inference.

In all such cases, one may want to treat the different phylogenetic trees as different estimations of the same underlying true evolutionary tree and then a *consensus method* can be used to construct a *consensus tree* that represents those parts of the evolutionary history on which the different phylogenetic trees agree, in some sense.

Although consensus methods can be formulated for rooted phylogenetic trees, such methods are less commonly used, probably because most tree inference methods produce unrooted trees, and also perhaps because the location of the roots in the input trees may affect the resulting consensus tree too much. We first discuss the two most important consensus methods for unrooted phylogenetic trees, namely the *strict consensus* and the *majority consensus*. We then describe the *Adams consensus*, which operates on rooted phylogenetic trees. Although the latter is rarely used in practice, it is of interest to us because it has a very natural interpretation in the context of rooted phylogenetic networks, as we will see in Section 6.6.

Let $\mathcal{X}$ be a set of taxa and assume that we are given a collection of phylogenetic trees $\mathcal{T} = (T_1, \ldots, T_k)$ on $\mathcal{X}$. Let $\mathcal{S}_{strict}(\mathcal{T})$ denote the set of all splits that occur in every input tree, that is, we define

$$\mathcal{S}_{strict}(\mathcal{T}) = \bigcap_{T \in \mathcal{T}} \mathcal{S}(T). \tag{3.45}$$

We call this the set of *strict consensus splits* of the collection of phylogenetic trees $\mathcal{T}$. The *strict consensus tree* is the (unique) phylogenetic tree T_{strict} for which $\mathcal{S}(T_{strict}) = \mathcal{S}_{strict}(\mathcal{T})$ holds. It can be obtained from any tree in $\mathcal{T}$ by contracting every edge whose split is not contained in $\mathcal{S}_{strict}(\mathcal{T})$.

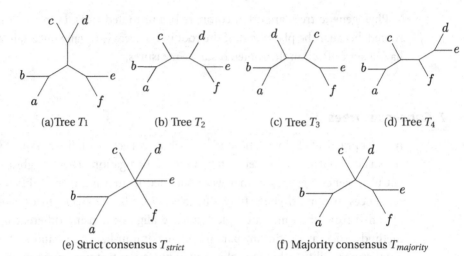

(a) Tree T_1 (b) Tree T_2 (c) Tree T_3 (d) Tree T_4

(e) Strict consensus T_{strict} (f) Majority consensus $T_{majority}$

Figure 3.26 For the collection of four phylogenetic trees $T = (T_1, \ldots, T_4)$ shown in (a–d) we display the strict consensus tree T_{strict} in (e) and the majority consensus tree $T_{majority}$ in (f).

In practice, the strict consensus tree may be quite unresolved, often being very close to the *star tree*, which consists of one central node incident to all leaf edges. The *majority consensus* is usually more informative.

Let $\mathcal{S}_{majority}(T)$ denote the set of *majority consensus splits*, that is, the set of all splits that occur in more than half of all trees in T, formally defined as

$$\mathcal{S}_{majority}(T) = \left\{ S : |\{T \in T : S \in \mathcal{S}(T)\}| > \frac{k}{2} \right\}. \qquad (3.46)$$

By the pigeon-hole principle, any two splits S_1 and S_2 that are contained in $\mathcal{S}_{majority}(T)$ must also occur in some input tree together, and thus must be *compatible* as defined in Section 5.3. Thus, by Theorem 5.3.2, there exists a unique phylogenetic tree $T_{majority}(T)$ corresponding to $\mathcal{S}_{majority}$, which is called the *majority consensus tree*.

In Figure 3.26 we show four unrooted phylogenetic trees on $\mathcal{X} = \{a, b, c, d, e, f\}$ and the corresponding strict and majority consensus trees.

We now discuss the Adams consensus. An important concept for the analysis of rooted phylogenetic trees is that of a *cluster* on $\mathcal{X}$, which is defined as any subset of $\mathcal{X}$, excluding both the empty set $\emptyset$ and the full set $\mathcal{X}$. This concept was briefly introduced in Section 1.6 and is discussed in more detail in Chapter 6.

Any edge $e = (u, v)$ in a rooted phylogenetic tree T defines a cluster C_e on $\mathcal{X}$ that consists of all taxa assigned to descendants of v. We say that a cluster C is *represented* by T, if $C = C_e$ for some edge e contained in T. We define the *lowest common ancestor* of a set of taxa $A \subset \mathcal{X}$ to be the lowest node u in T such that all leaves labeled by taxa in A are descendants of u. (Recall that a node w is *lower* than a node v, if w is a descendant of v).

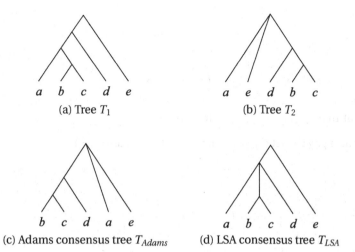

(a) Tree T_1 (b) Tree T_2

(c) Adams consensus tree T_{Adams} (d) LSA consensus tree T_{LSA}

Figure 3.27 For the two phylogenetic trees T_1 and T_2 shown in (a) and (b), we display the Adams consensus tree T_{Adams} in (c) and the LSA consensus tree T_{LSA} in (d).

For any two sets of taxa A and B on $\mathcal{X}$ we say that A *is nested in* B (in T), if the lowest common ancestor of A is a proper descendant of the lowest common ancestor of B. (Note that a is called a proper descendant of b if it is a descendant and $a \neq b$.) We say that a cluster A is *maximal* in a tree T, if the tree T does not represent any other cluster B that properly contains A.

For a collection of rooted phylogenetic trees $\mathcal{T} = (T_1, \ldots T_k)$ the *Adams consensus tree* T_{Adams} is defined as the unique rooted phylogenetic tree that satisfies the following two conditions [1, 2]:

- If A and B are any two sets on $\mathcal{X}$ such that A is nested in B for every tree $T \in \mathcal{T}$, then A is nested in B in T_{Adams}.
- If A and B are any two clusters on $\mathcal{X}$ that are represented by T_{Adams}, such that A is nested in B in T_{Adams}, then A is nested in B in every tree $T \in \mathcal{T}$.

Although this definition is a little unintuitive, the Adams consensus tree is easy to compute. The first step is to determine, for each input tree T_i ($i = 1, \ldots, k$), the set M_i of all clusters represented by T_i that are maximal in T_i. The second step it to consider all possible choices of one cluster $C_i \in M_i$ per input tree such that the intersection $C = C_1 \cap \cdots \cap C_k$ is non empty. Each such non-empty set C defines a cluster of T_{Adams}. We then recursively apply the algorithm to the set of all input trees restricted to C.

In Figure 3.27 we show the Adams consensus T_{Adams} of two trees T_1 and T_2. This is compared with the *LSA consensus tree* T_{LSA}, which is obtained using a new consensus method that we have developed in the context of rooted phylogenetic

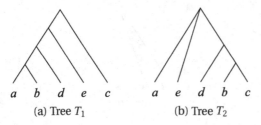

(a) Tree T_1 (b) Tree T_2

Figure 3.28 Two rooted phylogenetic trees T_1 and T_2 used in Exercise 3.17.1.

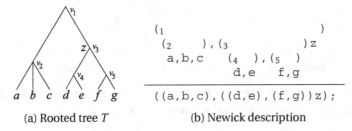

(a) Rooted tree T (b) Newick description

Figure 3.29 For the rooted phylogenetic tree T on $\mathcal{X} = \{a, \ldots, g\}$ in (a) we show how to construct the corresponding Newick string in (b). In this example, one internal node is labeled, namely node v_3 has label z. From top to bottom, the outer pair of brackets (labeled 1) correspond to the internal node v_1, the next two pairs of brackets (labeled 2 and 3) correspond to v_2 and v_3, and the last two pairs (labeled 4 and 5) correspond to v_4 and v_5. The labels a, $\ldots$, g correspond to the leaves labeled by the taxa $a, \ldots, g$. The final Newick description is written in the bottom line.

networks. We discuss the LSA consensus method in Section 6.6.1 and state that the two trees obtained by these two different methods are always very similar in the sense that they are *compatible* with each other.

Exercise 3.17.1 (Adams consensus for two trees) *Compute the Adams-consensus for the two trees T_1 and T_2 shown in Figure 3.28.*

3.18 The Newick format

The *Newick format* for trees is used to describe the topology and edge lengths of a rooted phylogenetic tree in a single line of text. Let T be a rooted phylogenetic tree on $\mathcal{X}$. In this format, each internal node of T is represented by a pair of round brackets "( )" and every leaf of T is represented by its taxon label. To assign a label to an internal node, write the label directly after the closing bracket associated with the node. If a node v is an ancestor of some node w in T, then the brackets or label representing w will be contained between the pair of brackets that represent

v. The brackets or labels representing the children $w_1, \ldots, w_k$ of an internal node v are listed inside the brackets representing v and are separated by commas. To indicate the length $\omega(e)$ of the edge e leading to a node v, we write a colon and the value of $\omega(e)$ after the brackets or label representing v. The end of a tree description is indicated by a semi-colon.

Although the Newick format is inherently rooted, it can also be used for describing unrooted phylogenetic trees. To write an unrooted phylogenetic tree in Newick format, simply root the tree arbitrarily and then proceed as in the rooted case. When constructing an unrooted phylogenetic tree from a Newick string, first construct the rooted phylogenetic tree and then erase the root. Also, note that there is no need for the leaf labels used to be unique and so the Newick format can be used to describe multi-labeled trees, as well (as defined in Section 11.7).

Exercise 3.18.1 (Reading and writing Newick) *Design an algorithm for writing a rooted phylogenetic tree in Newick format. Design an algorithm that parses a Newick string and constructs the corresponding rooted phylogenetic tree.*

Chapter notes

Most of the content of this chapter is based on teaching notes from our course on *Algorithms in Bioinformatics* taught at Tübingen University in recent years. Those notes, in turn, are based on the cited papers, on Chapter 11 of "Molecular Systematics" [230], and, in addition, on the following books: "Phylogenetics" by Semple and Steel [217], "Inferring Phylogenies" by Felsenstein [77] and "Biological Sequence Analysis" by Durbin and colleagues [64].

4

Introduction to phylogenetic networks

In the previous chapter we give a brief introduction to phylogenetic trees. Phylogenetic networks provide an alternative to phylogenetic trees and may be more suitable for datasets whose evolution involve significant amounts of reticulate events caused by hybridization, horizontal gene transfer, recombination, gene conversion or gene duplication and loss [56, 61, 89, 201, 219, 231]. Moreover, even for a set of taxa that have evolved according to a tree-based model of evolution, phylogenetic networks can be usefully employed to explicitly represent conflicts in a dataset that may, for example, be due to mechanisms such as incomplete lineage sorting or to inadequacies of an assumed evolutionary model [125].

While rooted phylogenetic networks can, in theory, be used to explicitly describe evolution in the presence of reticulate events, their calculation is difficult and computational methods for doing so have not yet matured into practical and widely used tools [24, 98, 106, 127, 225, 237]. In contrast, there are a number of established tools for computing unrooted phylogenetic networks, which can be used to visualize incompatible evolutionary scenarios in phylogeny and phylogeography [9, 10, 11, 32, 52, 122, 125].

In practice, most currently available algorithms for computing phylogenetic networks are based on combinatorics and this book focuses on such approaches. Some approaches developed within a maximum-parsimony or maximum-likelihood framework can be found, for example, in [59, 106, 141, 142, 143, 228].

In this chapter, we give an introduction to the topic of phylogenetic networks, very briefly describing the fundamental concepts and summarizing some of the most important methods that are available for the computation of phylogenetic networks. In essence, this chapter provides an overview of the material that is presented in Parts II and III of the book.

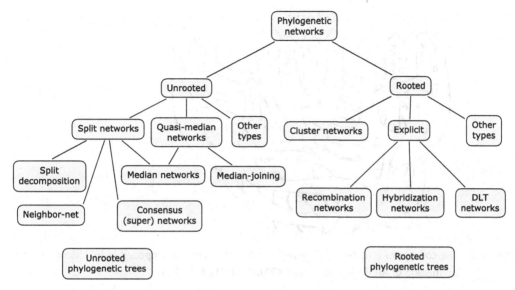

Figure 4.1   Overview of the main concepts mentioned in this chapter. First, we distinguish between unrooted (on the left) and rooted networks (on the right). While all phylogenetic networks mentioned on the left generalize unrooted phylogenetic trees, all mentioned on the right generalize rooted trees. Second, we distinguish between explicit networks (shown below the node labeled "Explicit" on the right) and abstract ones (all others).

4.1 Overview

Figure 4.1 shows the relationships between some of the concepts mentioned in this chapter. The aim is to give a brief introduction to some of the main types of phylogenetic networks. As with phylogenetic trees, a first major distinction is between *unrooted* and *rooted* phylogenetic networks. A second major distinction is between *explicit* and *abstract* phylogenetic networks, as we explain below.

4.2 What is a phylogenetic network?

In the literature, the term *phylogenetic network* is defined and used in a number of different ways, usually focusing on the specific type of network that an author happens to be interested in [7, 98, 165]. For example, a phylogenetic network is sometimes too narrowly defined as a rooted DAG whose leaves are labeled by taxa. In this book, we propose the following more general definition:

Definition 4.2.1 (Phylogenetic network)   *A phylogenetic network is any graph used to represent evolutionary relationships (either abstractly or explicitly) between a set of taxa that labels some of its nodes (usually the leaves).*

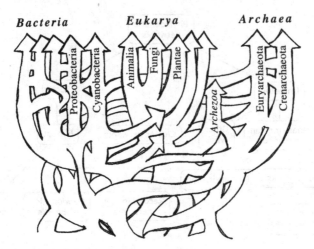

Figure 4.2   The concept of a rooted phylogenetic "network of life", as proposed in [61]. Reprinted by permission of AAAS: Science, 284(5423):2124-2128 ©1999.

The envisioned role of rooted phylogenetic networks in biology is to describe the evolution of life in a way that explicitly includes reticulate events. Ultimately, the main goal is to work out the details of a rooted phylogenetic *network of life*, such as the one proposed by Ford Doolittle [61], see Figure 4.2.

There are a number of different types of *unrooted* phylogenetic networks. In this book, we mainly focus on the important class of *split networks*, discussing the theory of split networks in Chapter 5 and describing a number of different algorithms for computing them in Part III of the book. A second important class of unrooted phylogenetic networks are *quasi-median networks*, which can be viewed as a generalization of split networks, and are discussed in Chapter 7. Algorithms for computing these and some other types of unrooted phylogenetic networks are presented in Part III of the book.

Rooted phylogenetic networks generalize rooted phylogenetic trees and such a network is usually defined as a rooted DAG whose leaves are labeled by taxa. We present the theory of clusters and rooted phylogenetic networks in Chapter 6 and describe a number of different algorithms for computing them in Part III of the book.

Phylogenetic networks can be used in two different ways: The first type of usage is as a tool for visualizing incompatible datasets in a fruitful manner, in which case we speak of an *abstract* phylogenetic network. The second type of usage is as a representation of a putative evolutionary history involving reticulate events, in which case the network is called *explicit*.

By definition, most (if not all) types of *unrooted* phylogenetic networks are abstract networks, as evolution in inherently rooted (and thus any unrooted phylogenetic *tree* is also abstract in this sense). However, *rooted* phylogenetic networks can be either abstract or explicit, depending on how they are constructed and interpreted.

The necessity of distinguishing between abstract and explicit networks was pointed out in [180]. They are called *implicit* and *explicit* in [123]. In [181], abstract and explicit networks are named *data-display* networks and *evolutionary* networks, respectively.

Phylogenetic networks can be computed from a wide range of data, including multiple sequence alignments, distance matrices, sets of trees, clusters, splits, rooted triplets or unrooted quartets.

In the literature, perhaps as many as 20 different names have been defined for different types of phylogenetic networks. A closer look reveals that some networks are named by the algorithms that compute them or by mathematical properties that define them, such as "neighbor-nets" or "median networks". Others are named by topological constraints that are imposed on them for computational reasons, such as "galled trees", "galled networks" or "level-k networks". Yet others are named by the types of evolutionary events which they model, such as "hybridization networks", "recombination networks" or "duplication-loss-transfer networks".

4.3 Unrooted phylogenetic networks

Let $\mathcal{X}$ be a set of taxa. An *unrooted phylogenetic network* N on $\mathcal{X}$ is any unrooted graph whose leaves are bijectively labeled by the taxa in $\mathcal{X}$. This is a very general definition and there are a number of special types of unrooted phylogenetic networks that are used in practice, the most important of which we consider in detail in this book.

4.3.1 Split networks

The foundation for split networks was laid in [9]. Let $\mathcal{X}$ be a set of taxa and assume that we are given a set of splits $\mathcal{S}$ on $\mathcal{X}$, usually together with a *weighting* that assigns a non-negative weight to each split, which may represent character changes distance, or may also have a more abstract interpretation. If the set of splits $\mathcal{S}$ is compatible, then it can be represented by an unrooted phylogenetic tree and each edge in the tree corresponds to exactly one of the splits. More generally, $\mathcal{S}$ can always be represented by a *split network*, which is an unrooted phylogenetic network with the property that every split S in $\mathcal{S}$ is represented by an array of parallel edges in N (see Section 5.5). An example is shown in Figure 4.3.

A split network N can be obtained from a number of different types of data. To be more precise, the algorithms mentioned below do not compute a split network directly, but rather they all compute a set of weighted splits $\mathcal{S}$. A split network N

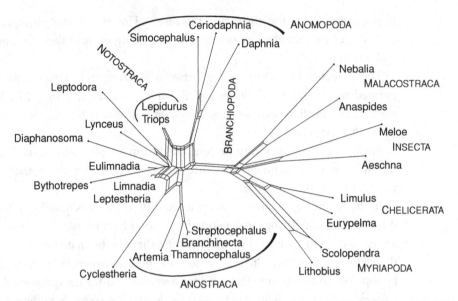

Figure 4.3   A split network on 25 species of Branchiopoda and outgroups, computed from 18S rDNA sequences using the neighbor-net algorithm, as reported in [239]. The authors compare this network to a maximum parsimony tree for the same dataset and discuss how the network exhibits conflicting signals that are not represented in the tree. Reprinted from BMC Evolutionary Biology 7:147 (2007) under the Creative Commons Attribution License.

is then computed using the *convex hull algorithm* or *circular network algorithm*, as described in Chapter 7. All splits-based algorithms discussed in this book are implemented in the program SplitsTree [125] (see Section 14.1).

Split networks from distances

There exist a number of methods for computing a set of weighted splits for a given distance matrix D on $\mathcal{X}$. The two most important are the *split decomposition method* [9] and the *neighbor-net method* [32], which are both described in Chapter 10.

The split decomposition method takes as input a distance matrix D on $\mathcal{X}$ and produces as output a set of weighted splits S on $\mathcal{X}$ that is *weakly compatible*, a property that ensures that the corresponding split network will not be too complicated (see Section 5.8). In practice, the split decomposition method is a very conservative method, in the sense that a split will only be present in the output if there is global support for it in the given dataset. For large or diverse datasets, the method tends to exhibit very low resolution and thus its use is limited to small datasets of less than 100 taxa, say.

The neighbor-net method takes as input a distance matrix D on $\mathcal{X}$ and produces as output a set of weighted splits S on $\mathcal{X}$ that is *circular*, which implies that it can be represented by an *outer-planar* split network (see Section 5.7), if used together with the circular network algorithm (see Section 7.21). The neighbor-net method is more popular than the split decomposition method because it is less conservative and so does not lose resolution on larger datasets. Moreover, the fact that the output of the method can always be represented by a planar split network and is thus easy to visualize adds to its attraction, see Figure 4.3.

Both network methods have the nice property that they produce a tree when given tree-like data.

Split networks from trees

Let $T = (T_1, \ldots, T_k)$ be a collection of unrooted phylogenetic trees on $\mathcal{X}$. These might be different gene trees, or trees for the same gene computed using different methods, or a set of trees obtained in a Bayesian analysis, for example. In Section 3.17, we discussed how to compute a strict consensus tree or majority consensus tree using the set of $\mathcal{S}_{strict}(T)$ or $\mathcal{S}_{majority}(T)$ of strict- or majority-consensus splits, respectively.

The set of majority-consensus splits is defined as the set of all splits that are present in more than 50 % of the input trees. By lowering the threshold to a proportion p of 50 % or less, one obtains a set of splits $\mathcal{S}_p(T)$ that will not necessarily be compatible. The split network N associated with $\mathcal{S}_p(T)$ is called a *consensus split network* and can be used to visualize conflicting signals in a set of trees [113], see Section 11.1.

In practice, in a collection of gene trees, the set of taxa that occurs in each tree will often differ between trees, simply because some gene sequences may not be not available for all taxa. To address this, methods have been developed to compute a *super split network* for a given set of unrooted phylogenetic trees T on overlapping, but non-identical taxon sets [126, 243], see Section 11.2. An application is shown in Figure 11.3.

Split networks from sequences

Assume that we are given a multiple sequence alignment M on $\mathcal{X}$. Any column in the alignment that contains exactly two different character-states defines a split of the taxon set, and the set of splits $\mathcal{S}(M)$ obtainable in this way can be represented by a split network N, computed using the convex hull algorithm.

If we label each edge of the network N by the columns in the alignment M that correspond to the split represented by the edge, then the resulting split network is called a *median network* [11]. This construction is suitable for datasets that have very few differences in them. Hence, median networks are mainly used in phylogeography and population studies. Because parallel mutations can lead to complicated structures in such a network, the concept of a *reduced* median network was also introduced [11], in which one attempts to simplify the network by

■ Northeast Asia domestic pig   ■ Domestic pig in region MDYZ   ■ Domestic pig in South China   ☐ Other
▨ Northeast Asia wild boar   ▨ Wild boar in region MDYZ   ▨ Wild boar in South China   ▨ Feral pigs
■ Domestic pig in region UMYR   ▨ Domestic pig in the Mekong region ■ Domestic pig in region URYZ ▨ Japanese domestic pig and ancient DNA
■ Domestic pig in region DRYR   ■ Wild boar in the Mekong region ▨ Wild boar in region URYZ   • Coalescent root type of haplogroup D1

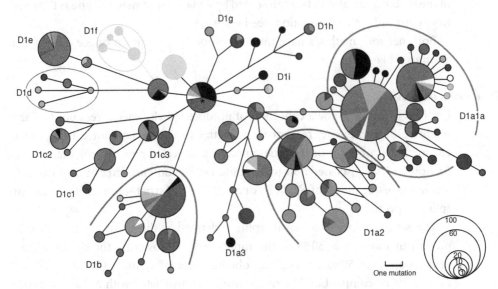

Figure 4.4 A reduced median network reported for a set of over 700 pigs, wild boar and other closely related animals sampled across China, South East Asia and India, based on mtDNA [245]. This network was constructed to investigate the question whether pigs were independently domesticated in multiple places. Reprinted from Genome Biology 8:R245 (2007) under the Creative Commons Attribution License.

postulating appropriate parallel mutation events. An example of a reduced median network is shown in Figure 4.4.

Split networks from quartets

Mathematicians are interested in developing methods that infer a phylogenetic tree or network from basic building blocks. In the computation of an unrooted tree or split network, these are phylogenetic trees on sets of four taxa, sometimes called *quartet trees*. One such method is the *quartet-net method*, or *QNet*, for short [91]. It takes as input a set $\mathcal{Q}$ of weighted quartet topologies on $\mathcal{X}$ and produces as output a set of weighted splits $\mathcal{S}$ on $\mathcal{X}$ that is circular, and thus can be represented by an outer-planar split network. This method is described in Section 12.3.

4.3.2 Quasi-median networks

As we mentioned in the previous section, a median network can be used to visualize a set of binary characters on a set of taxa $\mathcal{X}$. The concept of a *quasi-median network* was introduced as a type of network that can be used to represent multi-state characters.

A quasi-median network is a generalization of the concept of a split network and is discussed in detail in Section 9.5. In general the quasi-median network for a multiple sequence alignment M of DNA sequences on $\mathcal{X}$ is too large and too complicated to be of practical interest. At the other extreme, a *minimum spanning network* can be used to represent the differences between the sequences in M. This type of network is often of limited interest, too, because it contains one node per taxon and no additional nodes (see Figure 10.1 for an application).

The *median-joining* algorithm [10] constructs an informative subnetwork of the full quasi-median network, guided by the minimum spanning network, thus overcoming the drawbacks of both approaches (see Section 9.6). The median-joining method is best suited for very closely related sequences that have evolved without recombinations and is widely used in phylogeography and population studies, usually based on mtDNA or the Y chromosome. An application is shown in Figure 9.12.

Implementations of the median network and median-joining algorithms are provided by the programs Network and SplitsTree (see Chapter 14).

4.3.3 Other types

There are a number of other types of unrooted phylogenetic networks. We briefly describe two of them.

Haplotype networks

A *haplotype network* is an unrooted phylogenetic network in which nodes represent different haplotypes within a group of (usually very closely related) taxa and edges join those sequences or haplotypes that are very similar. The edges are usually labeled by the positions at which the joined haplotypes differ.

Both the median network computation and the median-joining algorithm can be used to compute a haplotype network. Another popular approach is the TCS approach [233] (see Section 14.3). It is based on the concept of *statistical parsimony* and aims at producing a haplotype network in which two haplotypes are joined by an edge if and only if a quantity called the "probability of parsimony" exceeds 95% for the edge.

Reticulograms

A reticulogram is an unrooted phylogenetic tree to which a set of auxiliary edges has been added. A reticulogram is obtained from a distance matrix D on $\mathcal{X}$ using the T-Rex software, which first computes a phylogenetic tree on $\mathcal{X}$ (using a method such as neighbor-joining) and then repeatedly adds *short-cut* edges to the graph until the distances between taxa in the graph shows a good fit to the distances in the original input matrix D [171], see Section 10.5.

4.4 Rooted phylogenetic networks

Let $\mathcal{X}$ be a set of taxa. A *rooted phylogenetic network N* on $\mathcal{X}$ is a rooted DAG whose set of leaves is bijectively labeled by the taxa in $\mathcal{X}$. Any node of indegree ≥ 2 is called a *reticulate* node and all others are called *tree* nodes. Any edge leading to a reticulate node is call a *reticulate* edge and all others are called *tree* edges. This is a very general definition and additional requirements can be made, for example, that the network is *bicombining*, in other words, that all reticulate nodes have indegree 2.

How to interpret such a rooted phylogenetic network mathematically? Perhaps the most important feature of a rooted phylogenetic tree or network is the set of clusters that the network represents, as clusters suggest putative monophyletic groups and thus provide hypotheses about the evolutionary relatedness of the taxa under consideration. Hence, in this book, we treat the calculation of rooted phylogenetic networks in a "cluster-centric" manner and usually interpret rooted phylogenetic networks as representations of sets of clusters.

Exactly how does a rooted phylogenetic network N on $\mathcal{X}$ *represent* a cluster? There are two different answers to this question. We say that a network N represents a given cluster C on $\mathcal{X}$ in the *hardwired* sense, if there exists a tree edge e in N such that the set of labels of leaves below e equals C. Alternatively, we say that N represents C in the *softwired* sense, if there exists a rooted phylogenetic tree T that is contained in N and represents C (in the hardwired sense). This is a very important distinction and we discuss it in detail in Section 6.8.

4.4.1 Clusters and networks

Assume that we are given a set of clusters C on $\mathcal{X}$. A *cluster network N* for C is a rooted phylogenetic network that represents the set of clusters on $\mathcal{X}$ in the *hardwired* sense [121]. It can be computed efficiently using the cluster-popping algorithm (see Section 6.4). The number of edges that it contains is at most quadratic in the number of given clusters. A cluster network is an abstract phylogenetic network that can be used, for example, to provide a combined visualization of a whole set of rooted phylogenetic trees. In Section 6.7 we show that a cluster network N that represents all clusters of a given set of trees also contains all the trees themselves, if they are bifurcating, and otherwise it contains resolutions of them.

A number of new methods aim at constructing a rooted phylogenetic network N that represents a set of clusters in the *softwired* sense, motivated by the assumption that the set of clusters that a network represents is its most important feature, as argued above. In Chapter 8, we discuss some of these new approaches, which aim at computing *galled trees, galled networks* [131] and *level-k networks* [136]. The latter algorithm shows particular promise of becoming a general tool for computing rooted phylogenetic networks from different types of data.

All cluster-based methods described in this book are implemented in the program Dendroscope [120].

4.4.2 Hybridization networks

Assume that we are given a set of taxa $\mathcal{X}$ that have evolved under a model of evolution that includes both speciation events and descent-with-modification, as usual, and, in addition, hybridization events. Then the evolutionary history of the taxa in $\mathcal{X}$ can be represented by a rooted phylogenetic network N on $\mathcal{X}$ whose tree nodes correspond to speciation events and whose reticulate nodes correspond to putative hybridization events [165]. (Descent-with-modification events, such as mutations, insertions or deletions always occur along the *edges* of a phylogenetic tree or network). A rooted phylogenetic network that is interpreted in this way is called a *hybridization network*, see Figure 4.5.

We may attempt to determine such a hybridization network computationally when given two or more gene trees on $\mathcal{X}$ whose topologies differ significantly and we suspect that these differences are due to hybridization events. The corresponding computational problem can be formulated as follows. Given a set $\mathcal{T}$ of two or more rooted phylogenetic trees on $\mathcal{X}$, determine a rooted phylogenetic network N that contains all trees in $\mathcal{T}$ and has a *minimum* number of reticulate nodes. This is known to be a computationally hard problem [27, 241].

In Section 11.5 we discuss an algorithm that addresses the problem of computing a hybridization network N for an input set consisting of exactly two bifurcating trees T_1 and T_2 [24]. In practice, the algorithm appears to run reasonably fast in many cases. No comparable algorithm exists, at present, for solving the problem on more than two input trees.

4.4.3 Recombination networks

Assume that we are given a set of taxa $\mathcal{X}$ that have evolved under a model of evolution that includes both speciation events and descent-with-modification, as usual, and, in addition, recombination events. Then, the evolutionary history of the taxa in $\mathcal{X}$ can be represented by a rooted phylogenetic network N on $\mathcal{X}$ whose tree nodes correspond to speciation events and whose reticulate nodes correspond to recombination events. In addition, we require that the following two labellings are given [89, 98, 128, 224]:

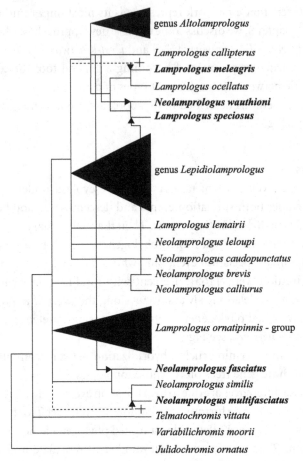

Figure 4.5   A rooted phylogenetic network on fish reported in [153]. In this hybridization network, black lines and triangles indicate a rooted phylogenetic tree inferred on AFLP data from the nuclear genome. Vertical arrows indicate suspected cases of hybridization and dashed lines indicate putative missing lineages, inferred by a comparison to phylogenetic trees based on mitochondrial genes. Reprinted from BMC Evolutionary Biology 7:7 (2009) under the Creative Commons Attribution License.

(i) a labeling of all nodes by sequences, and

(ii) a labeling of all tree edges by positions in the sequences at which mutations occur.

These labellings must be compatible in the sense that the sequences assigned to tree nodes of the network differ exactly by the indicated mutations, while the sequences assigned to reticulate nodes must be obtainable from the sequences assigned to the parents nodes by suitable recombinations, see Definition 9.8.1. A rooted phylogenetic network that is augmented and interpreted in this way is called a *recombination network*.

In this book, we describe two different approaches to the problem of computing a recombination network. An early method [106] is based on the idea of assigning a rooted phylogenetic tree to each position of a given multiple sequence alignment M on $\mathcal{X}$ in the most parsimonious way, see Section 9.8.1.

More recently, Gusfield and colleagues have established a different approach [98]. To obtain a problem that is computationally tractable, they restrict their attention to recombination networks that have the *galled tree* property. The approach is described in detail in Section 9.9. An application is shown in Figure 9.22.

Under the "coalescent-with-recombination" model of population genetics, a description of the history of n sampled sequences going backward in time gives rise to a graph that is called an *ancestral recombination graph* (ARG) [90, 108]. This graph is used to perform statistical analysis of the inheritance and prevalence of genes in populations and its exact topology is usually of little interest. Ancestral recombination graphs are beyond the scope of this book.

4.4.4 DLT networks

Assume that we are given a set of taxa $\mathcal{X}$ that have evolved under a model of evolution that includes both speciation events and descent-with-modification, as usual, and, in addition, gene duplication, loss and horizontal-transfer events [56, 197]. An example is shown in Figure 4.6.

The associated computational problem can be formulated as follows: Given a gene tree T and a corresponding species tree T_{sp}, reconcile all differences between the two trees by postulating an appropriate *duplication–loss–transfer* (DLT) scenario. Such a scenario provides a mapping of the gene tree onto the species tree that implies certain duplication, loss and transfer events. In Section 11.8 we present an approach that addresses this problem and seeks to minimize the number of implied duplication and transfer events [100].

4.4.5 Other types

There are a number of other types of rooted phylogenetic networks. We now briefly discuss three of them.

Reassortment networks

Many viruses are organized into segments of sequence and evolve both by descent-with-modification and also by *reassortment*, a process by which viruses that have co-infected a host exchange segments of their genomes. Reassortment is an important mechanism. For example, a possible route to infection of humans by avian strains of influenza A is that swine get co-infected by avian and human viruses and

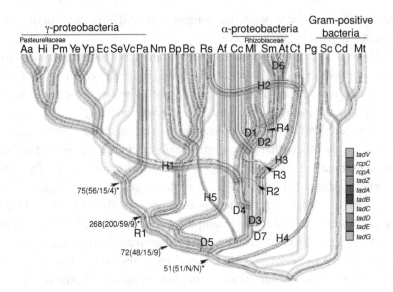

Figure 4.6   A rooted phylogenetic network reported in [197]. The aim is to provide a plausible evolutionary scenario for the tad (tight adherence) locus for colonization of surfaces and biofilm
formation by the human pathogen *Actinobacillus actinomycetemcomitans*. Here, tubes represent the species taxonomy. Embedded trees represent phylogenies of different genes.
Two-letter abbreviations are used to indicate the names of different bacterial species. Single letters indicate different types of events, namely H, horizontal gene transfer events; R,
recombination and gene replacement; D, duplications. Losses are not indicated for simplicity. Reprinted by permission from Macmillan Publishers Ltd: Nature Genetics 34:193-198
©2003.

these reassort to produce a new virus carrying both avian- and human-adapted
genes [47].

A *reassortment network* is a directed graph in which nodes represent viral isolates
and edges represent the evolutionary history of the viruses, including reassortment
events. Edge weights reflect the edit costs of reassortment and mutation events.
Such a graph is organized in layers that correspond to evolutionary stages such as
the seasons in which the viruses were isolated [21].

Networks from multi-labeled trees

Gene duplication is a common event in evolution and so many genes are present in
multiple copies in a genome. When some taxa are represented by multiple copies
of a gene in a phylogenetic tree, then the tree is called *multi-labeled*. To analyze
the duplication history of a gene, it may be helpful to map such a multi-labeled
phylogenetic tree onto a rooted phylogenetic network so as to see which parts of the
tree are similar and where are there differences [117]. Algorithms for constructing

such a network are discussed in Section 11.7 and an application is presented in Figure 11.16.

Networks from rooted triples

As already mentioned above, mathematicians are interested in developing methods that infer a phylogenetic tree or network from basic building blocks. In the computation of a rooted tree or network, these are rooted phylogenetic trees on three taxa, which are sometimes called *rooted triples*. In this context, the input is a set R of rooted triples on $\mathcal{X}$ and the goal is to compute a rooted phylogenetic network N that contains all the rooted triples in R and is optimal in some sense. One possible optimality criterion is to minimize the *level* of the network N, which is defined as the maximum number of reticulation nodes contained in any biconnected component of the network. We give an introduction to this type of approach in Section 12.1.

4.4.6 Combining view versus transfer view

A rooted phylogenetic network N on $\mathcal{X}$ can be drawn and interpreted in two different ways. In a *combining view* of N, all edges entering a particular reticulate node are rendered and interpreted as reticulate edges. Such a view is appropriate for rooted phylogenetic networks in which the reticulate nodes represent recombinations or hybridizations, say.

In a *transfer view*, for each reticulate node in N exactly one in-edge is drawn as a tree edge and all other in-edges are interpreted as *transfer edges*. This type of view is used, for example, when a rooted phylogenetic network N is to be interpreted as a species tree that has been augmented by some additional edges that represent putative horizontal gene transfer events, as in Figure 4.6. The two types of views are compared in Figure 13.12.

Exercise 4.4.1 (Combining view)   *The hybridization network in Figure 4.5 is shown in a transfer view. Draw it in a combining view.*

4.5 The extended Newick format

As described in Section 3.18, the Newick format is widely used for describing the topology and edge lengths of a rooted phylogenetic tree in a single line of text. The *Newick format* can be extended so as to describe a rooted phylogenetic network N on $\mathcal{X}$. Here we describe the extension that is implemented in the program Dendroscope [120], see also [43]. The main idea is to modify the network N so that it becomes a rooted phylogenetic tree, by cutting all but one in-edge of each

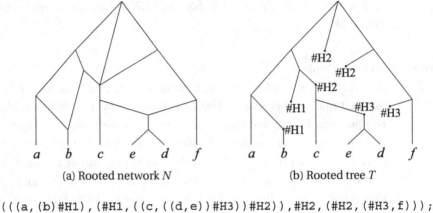

(a) Rooted network N (b) Rooted tree T

```
(((a, (b)#H1),(#H1,((c,((d,e))#H3))#H2)),#H2,(#H2,(#H3,f)));
```

(c) Newick description

Figure 4.7   (a) A rooted phylogenetic network N on $\mathcal{X} = \{a, \ldots, f\}$. (b) A multi-labeled phylogenetic tree T on $\mathcal{X}' = \{a, \ldots, f, \#H1, \#H2, \#H3\}$ that can be used to determine the Newick description of N. (c) The resulting Newick string.

reticulate node, and then labeling the new leaf nodes by additional formal taxa so as to indicate which nodes belong together.

In more detail, assume that the network N has k reticulate nodes $r_1, \ldots, r_k$. Assign a formal label $\#Hi$ to each reticulate node r_i, for all i, and set $\mathcal{X}' = \mathcal{X} \cup \{\#H1, \ldots, \#Hk\}$. For each reticulate node r_i choose one incoming edge e to keep. For all other edges f leading to r_i, replace f by a new edge to a new leaf labeled $\#Hi$. The resulting modification of N is a rooted phylogenetic tree on $\mathcal{X}'$ that is possibly multi-labeled and possibly has internal node labels. It can be written in Newick format as described in Section 3.18.

To construct the rooted phylogenetic network N encoded by a given Newick string, first create the represented rooted phylogenetic tree T on $\mathcal{X}'$ as described in Section 3.18. Then convert the tree into a rooted phylogenetic network by merging any two nodes that are labeled by the same formal taxon $\#Hi$, where i is a number. In Figure 4.7 we show a rooted phylogenetic network N on $\mathcal{X}$, the corresponding multi-labeled tree T on $\mathcal{X}'$ and the resulting Newick string.

How to specify that a rooted phylogenetic network N should be shown in a transfer view? Assume that N has k reticulate nodes $r_1, \ldots, r_k$. For each reticulate node r_i in N, exactly one of the occurrences of the formal label $\#Hi$ must be replaced by $\#\#Hi$ to indicate which in-edge of r_i is to be drawn as a tree edge.

Exercise 4.5.1 (Newick for combining and transfer views)   *Write down the Newick string for the hybridization network shown in Figure 4.5, first for the combining view and then for the transfer view. Enter the two strings into the Dendroscope program (see Section 14.4) and compare the resulting visualizations.*

4.6 Which types of networks are currently used in practice?

For which of the types of phylogenetic networks mentioned above do there exist established and robust methods and software for their computation in evolutionary studies?

For unrooted phylogenetic networks, most of the methods mentioned are routinely used in phylogenetic analysis or phylogeography, in particular, neighbor-net, consensus split (super) networks and median-joining, given distances, trees or sequences, respectively.

This is not the case for rooted phylogenetic networks. While a number of algorithms have been described for computing rooted phylogenetic networks, there are some problems to overcome. First, for many of the algorithms there exist only proof-of-concept implementations that are not designed to be used as tools in real studies. Second, the computational problems are often hard and the algorithms have impractical running times. Third, the calculation of rooted phylogenetic networks must be more closely linked to detailed biological models of reticulate evolution so as to produce more plausible results.

At present, none of the existing methods for computing a rooted phylogenetic method is widely or routinely used as a tool to help understand the evolutionary history of a given set of taxa in terms of mutations, speciations and specific types of reticulate events. While rooted phylogenetic networks are conceptually very appealing, the development of suitable methods for their computation remains a formidable challenge.

Part II

Theory

In this second part of the book we develop the theoretical foundations for phylogenetic networks. First we study splits and unrooted phylogenetic networks (Chapter 5). The mathematical and computational aspects of such networks have been worked out in detail over the last 20 years. The existing algorithms are routinely used in numerous publications in systematic biology. Then we turn to clusters and rooted phylogenetic networks (Chapter 6). Their theory is still under development and widely used methods for the computation of rooted phylogenetic networks have yet to emerge.

5

Splits and unrooted phylogenetic networks

The concept of a *split* plays an important role in the mathematics of phylogeny. It is motivated by the simple, but crucial, observation that every edge *e* in an unrooted phylogenetic tree *T* defines a bipartition of the underlying taxon set $\mathcal{X}$ into two non-empty and disjoint subsets, *A* and *B*, known as a *split*. The splits of an unrooted phylogenetic tree uniquely define the topology of the tree and splits are used, for example, to compare different trees or to compute consensus trees. Any set of splits that is *compatible* corresponds to a phylogenetic tree and so one possible way to generalize from trees to networks is to consider sets of splits that are *incompatible*.

Splits provide the basis of unrooted phylogenetic trees and a large class of unrooted phylogenetic networks, namely *split networks*, just as clusters provide the basic building blocks for rooted phylogenetic trees and networks (see Chapter 6). The foundation for the theory of split networks was laid down in a seminal paper by Bandelt and Dress [9].

5.1 Overview

Figure 5.1 shows the relationships between some of the main concepts introduced in this chapter. The focus is on *splits*, and on *split networks* as an important type of *unrooted phylogenetic networks*. We compare splits to clusters, since the two concepts are closely related to each other. We describe the *Buneman graph* construction that provides a canonical split network for an arbitrary set of splits. We shall see that splits can be *compatible*, *circular* or *weakly compatible*. A nice feature of circular splits is that they can be represented by split networks that are *outer-labeled planar*. Weakly compatible splits are of interest because they can be efficiently calculated using the *split decomposition* algorithm. We also give a brief introduction to *T-theory*, a related mathematical theory.

Tree inference approaches (described in Chapter 3) can be thought of as methods for computing compatible sets of splits. The *split decomposition method* provides

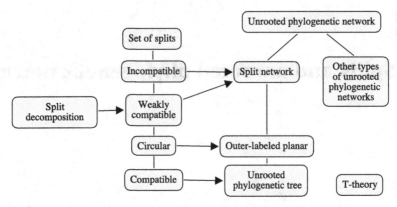

Figure 5.1   Overview of the main concepts introduced in this chapter. On the left, we list the different properties that a set of splits can have, in order of decreasing generality, and in the middle, we list the corresponding types of split networks.

one way of computing a set of incompatible splits on $\mathcal{X}$. Other methods for computing a set of splits that is not necessarily compatible are presented in Part III of the book.

5.2 Splits

Splits and clusters are closely related concepts. While clusters *group* taxa to emphasize their common features, splits *divide* taxa to emphasize their distinctive features. Here is the formal definition of a split:

Definition 5.2.1 (Split)   *A split $S = A \mid B$ is a bipartition of a set of taxa $\mathcal{X}$ into two non-empty subsets A, B with $A \cap B = \emptyset$ and $A \cup B = \mathcal{X}$. We also use the notation $\frac{A}{B}$ and $\frac{B}{A}$ equivalently, to improve readability. A weighted split is a split S that has been assigned a weight $\omega(S) \geq 0$.*

Note that there is no ordering in a split and so $A \mid B$ equals $B \mid A$. Consider a split $S = A \mid B$ on $\mathcal{X}$. We call A and B the two *split parts* of S. The *size* of a split $S = A \mid B$ is defined as the minimal cardinality of the two parts A and B. More formally, $\text{size}(S) = \min\{|A|, |B|\}$. A split of size one is called a *trivial split*. For any taxon $x \in \mathcal{X}$, we use $S(x)$ to denote the split part that contains x and we use $\bar{S}(x)$ to denote the other part.

Let $\mathcal{S}$ be a set of splits on $\mathcal{X}$ and let $\mathcal{X}' \subset \mathcal{X}$ be a subset of taxa. We define the set of splits *induced* by $\mathcal{X}'$, or the *restriction* of $\mathcal{S}$ to the subset $\mathcal{X}'$, as

$$\mathcal{S}|_{\mathcal{X}'} = \left\{ \frac{A \cap \mathcal{X}'}{B \cap \mathcal{X}'} : \frac{A}{B} \in \mathcal{S}, \ A \cap \mathcal{X}' \neq \emptyset \text{ and } B \cap \mathcal{X}' \neq \emptyset \right\}. \tag{5.1}$$

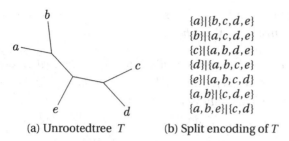

$$\{a\}|\{b,c,d,e\}$$
$$\{b\}|\{a,c,d,e\}$$
$$\{c\}|\{a,b,d,e\}$$
$$\{d\}|\{a,b,c,e\}$$
$$\{e\}|\{a,b,c,d\}$$
$$\{a,b\}|\{c,d,e\}$$
$$\{a,b,e\}|\{c,d\}$$

(a) Unrootedtree T (b) Split encoding of T

Figure 5.2 (a) An unrooted phylogenetic tree T on $\mathcal{X} = \{a, \ldots, e\}$. (b) The seven splits represented by T.

As already mentioned, any edge e of a phylogenetic tree T defines a split of the underlying taxon set $\mathcal{X}$ as follows: Deletion of e produces precisely two subtrees, T' and T'', say; the split defined by e is then the split $A \mid B$, where A and B are the taxon labels occurring in T' and T'' respectively. We use $\sigma_T(e) = A \mid B$ to denote the *split represented by* e. Because every leaf in a phylogenetic tree T is labeled by some taxon, it follows that both A and B are non-empty. Further, as each taxon occurs precisely once in T, it follows that $A \cap B = \emptyset$ and $A \cup B = \mathcal{X}$. If the edges of the tree have lengths or weights, then these can be assigned to the corresponding splits, as well.

If T does not contain any unlabeled nodes of degree two, then any two different edges e and f always represent two different splits, that is, $\sigma_T(e) \neq \sigma_T(f)$ must hold. The only situation in which a phylogenetic tree can contain an unlabeled node of degree 2 is when it has a root with outdegree two. In this case, the two edges e and f that originate at the root ρ give rise to the same split $\sigma_T(e) = \sigma_T(f)$, but to two complementary clusters.

A common construction to avoid this special case at the root ρ is to attach an additional leaf to ρ that is labeled by a special taxon o, which we call a (formal) *outgroup*. Then the root of a phylogenetic tree is specified as the node to which the leaf edge of o attaches. All phylogenetic trees that are discussed in this chapter are unrooted. However, by using this *outgroup trick* much of what we discuss concerning unrooted trees can also be adapted to rooted phylogenetic trees.

Let T be an unrooted phylogenetic tree on $\mathcal{X}$. We define the *split encoding* $\mathcal{S}(T)$ to be the set of all splits represented by T, that is,

$$\mathcal{S}(T) = \{\sigma(e) \mid e \text{ is an edge in } T\}. \tag{5.2}$$

The term *encoding* is justified by the observation that the tree T can be uniquely reconstructed from $\mathcal{S}(T)$, as we see in the next section.

Figure 5.2 shows an unrooted phylogenetic tree that has seven edges and thus gives rise to seven different splits. We can determine all splits associated with any given unrooted phylogenetic tree T in quadratic time using the following algorithm:

Algorithm 5.2.2 (Splits from tree) *The set $\mathcal{S}(T)$ of all splits associated with an unrooted phylogenetic tree T on $\mathcal{X}$ can be computed as follows:*

(i) *Choose a start leaf ρ and assume that all edges of T are directed away from ρ.*
(ii) *In a postorder traversal of T, for each node v compute the set $L(v)$ of taxon labels that are encountered in the subtree rooted at v.*
(iii) *For each edge $e = (u, v)$ of T, add the split $\sigma(e) = \frac{L(v)}{\mathcal{X}-L(v)}$ to $\mathcal{S}(T)$.*

Exercise 5.2.3 (Splits on a tree) *Consider the following set of splits*

$$\mathcal{S} = \left\{ \frac{\{a\}}{\{b,c,d,e\}}, \frac{\{b\}}{\{a,c,d,e\}}, \frac{\{c\}}{\{a,b,d,e\}}, \frac{\{d\}}{\{a,b,c,e\}}, \frac{\{e\}}{\{a,b,c,d\}}, \frac{\{a,b\}}{\{c,d,e\}}, \frac{\{a,e\}}{\{b,c,d\}} \right\}. \tag{5.3}$$

Does there exist an unrooted phylogenetic tree T on $\mathcal{X} = \{a, \ldots, e\}$ whose split encoding equals $\mathcal{S}$?

5.3 Compatibility and incompatibility

Suppose we are given an arbitrary set of splits $\mathcal{S}$ on $\mathcal{X}$. We would like to know whether $\mathcal{S}$ can be represented by some unrooted phylogenetic tree T, that is, does there exist some tree T with $\mathcal{S} = \mathcal{S}(T)$? The answer is given by the concept of compatibility [37].

Definition 5.3.1 (Compatibility of splits) *Two splits $S_1 = A_1 \mid B_1$ and $S_2 = A_2 \mid B_2$ on $\mathcal{X}$ are called* compatible, *if one of the following four possible intersections of their split parts is empty:*

$$A_1 \cap A_2, \ A_1 \cap B_2, \ B_1 \cap A_2 \text{ or } B_1 \cap B_2. \tag{5.4}$$

Otherwise, the two splits are called incompatible. *A set of splits $\mathcal{S}$ is called* compatible *if all pairs of splits in $\mathcal{S}$ are compatible.*

The following result states that the local condition that every pair of splits in $\mathcal{S}$ is compatible suffices to ensure that we have the global property that all splits in $\mathcal{S}$ can be realized simultaneously on a single unrooted phylogenetic tree:

Theorem 5.3.2 (Compatibility Theorem) *Let $\mathcal{S}$ be a set of splits on $\mathcal{X}$ and assume that $\mathcal{S}$ contains all trivial splits on $\mathcal{X}$. There exists a unique unrooted phylogenetic tree T that realizes $\mathcal{S}$, that is, with $\mathcal{S}(T) = \mathcal{S}$, if and only if $\mathcal{S}$ is compatible.*

This result was first formulated and proved in [37]. One way to see that the result holds is simply to apply the outgroup trick and then the result follows from the analogous result for clusters and rooted phylogenetic trees, which is shown in Section 6.4.

$\{a,b,c\}|\{d,e,f\}$,
$\{a,b,e,f\}|\{c,d\}$,
$\{a,b,f\}|\{c,d,e\}$,
$\{a,c\}|\{b,d,e,f\}$

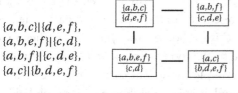

(a) Set of splits   (b) Incompatibility graph

Figure 5.3   (a) A set of four splits S on $\mathcal{X} = \{a, \ldots, f\}$. (b) The corresponding incompatibility graph $IG(S)$, which has four nodes and four edges.

In the above theorem we require that the set of splits S contains all trivial splits on $\mathcal{X}$. This assumption can be dropped if we use $\mathcal{X}$-trees rather than phylogenetic trees in the theorem.

It is often useful to represent the incompatibilities among a set of splits S by a graph that is defined as follows [9] (see Figure 5.3):

Definition 5.3.3 (Incompatibility graph) *The incompatibility graph $IG(S)$ of a set of splits S is the graph (V, E) that has node set $V = S$ and edge set $E = \big\{\{S_1, S_2\} \mid S_1$ and S_2 are incompatible$\big\}$.*

A split $S \in S$ is compatible with all other splits in S, if and only if it is an isolated node in the incompatibility graph. In consequence, a set of splits S is compatible if and only if the incompatibility graph $IG(S)$ has no edges.

5.4 Splits and clusters

Splits are the unrooted counterparts of clusters and the two concepts are closely related. As discussed in Section 3.16 and Chapter 6, clusters can be represented by rooted phylogenetic trees or networks. In this section, we discuss how to transform splits to clusters and vice versa.

To obtain a set of clusters C from a set of splits S on $\mathcal{X}$, we must first choose an *outgroup* taxon $o \in \mathcal{X}$. Then, for each split $S = \frac{A}{B}$ in S, we define the cluster C associated with S to be the split part that does not contain o, that is, we set $C = \bar{S}(o)$. We usually also consider $\{o\}$ as a cluster, to ensure that all trivial clusters are present in C.

Exercise 5.4.1 (Splits to clusters) *Prove that this assignment preserves incompatibilities, in other words, that S is compatible if and only if C is compatible (see Definition 6.2.2).*

Exercise 5.4.2 (Splits to clusters example) *Using c as an outgroup, list all clusters for the set of splits shown in Figure 5.3(a).*

Let S be a set of splits on $\mathcal{X}$. If S is compatible, then there exists an unrooted phylogenetic tree T that represents S. In this case, an alternative method to define the associated set of clusters $\mathcal{C}$ is to choose a root ρ for T and then to let $\mathcal{C}$ be the set of all clusters represented by the rooted version of T. If we choose a node of T to be the root, then there is a simple one-to-one correspondence between the splits and clusters. On the other hand, if the root is chosen so as to subdivide some edge e of T, then the split $S = \frac{A}{B}$ associated with e gives rise to precisely two clusters, namely A and B.

Now let us look at the opposite problem of defining a set of splits S for a given set of clusters $\mathcal{C}$ on $\mathcal{X}$. For a given cluster C, we could simply define the associated split S as C versus the complement of C, that is, as $S = \frac{C}{(\mathcal{X}-C)}$. Unfortunately, this assignment does not preserve incompatibilities. For example, the clusters $\{a, b\}$ and $\{b, c\}$ on $\mathcal{X} = \{a, b, c\}$ are incompatible, whereas the two splits associated in the manner suggested, $\frac{\{a,b\}}{\{c\}}$ and $\frac{\{a\}}{\{b,c\}}$, are not. To address this problem, we add a new (formal) *outgroup* taxon $o \notin \mathcal{X}$ that is then always placed in the split part that contains the complement of a cluster. In other words, for every cluster C we define the associated split as $S = \frac{C}{(\mathcal{X}-C)\cup\{o\}}$ on $\mathcal{X}' = \mathcal{X} \cup \{o\}$.

In the special case that the set of clusters $\mathcal{C}$ is compatible, and thus corresponds to some rooted phylogenetic tree T, we can obtain the set of associated splits directly from T as $S(T)$, after first unrooting the tree.

Exercise 5.4.3 (Number of splits and clusters) *How many different splits and clusters are possible on a set $\mathcal{X}$ of n taxa?*

5.4.1 Optimal compatible subsets

Let S be a set of splits on $\mathcal{X}$. If S is incompatible, then there are two basic computational problems that are sometimes of interest. The first problem is to remove a minimum number of splits such that the remaining set of splits is compatible:

Problem 5.4.4 (Maximum compatibility problem) *Determine a maximum-size subset of splits $S' \subset S$ that is compatible.*

The second problem is to remove a minimum number of taxa such that the set of splits induced on the remaining taxa is compatible:

Problem 5.4.5 (Maximum compatible subset problem) *Determine a maximum-size subset of taxa $\mathcal{X}' \subset \mathcal{X}$ such that the set of splits $S|_{\mathcal{X}'}$ induced on $\mathcal{X}'$ is compatible.*

It follows from the NP-completeness of the two analogous problems formulated for clusters in Section 6.2.1 that these two problems are NP-complete.

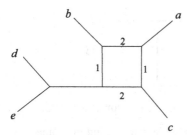

Figure 5.4   A split network N on $\mathcal{X} = \{a, \dots, e\}$. Every leaf edge, and, in this example, also the edge separating the taxa d and e from the rest of the taxa, each represents a split in the same way that they would in an unrooted phylogenetic tree. The two edges labeled 1 together represent the split $\frac{\{a,b\}}{\{c,d,e\}}$, whereas the two edges labeled 2 represent the split $\frac{\{a,c\}}{\{b,d,e\}}$.

5.5 Split networks

As we have seen, any set of compatible splits (containing all trivial splits) corresponds to an unrooted phylogenetic tree on $\mathcal{X}$. In this section, we introduce a mathematical generalization of the concept of an unrooted phylogenetic tree, called a *split network*, which can be used to represent an arbitrary, in particular, incompatible, set of splits.

In a split network, we use one or more edges to represent a split. The set of edges used to represent a given split S has the property that deletion of all these edges produces exactly two connected components, and, as in the case of phylogenetic trees, the two parts of S are given by the sets of taxa that occur as labels of one component, or the other, respectively. For an example, see Figure 5.4.

First we define the concept of a split graph. Think of this as the underlying graph-theoretical concept, just like phylogenetic trees are based on the underlying graph-theoretical concept of a tree, as defined in Chapter 1. A split network N is then obtained from such a graph by specifying a labeling of the leaves by taxa and a labeling of the edges by splits.

Below, we define a split graph as a finite, connected graph, together with an edge coloring whose most important property is that deleting all edges of a fixed color produces precisely two connected components. In a split network, all edges of a given color represent one split S and things are arranged such that the two connected components separated by the edges are each labeled by the taxa of one of the two parts of the split.

To develop the necessary formal concepts, let $G = (V, E)$ be a finite connected graph. Further, let K denote a finite set of labels, which we call *colors*. An *edge coloring* is a mapping $\sigma : E \to K$ that has the property that no two adjacent edges are given the same color. Such a mapping is called *surjective*, if it uses all colors in K.

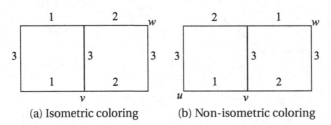

(a) Isometric coloring   (b) Non-isometric coloring

Figure 5.5   (a) A split graph with an isometric coloring of its edges using the colors $K = \{1, 2, 3\}$. All shortest paths between any two nodes, for example v and w, contain the same set of colors. (b) The coloring indicated here is not isometric; for example, one shortest path between v and w has colors 1 and 3, whereas the other one has colors 2 and 3. Additionally, there exists a shortest path connecting the nodes u and w that is not properly colored, as it contains two edges with color 1.

Consider an undirected path,

$$P = (u_0, e_1 = \{u_0, u_1\}, u_1, e_2 = \{u_1, u_2\}, \ldots, e_t = \{u_{t-1}, u_t\}, u_t), \quad (5.5)$$

containing exactly $\text{len}(P) = t$ different edges. We denote the set of colors that occur in P by $\sigma(P)$, defined as

$$\sigma(P) = \{\sigma(e_1), \ldots, \sigma(e_t)\} \quad (5.6)$$

and call P *properly colored*, if all edges in P have different colors, that is, if $|\sigma(P)| = t$. We call an edge coloring σ of G an *isometric coloring*, if for any two nodes, all shortest paths between them are properly colored and use exactly the same set of colors (see Figure 5.5).

Definition 5.5.1 (Split graph)   *A split graph consists of a finite, simple, connected, bipartite graph $G = (V, E)$, together with an edge coloring $\sigma : E \to K$ that is surjective and isometric.*

With this definition, we get the crucial property that deletion of all edges of any given color produces precisely two connected components [62]:

Theorem 5.5.2 (Deletion of split produces two components)   *Let $(G = (V, E), \sigma : E \to K)$ be a split graph. For any color $c \in K$ we have: the graph G_c obtained by deleting all edges of color c consists of precisely two connected components, which we denote by $G_c^0 = (V_c^0, E_c^0)$ and $G_c^1 = (V_c^1, E_c^1)$.*

Proof   For two nodes v and w we use $d(v, w)$ to denote the minimal number of edges in any path from v to w, and we use $\sigma(v, w)$ to denote the set of colors that occur in every shortest path from v to w. Note that σ is an isometric coloring, so $d(v, w) = |\sigma(v, w)|$ for all $v, w \in V$.

First we show that *any* path P from v to w uses all colors from $\sigma(v, w)$, by induction on $\mathrm{len}(P) \geq d(v, w)$.

To start the induction, assume $\mathrm{len}(P) = d(v, w)$. In this case P is a shortest path from v to w and so the claim follows directly from the definition of $\sigma(v, w)$.

Now, assume that $\mathrm{len}(P) > d(v, w)$ and so $P = (v, e_1, \ldots, w', e_t, w)$, for some $t > d(v, w)$. Because G is bipartite, we cannot have $d(v, w) = d(v, w')$ and so we must either have

$$d(v, w) = d(v, w') - 1 \quad \text{or} \quad d(v, w) = d(v, w') + 1. \tag{5.7}$$

In the first case every shortest path from v to w can be extended to a shortest path from v to w', implying

$$\sigma(v, w) \subseteq \sigma(v, w'). \tag{5.8}$$

Now, because $\mathrm{len}(v, e_1, \ldots, e_{t-1}, w') < \mathrm{len}(P)$, we can apply the induction hypothesis and obtain

$$\sigma(v, w) \subseteq \sigma(v, w') \subseteq \sigma(v, e_1, \ldots, e_{t-1}, w') \subseteq \sigma(P), \tag{5.9}$$

as claimed.

In the other case, we have

$$\sigma(v, w) = \sigma(v, w') \cup \{\sigma(e_t)\}. \tag{5.10}$$

Moreover, because $\mathrm{len}(v, e_1, \ldots, e_{t-1}, w') < \mathrm{len}(P)$, we also have

$$\sigma(v, w') \subseteq \sigma(v, e_1, \ldots, e_t, w'). \tag{5.11}$$

Putting these two observations together, we see that

$$\sigma(v, w) = \sigma(v, w') \,\dot{\cup}\, \{\sigma(e_t)\} \subseteq \sigma(v, e_1, \ldots, e_{t-1}, w') \,\dot{\cup}\, \{\sigma(e_t)\} = \sigma(P), \tag{5.12}$$

as claimed.

The proved statement implies that any two nodes v and w are in the same connected component of G_c if and only if $c \notin \sigma(v, w)$ holds. This implies that G_c must have *at least* two components, as σ is surjective. In the remainder of the proof, we show that G_c has *at most* two components.

Consider an edge $e = \{w', w\}$ that has color c and assume that $P = (v, e_1, \ldots, e_t, w')$ is a shortest path from some node v to w', then, as above, there are two possible cases: (1) If $d(v, w) = d(v, w') + 1$, then v and w' must lie in the same component of G_c, because $(v, e_1, \ldots, e_t, w', e, w)$ is a shortest path from v to w, which in turn implies (by the isometric property) that $c = \sigma(e) \notin \{\sigma(e_1), \ldots, \sigma(e_t)\}$ holds. (2) If $d(v, w) = d(v, w') - 1$, then any shortest path

$$S_1 = \{a,b,c\}|\{d,e,f\}$$
$$S_2 = \{a,e,f\}|\{b,c,d\}$$
$$S_3 = \{a,f\}|\{b,c,d,e\}$$
$$S_4 = \{a,b,f\}|\{c,d,e\}$$

(a) Split network N (b) Non-trivial splits   (c) Split $S_2 = \frac{\{a,e,f\}}{\{b,c,d\}}$

Figure 5.6   (a) A split network N representing all trivial splits on $\mathcal{X} = \{a, \ldots, f\}$ and the four non-trivial splits listed in (b). All the edges representing a particular non-trivial split are labeled by that split. However, the labeling of edges by splits is usually omitted, as shown in (c). Edges representing the same split, such as the three edges shown in bold lines representing S_2, are drawn parallel and with the same length.

$(v, e'_1, \ldots, e'_{t-1}, w)$ from v to w gives rise to a shortest path $(e'_1, \ldots, e'_{t-1}, e)$ from v to w'. This implies, as above, that $c = \sigma(e) \notin \{\sigma(e'_1), \ldots, \sigma(e'_{t-1})\} = \sigma(v, w)$ holds.

So, in G_c every node v is connected either to w or to w', but not to both of these nodes. This implies that G_c has at most two components.   □

A split network N on $\mathcal{X}$ is obtained from a split graph by labeling its nodes by a taxon set $\mathcal{X}$ and labeling its edges by a set of splits $\mathcal{S}$ on $\mathcal{X}$ [62]:

Definition 5.5.3 (Split network)   *Let $\mathcal{S}$ be a set of splits on $\mathcal{X}$. A split network $N = (V, E, \sigma, \lambda)$ that represents $\mathcal{S}$ is given by a split graph $(G = (V, E), \sigma : E \to \mathcal{S})$ and a node labeling $\lambda : \mathcal{X} \to V$, with the property that for every split $S = A \mid B$ in $\mathcal{S}$ we have:*

$$A = \bigcup_{v \in V_S^0} \lambda^{-1}(v) \quad and \quad B = \bigcup_{v \in V_S^1} \lambda^{-1}(v), \tag{5.13}$$

(or vice versa), in other words, deletion of all edges of color S produces a graph consisting of precisely two connected components, one containing all nodes labeled with elements of A and the other containing all nodes labeled with elements of B.

Formally, all edges representing a given split S are labeled by that split. In drawings, however, the labeling or coloring of the edges is usually omitted and it is customary, instead, to draw all edges representing the same split as parallel lines of the same length, so as to help make clear which edges belong together, see Figure 5.6. Such a representation is always possible, see Section 13.8.

Note that any unrooted phylogenetic tree T can be regarded as a split network: Define $\sigma : E \to \mathcal{S}$ as $\sigma(e) := \sigma_T(e)$, then it is obvious that deletion of all edges

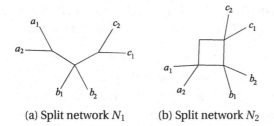

(a) Split network N_1 (b) Split network N_2

Figure 5.7   The two split networks N_1 and N_2 shown in (a) and (b) both represent the same set of splits S, consisting of the two non-trivial splits $\frac{\{a_1,a_2\}}{\{b_1,b_2,c_1,c_2\}}$ and $\frac{\{a_1,a_2,b_1,b_2\}}{\{c_1,c_2\}}$, together with all trivial splits on $\mathcal{X} = \{a_1, a_2, b_1, b_2, c_1, c_2\}$. Because S is compatible, there exists a corresponding split network that is a tree, namely N_1. The split network N_2 also represents S, but is not a tree.

(that is, in fact, only one edge) of color $S = \sigma_T(e)$ produces two connected components which correspond to the two parts of S. So in this sense, split networks are a generalization of unrooted phylogenetic trees. Moreover, we have:

Lemma 5.5.4 (Split networks and compatiblity)   *A set of splits S on $\mathcal{X}$ is compatible if and only if there exists a split network N representing S that is a tree.*

Figure 5.7 demonstrates that the split network associated with a set of splits is not uniquely defined and also that a split network representation of a compatible set of splits need not necessarily be a tree, although a tree representation for a compatible set of splits always exists. We would like to point out, however, that both algorithms described in Chapter 7 for constructing a split network for a given set of splits produce a tree, when run with a compatible set of splits as input.

Exercise 5.5.5 (Draw a split network)   *Consider the set of splits S on $\mathcal{X} = \{a, \ldots, g\}$ given by*

$$\frac{\{a, b\}}{\{c, d, e, f, g\}}, \quad \frac{\{a, b, c\}}{\{d, e, f, g\}}, \quad \frac{\{a, b, e, f, g\}}{\{c, d\}}, \quad \frac{\{a, f, g\}}{\{b, c, d, e\}}, \quad (5.14)$$

and all trivial splits on $\mathcal{X}$. Draw a split network N that represents S.

5.6 The canonical split network

Let S be a set of splits on $\mathcal{X}$. The definition of a split network (Definition 5.5.3) gives no indication of how one might obtain such a split network for S. To address this, in this section we discuss how to define a *canonical split network* N that represents S, which is also called the *Buneman graph* [17].

Definition 5.6.1 (Projection) *Let $S = \{S_1, \ldots, S_k\}$ be a set of splits on $\mathcal{X}$. We call a vector $p = (D_1, \ldots, D_k)$ of length k a projection of S, if it has the two following properties:*

(i) *The i-th component of p consists of one of the two split parts of the i-th split $S_i = \frac{A_i}{B_i}$ in S, that is, we have either $D_i = A_i$ or $D_i = B_i$, for all $i = 1, \ldots, k$.*

(ii) *Any two components of p have a non-empty intersection, that is, we have $D_i \cap D_j \neq \emptyset$ for all $i, j = 1, \ldots, k$.*

The core of a projection $p = (D_1, \ldots, D_k)$ is defined as the set $\overset{\circ}{p} = \bigcap_{i=1}^{k} D_i$ of all such taxa that are contained in all components of p.

We have the following easy result:

Lemma 5.6.2 (Every taxon in exactly one core) *Let S be a set of splits on $\mathcal{X}$. Then every taxon x in $\mathcal{X}$ is contained in the core $\overset{\circ}{p}$ of exactly one projection p of S.*

Proof Let $S = \{S_1, \ldots, S_k\}$ be a set of splits on $\mathcal{X}$. Consider any taxon x in $\mathcal{X}$. For $i = 1 \ldots, k$ let D_i be the split part of split S_i that contains x. Then x is contained in every component of $p = (D_1, \ldots, D_k)$ and so p is a projection and x is contained in $\overset{\circ}{p}$.

Now, assume that there exists a second projection $q \neq p$ whose core also contains x. Then p and q differ on at least one component, say the i-th component. The i-th component of p contains one part of the split S_i and q contains the other part. As the two split parts are disjoint, they cannot both contain x and so x cannot be contained in both cores, implying that such a second projection q does not exist. □

We can now define the canonical split network for a given set of splits:

Definition 5.6.3 (Buneman graph) *Let S be a set of splits on $\mathcal{X}$. The canonical split network or Buneman graph associated with S is the split network $N = (V, E, \sigma, \lambda)$ that is defined as follows [17]:*

(i) *The node set V is given by the set of all projections $\mathcal{P}(S)$ of S.*

(ii) *Any two nodes p and p' are connected by an edge e of color $\sigma(e) = i$ in E if and only if p and p' differ precisely in their i-th component.*

(iii) *For each taxon x in $\mathcal{X}$, we set $\lambda(x) = p$, where p is the unique projection whose core contains x, i.e. with $x \in \overset{\circ}{p}$.*

As an example, consider the set of splits

$$S = \{S_1, S_2, S_3, S_4\} = \left\{ \frac{\{a\}}{\{b, c, d, e\}}, \frac{\{a, b, c\}}{\{d, e\}}, \frac{\{a, b, d\}}{\{c, e\}}, \frac{\{a, c, e\}}{\{b, d\}} \right\} \quad (5.15)$$

Figure 5.8   The canonical split network (Buneman graph) N for a set of splits $\mathcal{S}$ on $\mathcal{X} = \{a, b, c, d, e\}$ described in the text.

on the taxon set $\mathcal{X} = \{a, b, c, d, e\}$. There are exactly seven projections for $\mathcal{S}$ and these provide the set of nodes for the canonical split network N associated with $\mathcal{S}$:

$$
\begin{aligned}
p_1 &= (& \{a\}, & \quad \{a, b, c\}, & \quad \{a, b, d\}, & \quad \{a, c, e\} &), \\
p_2 &= (& \{b, c, d, e\}, & \quad \{a, b, c\}, & \quad \{a, b, d\}, & \quad \{a, c, e\} &), \\
p_3 &= (& \{b, c, d, e\}, & \quad \{a, b, c\}, & \quad \{a, b, d\}, & \quad \{b, d\} &), \\
p_4 &= (& \{b, c, d, e\}, & \quad \{a, b, c\}, & \quad \{c, e\}, & \quad \{a, c, e\} &), \\
p_5 &= (& \{b, c, d, e\}, & \quad \{d, e\}, & \quad \{a, b, d\}, & \quad \{a, c, e\} &), \\
p_6 &= (& \{b, c, d, e\}, & \quad \{d, e\}, & \quad \{a, b, d\}, & \quad \{b, d\} &), \quad \text{and} \\
p_7 &= (& \{b, c, d, e\}, & \quad \{d, e\}, & \quad \{c, e\}, & \quad \{a, c, e\} &).
\end{aligned}
\tag{5.16}
$$

The cores of these projections are: $\mathring{p}_1 = \{a\}$, $\mathring{p}_2 = \emptyset$, $\mathring{p}_3 = \{b\}$, $\mathring{p}_4 = \{c\}$, $\mathring{p}_5 = \emptyset$, $\mathring{p}_6 = \{d\}$, and $\mathring{p}_7 = \{e\}$. Thus, the mapping of taxa to nodes is $\lambda(a) = p_1, \lambda(b) = p_3$, $\lambda(c) = p_4$, $\lambda(d) = p_6$ and $\lambda(e) = p_7$. The set of edges is given by all pairs of nodes that disagree on precisely one component: $e_1 = \{p_1, p_2\}$, $e_2 = \{p_2, p_3\}$, $e_3 = \{p_2, p_4\}$, $e_4 = \{p_2, p_5\}$, $e_5 = \{p_3, p_6\}$, $e_6 = \{p_4, p_7\}$, $e_7 = \{p_5, p_6\}$ and $e_8 = \{p_5, p_7\}$. Each edge is colored by the component on which the two incident nodes disagree on. In this example, the coloring of edges is $\sigma(e_1) = 1, \sigma(e_2) = \sigma(e_7) = 4$, $\sigma(e_3) = \sigma(e_8) = 3$ and $\sigma(e_4) = \sigma(e_5) = \sigma(e_6) = 2$. The resulting split network is shown in Figure 5.8.

Exercise 5.6.4 (Buneman graph and compatibility)   *Show that the Buneman graph of a compatible set of splits $\mathcal{S}$ is an unrooted phylogenetic tree.*

Lemma 5.6.5 (Buneman graph is a split network)   *Let $\mathcal{S}$ be a set of splits on $\mathcal{X}$. The Buneman graph for $\mathcal{S}$ is a split network for $\mathcal{S}$.*

Proof   We must show that the Buneman graph is a split graph as defined in Definition 5.5.1 and then the claim follows with the help of Theorem 5.5.2.

Let $\mathcal{S} = \{S_1, \ldots, S_k\}$ be a set of splits on $\mathcal{X}$ and let N be the corresponding Buneman graph.

We first show that the edge coloring on N is surjective. Consider any number $i = 1, \ldots, k$. We must show that N contains an edge that has color i, that is, there exist two projections p and q that differ only on their i-th component. Let p_0 be some projection of S. Let $A = \{D_j \mid D_j \subset D_i\}$ be the set of components of p_0 that are subsets of D_i, not including D_i, and let $B = \{D_j \mid D_j \not\subset D_i\}$ be the set of components that are not subsets of D_i. Define $\bar{A} = \{\mathcal{X} - D_j \mid D_j \in A\}$ as the set of complements of all the components in A. By definition of A and B we have the following: Each set in $\bar{A}$ intersects every component in B. Moreover, each set in $\bar{A}$ intersects all other sets in $\bar{A}$. Finally, D_i and $\mathcal{X} - D_i$ intersects all sets in $\bar{A}$ and B. Hence, the vector p obtained by replacing each component of p_0 that is listed in A by its complement is a projection. Similarly, the vector q obtained by replacing each component of p_0 that S in the set $A \cup \{D_i\}$ by its complement is a projection. As p and q differ only by their i-th component, we are done.

Our goal is to show that the the graph is connected, bipartite and the edge coloring is isometric, by proving the following statement: For any two nodes p and q that differ on exactly d components, the length of any shortest path P (and such a path always exists) from p to q is d and each of the indices $i_1, \ldots, i_d$ on which p and q differ occurs as an edge color in P.

The proof is by induction on the number d of components on which p and q differ. If $d = 1$, then there is indeed a path from p to q consisting of a single edge e and that edge is colored by the index of the component on which the two nodes differ.

Now, assume that $d > 1$ and that any two nodes that differ on $d - 1$ components are connected by a path of length $d - 1$ colored by the indices of the components on which the two nodes differ. Let p and q be two nodes that differ on exactly d components.

Let D_i be a split part of some split S_i that occurs as the i-th component of p but not of q, and assume that D_i has minimum size with this property. Consider the vector p' that is obtained from p by by replacing D_i by the complementary split part $\mathcal{X} - D_i$.

As D_i is minimal, none of the other components of p can be completely contained in D_i and so they all must contain at least one element that is not contained in D_i. This implies that p' is a valid projection and hence a node of the Buneman graph.

Because p and q disagree on d components and because $\mathcal{X} - D_i$ is the i-th component of q, it follows that p' and q disagree on only $d - 1$ components. By assumption, all shortest paths from p' to q have length $d - 1$ and the edges are colored by the $d - 1$ indices of the components on which the two nodes differ, none colored by the index of D_i since p' and q do not differ on D_i.

Because p and p' differ by only one component, they are connected by an edge in N, which can be used to extend any shortest path from p' to q by one edge to obtain a path from p to q that has d edges and d different colors. Hence, there *exists* a path from v to w with the desired properties.

It remains to be shown that *all* shortest paths from v to w have the desired properties. To this end, consider an arbitrary path P of length d from p to q and let p' be the second node in the path. By assumption, p' is connected to q by a path containing $d - 1$ edges, each colored by a different index and none colored by the index of the component on which p and p' differ. Hence, P has d different colors.

Let us demonstrate now that N is bipartite. Chose a fixed taxon x in $\mathcal{X}$. The *parity* of a projection p is set to 0, if the number of components of p that contain the taxon x is even, and is set to 1, otherwise. Using this measure, we can partition the set of nodes V of the Buneman graph in two disjoint subsets V_1 and V_2 such that for every edge $e \in E$ one of the endpoints lies in V_1 and the other endpoint lies in V_2, in the following way: Let V_1 and V_2 be the set of nodes with odd and even parity, respectively. Then, for each pair of nodes u, v in V_1 (or in V_2), there does not exist an edge connecting them since they differ on more than one component. This completes the proof that the Buneman graph is a split graph as defined in Definition 5.5.1.

To prove that the Buneman graph is a split network, we need to show for the node labeling $\lambda : \mathcal{X} \to V$ and for every split $S = A \mid B$ in $\mathcal{S}$ that the deletion of all edges of color S produces a graph consisting of precisely two connected components, one containing all nodes labeled with elements of A and the other containing all nodes labeled with elements of B (see Definition 5.5.1). Theorem 5.5.2 ensures that, for each split S of a split set $\mathcal{S}$ represented by N, by deleting all edges of color S we obtain a graph N_S consisting of precisely two connected components N_S^0 and N_S^1. We need to prove that all nodes labeled with elements of A (or of B) are contained in N_S^0 (or N_S^1, respectively).

In the proof of Theorem 5.5.2, we showed that any two nodes p and q are contained in the same connected component of N_S if and only if $S \notin \sigma(p, q)$ holds. For every pair of nodes p and q such that $\lambda^{-1}(p) = a$ and $\lambda^{-1}(q) = b$, with a in A and b in B, it holds that p and q differ on the split S. As demonstrated above, this implies that $S \in \sigma(p, q)$, so p and q are not in the same connected component of N_S^0 and N_S^1. Since this holds for each pair of nodes p and q in V and for every split, this concludes the proof. $\qquad \square$

Let $\mathcal{S}$ be a set of splits on $\mathcal{X}$. How many nodes and edges might the canonical split network contain? If the incompatibility graph $IG(\mathcal{S})$ contains a clique of size k, then any one of the 2^k possible choices of split parts gives rise to a node in the network and thus the number of nodes and edges of the network is exponential in the number of splits in the worst case.

In Chapter 7 we discuss how to compute the Buneman graph using the convex hull algorithm.

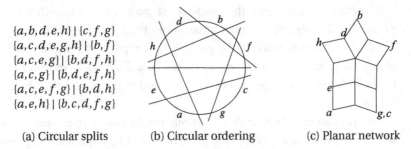

$$\{a,b,d,e,h\}\mid\{c,f,g\}$$
$$\{a,c,d,e,g,h\}\mid\{b,f\}$$
$$\{a,c,e,g\}\mid\{b,d,f,h\}$$
$$\{a,c,g\}\mid\{b,d,e,f,h\}$$
$$\{a,c,e,f,g\}\mid\{b,d,h\}$$
$$\{a,e,h\}\mid\{b,c,d,f,g\}$$

(a) Circular splits   (b) Circular ordering   (c) Planar network

Figure 5.9   (a) A set of six circular splits S on $\mathcal{X} = \{a, b, \ldots, h\}$. (b) An arrangement of the taxa around a circle such that every split $S = A \mid B \in S$ can be realized by a straight line through the circle that separates the two split parts A and B. A circular ordering is given by (a, g, c, f, b, d, h, e). (c) An outer-labeled planar split network representing S.

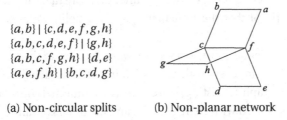

$$\{a,b\}\mid\{c,d,e,f,g,h\}$$
$$\{a,b,c,d,e,f\}\mid\{g,h\}$$
$$\{a,b,c,f,g,h\}\mid\{d,e\}$$
$$\{a,e,f,h\}\mid\{b,c,d,g\}$$

(a) Non-circular splits   (b) Non-planar network

Figure 5.10   (a) A set of four non-circular splits S on $\mathcal{X} = \{a, b, \ldots, h\}$. (b) A non-planar split network representing S.

5.7 Circular splits and planar split networks

One practical problem that arises when working with split networks is that the networks can be very complicated and thus difficult to visualize in a comprehensible way. Hence, a number of restricted classes of sets of splits have been introduced in an attempt to avoid overly complicated networks. The two most important are *circular* splits, which are the focus of this section, and *weakly compatible* splits, which we introduce in the next section.

Informally, a set of splits S on $\mathcal{X}$ is called *circular*, if the taxa in $\mathcal{X}$ can be placed around a circle in such a way that each split $S = \frac{A}{B}$ can be realized by a line through the circle that separates the plane into two half-planes, one containing all taxa in A and the other containing all taxa in B (see Figure 5.9). An example of a non-circular set of splits and the corresponding split network is shown in Figure 5.10. More formally [9]:

Definition 5.7.1 (Circular splits)   *A set of splits S on $\mathcal{X}$ is called* circular, *if there exists a linear ordering* $(x_1, \ldots, x_n)$ *of the elements of $\mathcal{X}$ for S such that each split*

$$
\begin{array}{cccccccc}
a & b & c & d & e & f & g & h \\
0 & 0 & 1 & 0 & 0 & 1 & 1 & 0 \\
0 & 1 & 0 & 0 & 0 & 1 & 0 & 0 \\
0 & 1 & 0 & 1 & 0 & 1 & 0 & 1 \\
0 & 1 & 0 & 1 & 1 & 1 & 0 & 1 \\
0 & 1 & 0 & 1 & 0 & 0 & 0 & 1 \\
0 & 1 & 1 & 1 & 0 & 1 & 1 & 0
\end{array}
\qquad
\begin{array}{cccccccc}
a & e & h & d & b & f & c & g \\
0 & 0 & 0 & 0 & 0 & 1 & 1 & 1 \\
0 & 0 & 0 & 0 & 1 & 1 & 0 & 0 \\
0 & 0 & 1 & 1 & 1 & 1 & 0 & 0 \\
0 & 1 & 1 & 1 & 1 & 1 & 0 & 0 \\
0 & 0 & 1 & 1 & 1 & 0 & 0 & 0 \\
0 & 0 & 0 & 1 & 1 & 1 & 1 & 1
\end{array}
$$

(a) Input matrix   (b) Permuted matrix

Figure 5.11 (a) An input matrix M for the Consecutive Ones problem that corresponds to the set of splits shown in Figure 5.9(a). (b) A solution obtained by permuting the columns of M in the order suggested by Figure 5.9(b).

$S \in \mathcal{S}$ *has the form*

$$
S = \frac{\{x_p, x_{p+1}, \ldots, x_q\}}{\mathcal{X} - \{x_p, x_{p+1}, \ldots, x_q\}},
\tag{5.17}
$$

for appropriately chosen $1 < p \le q \le n$.

We call such an ordering $(x_1, \ldots, x_n)$ a *circular ordering* for $\mathcal{S}$, as it holds that $(x_1, \ldots, x_n)$ is a circular ordering for $\mathcal{S}$ if and only if $(x_n, x_{n-1}, \ldots, x_1)$ and $(x_2, x_3, \ldots, x_n, x_1)$ both are. How to determine whether a set of splits $\mathcal{S}$ is circular? This is equivalent to a well-known problem in computer science:

Problem 5.7.2 (Consecutive Ones problem) *Let M be a binary matrix. Does there exist a permutation of the columns of the matrix M such that in each row, all ones in the row occur in a single consecutive run?*

It is straightforward to translate the problem of determining whether $\mathcal{S}$ is circular into an instance of the Consecutive Ones problem. Simply define a binary matrix M in which each row r corresponds to some split $S \in \mathcal{S}$ and every column c corresponds to some taxon $x \in \mathcal{X}$. Then set $M(r, c) = 1$ if and only if S separates x from the taxon represented by the first column of M.

Exercise 5.7.3 (Circular ordering and Consecutive Ones) *Prove the following statement: There exists a circular ordering of $\mathcal{X}$ for $\mathcal{S}$ if and only if a solution of the Consecutive Ones problem exists for M. Hint: construct a binary matrix M whose rows represent splits and whose columns represent taxa (see Figure 5.11).*

The Consecutive Ones problem can be solved in linear time [23]. However, if no circular ordering exists for a given set of splits $\mathcal{S}$ on $\mathcal{X}$, then one might try to determine a solution that is as good as possible. For our purposes, an appropriate optimization goal is to find an ordering of $\mathcal{X}$ that minimizes the number of runs of ones in the matrix M. This problem, known as the *Optimal Consecutive Ones* problem, is NP-hard.

A useful heuristic for finding an optimal circular ordering for a set of non-circular splits S on $\mathcal{X}$ is to define a distance matrix D on $\mathcal{X}$ by setting the distance between any two taxa a and b equal to the sum of all weights of the splits that separate them. This distance matrix D is then provided as input to the neighbor-net algorithm, described in Section 10.4, which outputs an ordering of $\mathcal{X}$, that is guaranteed to be circular, whenever the given input set S is circular [32].

Circular split systems are of particular interest because they can always be represented graphically in an appealing way, namely by a split network that is drawn in the plane without any edges crossing and with all labeled nodes appearing around the outside of the network (see Figure 5.9c), conforming to the following:

Definition 5.7.4 (Outer-labeled planar) *Let G be a graph in which some of the nodes are labeled. We call G outer-labeled planar, if there exists a drawing of G in the plane such that no two edges intersect and all labeled nodes lie on the outside of the graph.*

With this definition we have [62]:

Theorem 5.7.5 (Circular implies outer-labeled planar) *A set of splits S on $\mathcal{X} = \{x_1, \ldots, x_n\}$ is circular if and only if it can be represented by a split network N that is outer-labeled planar.*

Proof If S is circular, then we can position the elements of $\mathcal{X} = \{x_1, \ldots, x_n\}$ around a circle in the plane in such a way that every split $S \in \mathcal{S}$ is induced by a chord of the circle. Applying de Bruijn's "dualization principle" [29] to the set of these chords produces a "zonotopal graph" that can be labeled so as to form an outer-labeled planar split network for S [20].

Now assume that we are given a split network N that is embedded in the plane such that all labels occur on the outer boundary of the network. A circular ordering is given by the order in which the taxa are encountered when traveling once around the outside of the embedded network.   □

As shown in Section 10.4, the neighbor-net algorithm, which is a widely used method for computing a set of incompatible splits from a distance matrix, always produces a circular set of splits and thus can always be represented by an outer-labeled planar split network.

Exercise 5.7.6 (Splits from one or two trees) *Prove that the set of splits obtained from a single unrooted phylogenetic tree is always circular. Construct an example that shows that the set of splits taken from two unrooted phylogenetic trees is not necessarily circular.*

As we saw above, the number of nodes and edges of a split network N representing a general set of splits can be exponential in the number of splits. However, a set of circular splits can be represented by a split network that is of quadratic size only:

Lemma 5.7.7 (Circular splits have quadratic-size network) *If S is a set of m circular splits on $\mathcal{X}$, then the number of nodes and edges in an outer-labeled planar split network N representing S is at most quadratic in m.*

Proof Because S is circular, we can arrange all taxa on a circle in such a way that every split $S = \frac{A}{B}$ corresponds to some line l_S that passes through the circle and separates all taxa in A from all taxa in B. We assume that all such lines are in *general position*, that is, that no more than two lines intersect at any given point (see Figure 5.9b for an example). This can always be achieved by allowing ourselves to bend lines locally to avoid more than two lines meeting in a point. The collection of lines is then technically what is known as a *pseudo-line arrangement*. As indicated in the proof of Theorem 5.7.5, a split network N can be obtained by dualization of the complete (pseudo) line arrangement. In this process, each line l gives rise to $t + 1$ edges, where t is the number of intersection points that l shares with other lines in the arrangement. As t is at most $m - 1$, it follows that N contains only a quadratic number of edges and, thus also, of nodes.   □

Note that the canonical construction is less suited for circular split systems. Indeed, the canonical split network (Buneman graph) N associated with a set of splits S on $\mathcal{X}$ is not necessary planar when S is circular. This is a main reason why we make the distinction between the concept of a split network and a canonical split network. In Section 7.2, we present the circular network algorithm, which is guaranteed to produce an outer-labeled planar split network N when given a set of circular splits S on $\mathcal{X}$.

5.8 Weak compatibility

As discussed above, any *two* splits $S_1 = A_1 \mid B_1$ and $S_2 = A_2 \mid B_2$ on $\mathcal{X}$ are called compatible if one of the four possible intersections, $A_1 \cap A_2$, $A_1 \cap B_2$, $B_1 \cap A_2$ or $B_1 \cap B_2$ is empty. One possible mathematical generalization of compatibility is known as *weak compatibility* [9] and is defined as a relation between any *three* splits, requiring that certain intersections of the split parts are empty, as we discuss below.

Weak compatibility is an important concept for two reasons. First, the set of splits produced by the split decomposition method, which we describe in Section 5.9, is always weakly compatible. The same is also true for the (less widely known) parsimony splits method, which is presented in Section 9.3. Second, in practice the

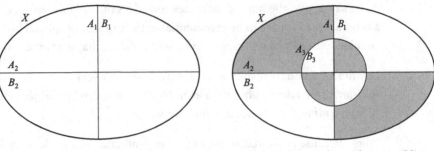

(a) Venn diagram for compatibility   (b) Venn diagram for weak compatiblity

Figure 5.12 (a) Two splits $S_1 = \frac{A_1}{B_1}$ and $S_2 = \frac{A_2}{B_2}$ on $\mathcal{X}$ are compatible, if and only if one of the four regions in this Venn diagram is empty. (b) Three splits $S_1 = \frac{A_1}{B_1}$, $S_2 = \frac{A_2}{B_2}$ and $S_3 = \frac{A_3}{B_3}$ on $\mathcal{X}$ are weakly compatible if and only if at least one of the gray regions and one of the white regions in this Venn diagram are empty.

resulting split networks are often quite close to being planar, as they usually have only a few edges crossing over each other and do not contain any "high-dimensional cubes", which may occur for completely unrestricted sets of splits.

Definition 5.8.1 (Weak compatibility) *A set of splits S on $\mathcal{X}$ is called* weakly compatible, *if any three distinct splits in S are weakly compatible. Three such splits $S_1 = \frac{A_1}{B_1}$, $S_2 = \frac{A_2}{B_2}$, and $S_3 = \frac{A_3}{B_3}$ are called* weakly compatible, *if*

(i) *at least one of the following four intersections is empty:*

$$A_1 \cap A_2 \cap A_3, \quad A_1 \cap B_2 \cap B_3, \quad B_1 \cap A_2 \cap B_3 \quad and \quad B_1 \cap B_2 \cap A_3,$$
(5.18)

and, symmetrically,

(ii) *at least one of the following four intersections is empty:*

$$B_1 \cap B_2 \cap B_3, \quad B_1 \cap A_2 \cap A_3, \quad A_1 \cap B_2 \cap A_3 \quad and \quad A_1 \cap A_2 \cap B_3.$$
(5.19)

In Figure 5.12 we show two Venn diagrams illustrating the definitions of compatibility and weak compatibility. If three splits are not weakly compatible, then it can happen that every possible intersection of any three split parts is non-empty and the split network required to represent the three splits consists of the eight edges of a cube.

Exercise 5.8.2 (Number of weakly compatible splits) *What is the maximum number of weakly compatible splits that is possible on a taxon set of size 4, 5 or 6?*

Consider two unrooted phylogenetic trees T_1 and T_2 on $\mathcal{X}$ and let S be the set of splits obtained from both trees. Is S weakly compatible? This question can easily be answered using Figure 5.12: Given any three splits S_1, S_2 and S_3, at least one

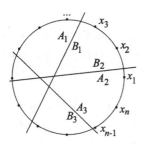

Figure 5.13 For a circular set of splits S with circular ordering $(x_1, \ldots, x_n)$, we represent three distinct splits $S_1 = \frac{A_1}{B_1}$, $S_2 = \frac{A_2}{B_2}$ and $S_3 = \frac{A_3}{B_3}$ by lines in a configuration in which every pair of lines intersects.

pair of splits, S_1 and S_2, say, must come from the same tree and is thus compatible. Hence, one of the four regions shown in (a) is empty. Comparison of (a) and (b) shows that every such region contains both a gray and a white region of the Venn diagram shown in (b). Therefore, one gray and one white region shown in (b) must be empty and so the three splits are weakly compatible. Moreover, we have that any set of circular splits is weakly compatible:

Lemma 5.8.3 (Circular implies weakly compatible)   *Let S be a set of splits on $\mathcal{X}$. If S is circular, then S is weakly compatible.*

Proof Let S be a circular set of splits on $\mathcal{X}$ and assume that we have placed the taxa around a circle in such a way that each split in S is represented by a line through the circle. Consider three distinct splits $S_1 = \frac{A_1}{B_1}$, $S_2 = \frac{A_2}{B_2}$ and $S_3 = \frac{A_3}{B_3}$. If all three lines representing these splits intersect each other, as show in Figure 5.13, then the intersections $A_1 \cap A_2 \cap A_3$ and $B_1 \cap B_2 \cap B_3$ are both empty. This implies that the three splits are weakly compatible, as the one intersection is colored gray and the other intersection is colored white in Figure 5.12(b). Otherwise, assume that (at least) two lines do not intersect, then the two corresponding splits are compatible and it follows that all three splits must be weakly compatible, using the same argument as applied above in the analysis of the situation of two trees.   □

5.9 The split decomposition

The first practical method proposed for computing a set of incompatible splits from evolutionary data was the split decomposition method, introduced in [9]. As we will see, the split decomposition method takes as input a distance matrix and produces as output a set of weighted splits that is weakly compatible. To motivate this approach, we first discuss the related *Buneman tree* method as proposed by Bandelt and Dress [9].

5.9.1 The Buneman tree

One way to determine whether a given set of splits S on X is compatible is to check for each quadruple of taxa $Q \subseteq X$ and every pair of splits S_1 and S_2 whether the splits induced by S_1 and S_2 on Q are compatible:

Lemma 5.9.1 (Quadruple condition for compatibility) *Two distinct splits S_1 and S_2 on X are compatible if and only if the set $\{S_1|_Q, S_2|_Q\}$ contains at most one non-trivial split for every quadruple Q on X.*

Proof Let $S_1 = \frac{A_1}{B_1}$ and $S_2 = \frac{A_2}{B_2}$ be two splits on X.

First consider the case that S_1 and S_2 are compatible. Assume that the claim is not true, then there exists a quadruple of distinct taxa $Q = \{w, x, y, z\}$ on X such that $S_1|_Q$ and $S_2|_Q$ are two distinct, non-trivial splits on Q. Moreover, assume, without loss of generality, that $\{w, x\} \subseteq A_1$ and $\{y, z\} \subseteq B_1$ hold. Because $S_1|_Q$ and $S_2|_Q$ are distinct, we can also assume, again without loss of generality, either that $\{w, y\} \subseteq A_2$ and $\{x, z\} \subseteq B_2$ hold or that $\{w, z\} \subseteq A_2$ and $\{x, y\} \subseteq B_2$ hold. In either case, we have $A_1 \cap A_2 \neq \emptyset$, $A_1 \cap B_2 \neq \emptyset$, $B_1 \cap A_2 \neq \emptyset$ and $B_1 \cap B_2 \neq \emptyset$, and therefore S_1 and S_2 are incompatible, a contradiction. Hence, in this case $S_1|_Q$ and $S_2|_Q$ cannot be two distinct, non-trivial splits on Q.

Now consider the case that S_1 and S_2 are incompatible. Then, there exist four distinct taxa $w, x, y, z \in X$ such that $w \in A_1 \cap A_2$, $x \in A_1 \cap B_2$, $y \in B_1 \cap A_2$ and $z \in B_1 \cap B_2$ hold. With $Q = \{w, x, y, z\}$, this implies $S_1|_Q = \frac{\{w, x\}}{\{y, z\}}$ and $S_2|_Q = \frac{\{w, y\}}{\{x, z\}}$, as required. $\square$

In other words, two splits S_1 and S_2 on X are compatible, unless there exists a quadruple Q on X that induces one non-trivial split $S_1|_Q$ on S_1 and a different non-trivial split $S_2|_Q$ on S_2. A non-trivial split $\frac{\{w, x\}}{\{y, z\}}$ on a quadruple $Q = \{w, x, y, z\}$ is called a *quartet topology*. The lemma suggests the following strategy for constructing a set of compatible splits on X:

- For each quadruple Q on X choose one of the three possible quartet topologies and denote it by $\hat{Q}$.
- Construct the set of all splits S that *respect* the chosen quartet topologies, that is, determine every split $S = \frac{A}{B}$ on X for which $S|_Q = \hat{Q}$ holds for all quadruples Q on X with $|Q \cap A| = |Q \cap B| = 2$.

Exercise 5.9.2 (Compatibility) *Show that the set of splits S computed in this way is compatible.*

Assume that we are given a distance matrix D on X. How to select a quartet topology for each quadruple $Q \subseteq X$? For a given quadruple $Q = \{w, x, y, z\} \subseteq X$, consider

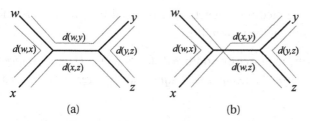

Figure 5.14 A phylogenetic tree on four taxa $\{w, x, y, z\}$. (a) Here we see that $d(w, y) + d(x, z) - d(w, x) - d(y, z)$ covers the length of the central edge e exactly twice, and thus that the Buneman index $\hat{\beta}_D(\frac{\{w,x\}}{\{y,z\}})$ equals the length of e. (b) Similarly, in this case $d(w, z) + d(x, y) - d(w, x) - d(y, z)$ does so, too.

the three possible sums of pairs of distances on Q:

$$d(w, x) + d(y, z), \quad d(w, y) + d(x, z) \quad \text{and} \quad d(w, z) + d(x, y). \quad (5.20)$$

We choose the quartet topology corresponding to the smallest of these three values. For example, if $d(w, x) + d(y, z)$ is smallest, then we set $\hat{Q} = \frac{\{w,x\}}{\{y,z\}}$. If more than one quartet topology takes on the smallest value, then no topology is chosen for the quadruple. The motivation for this choice is the desire to minimize the distances between taxa on the same side of a split.

To describe this more formally, consider any four taxa w, x, y and z with $\{w, x\} \cap \{y, z\} = \emptyset$, but not necessarily $w \neq x$ or $y \neq z$. We define the (quartet) *Buneman index* of $\hat{Q} = \frac{\{w,x\}}{\{y,z\}}$ with respect to D as:

$$\hat{\beta}_D\left(\tfrac{\{w,x\}}{\{y,z\}}\right) = \tfrac{1}{2}\left(\min\{d(w, y) + d(x, z), d(w, z) + d(x, y)\} \right.$$
$$\left. - d(w, x) - d(y, z) \right). \quad (5.21)$$

It follows from this definition that the Buneman index is positive for, at most, one quartet topology on any given quadruple; and this is the one that we choose. To understand the quartet Buneman index, note that it measures the length of the central edge in an edge-weighted phylogenetic tree on four taxa, as illustrated in Figure 5.14.

Now, assume we are given a distance matrix D on $\mathcal{X}$. We define the *Buneman index* of a split $S = \frac{A}{B}$ on $\mathcal{X}$ as the minimum quartet Buneman index of all the quadruples induced by S:

$$\beta_D(S) = \min\left\{ \hat{\beta}_D\left(\tfrac{\{w,x\}}{\{y,z\}}\right) \mid w, x \in A, \ y, z \in B \right\}. \quad (5.22)$$

Let $\mathcal{S}$ be the set of all splits on $\mathcal{X}$ that have a positive Buneman index. It follows from the definitions and Lemma 5.9.1 that $\mathcal{S}$ is compatible. In Section 5.9.4, we describe an algorithm that can be used to find all such splits efficiently. The corresponding unrooted phylogenetic tree is called the *Buneman tree* for D and the following result can be proven [9]:

Lemma 5.9.3 (Buneman tree on additive distances) *Let D be a distance matrix on X. The distances in the Buneman tree T equal the distances in D, if and only if D is additive.*

Although this result shows that the Buneman tree method computes the correct tree for any distance matrix D that satisfies the four-point condition, the method is rarely used in practice. This is because violations of the four-point condition usually lead to a loss of splits and the resulting tree is often very unresolved.

5.9.2 A four-point condition for weak compatibility

In the previous section, we saw that if a set of splits S on X has the property that it gives rise to at most one quartet topology for each quadruple Q on X, then this property suffices to ensure that the given set of splits S is compatible. In this section, we show that prescribing *up to two* of the three possible quartet topologies for every quadruple Q on X ensures that a given set of splits is *weakly* compatible. We start with the following observation [9]:

Lemma 5.9.4 (Quadruple condition for weak compatibility) *Three distinct splits S_1, S_2 and S_3 on X are weakly compatible if and only if the set $\{S_1|_Q, S_2|_Q, S_3|_Q\}$ contains at most two different non-trivial splits on Q for every quadruple Q on X.*

Exercise 5.9.5 (Proof) *Prove Lemma 5.9.4.*

This result suggests the following strategy for constructing a set of weakly compatible splits on X:

- For each quadruple Q on X we make a fixed choice of up to two of the three possible quartet topologies.
- Construct the set of all splits S that respect the chosen quartet topologies, that is, determine every split $S = \frac{A}{B}$ on X for which $S|_Q$ equals one of the chosen quartet topologies on Q, for all quadruples Q on X with $|Q \cap A| = |Q \cap B| = 2$.

Lemma 5.9.6 (Result is weakly compatibility) *The set of splits S computed by the described strategy is weakly compatible.*

Proof Let S be a set of splits on X computed using the described approach. To obtain a contradiction, assume that S is not weakly compatible. Then, there exist three splits $S_1 = \frac{A_1}{B_1}$, $S_2 = \frac{A_2}{B_2}$ and $S_3 = \frac{A_3}{B_3}$ such that either the four intersections $A_1 \cap A_2 \cap A_3$, $A_1 \cap B_2 \cap B_3$, $B_1 \cap A_2 \cap B_3$ and $B_1 \cap B_2 \cap A_3$ are all non-empty, or the four intersections $B_1 \cap B_2 \cap B_3$, $B_1 \cap A_2 \cap A_3$, $A_1 \cap B_2 \cap A_3$ and $A_1 \cap A_2 \cap B_3$ are all non-empty, or both. Assume, without loss of generality, that the first case holds. Then we can choose four distinct taxa $w \in A_1 \cap A_2 \cap A_3$, $x \in A_1 \cap B_2 \cap B_3$, $y \in B_1 \cap A_2 \cap B_3$ and $z \in B_1 \cap B_2 \cap A_3$ and set $Q = \{w, x, y, z\}$. We have $S_1|_Q = \frac{\{w,x\}}{\{y,z\}}$, $S_2|_Q = \frac{\{w,y\}}{\{x,z\}}$ and $S_3|_Q = \frac{\{w,z\}}{\{x,y\}}$. This is a contradiction to the construction of S. $\square$

5.9.3 Weak compatibility and D-splits

As we see later, there are two methods that are based on the above approach, namely the split decomposition method described in Section 5.9.4 of this chapter and the parsimony splits method, presented in Section 9.3. In preparation of the split decomposition method, our aim now is to introduce the concept of a *D-split* and to prove the key result that *D*-splits are always weakly compatible.

Assume we are given a distance matrix D on $\mathcal{X}$. How to select up to two quartet topologies for each quadruple $Q \subseteq \mathcal{X}$? For a given quadruple $Q = \{w, x, y, z\} \subseteq \mathcal{X}$, consider the three possible sums of distances on Q:

$$d(w, x) + d(y, z), \ d(w, y) + d(x, z) \text{ and } d(w, z) + d(x, y). \tag{5.23}$$

Generalizing the method discussed in the previous section, we choose all (that is, zero, one or two) quartet topologies for which the corresponding sum is smaller than the largest sum. More precisely, if the three sums are all different, or if two are equal and smaller than the third, then we choose the two topologies associated with the two smaller sums. If two are equal and larger than the third, then we choose the topology associated with the third value. If all three values are the same, then none of the three topologies is chosen. For example, if $d(w, x) + d(y, z) < d(w, y) + d(x, z) = d(w, z) + d(x, y)$, then we (only) choose the quartet topology $\frac{\{w,x\}}{\{y,z\}}$.

Consider any four taxa w, x, y and z with $\{w, x\} \cap \{y, z\} = \emptyset$, but not necessarily $w \neq x$ or $y \neq z$. We define the (quartet) *isolation index* of $\hat{Q} = \frac{\{w,x\}}{\{y,z\}}$ with respect to D as:

$$\hat{\alpha}_D\left(\frac{\{w,x\}}{\{y,z\}}\right) = \frac{1}{2}\left(\max\left\{ d(w, x) + d(y, z), d(w, y) \right.\right.$$
$$\left.\left. + d(x, z), d(w, z) + d(x, y) \right\} - d(w, x) - d(y, z)\right). \tag{5.24}$$

Note that the quartet isolation index is always non-negative, because the quantity that is subtracted also occurs as an argument of the maximum function. Also, because one of the three sums has to be largest, the quartet isolation index can only be positive for at most two quartet topologies, and these are the ones that we choose.

We define the *isolation index* $\alpha_D(S)$ of a split $S = \frac{A}{B}$ on $\mathcal{X}$ as the minimum isolation index of all the quartets induced by S:

$$\alpha_D(S) = \min\{\hat{\alpha}_D\left(\tfrac{\{w,x\}}{\{y,z\}}\right) \mid w, x \in A, \ y, z \in B\} \geq 0. \tag{5.25}$$

A split S whose isolation index $\alpha_D(S)$ is greater than 0 is called a *D-split*. By construction, we have:

Lemma 5.9.7 (D-splits are weakly compatible)   *Let D be a distance matrix on $\mathcal{X}$. The set of D-splits on $\mathcal{X}$ is weakly compatible.*

Let $\mathcal{X}$ be a set of taxa. A split S on a proper subset of $\mathcal{X}$ is called a *partial split* on $\mathcal{X}$. A split $S' = \frac{A'}{B'}$ is called an *extension* of a partial split $S = \frac{A}{B}$, if $A \subseteq A'$ and $B \subseteq B'$.

An important observation is that the weight of any extension S' of a D-split S never exceeds the weight of the extended split, that is, $\alpha_D(S') \le \alpha_D(S)$. Moreover, the sum of weights of all full extensions of a given split never exceeds the weight of that split:

Theorem 5.9.8 (Extension decreases weight)   *Let D be a distance matrix on $\mathcal{X}$. Let S be a partial D-split and let $\langle S \rangle$ denote the set of all full D-splits that extend S. We have:*

$$\sum_{S' \in \langle S \rangle} \alpha_D(S') \le \alpha_D(S). \tag{5.26}$$

The proof of this key result is based on the analysis of the case in which a partial split is extended by a single taxon in the two possible ways:

Lemma 5.9.9 (Extension lemma)   *Let D be a distance matrix on $\mathcal{X} = \{x_1, \ldots, x_n\}$ and let $S = \frac{A}{B}$ be a D-split on the first $n - 1$ taxa $\{x_1, \ldots, x_{n-1}\}$. For all $a_1, a_2 \in A$, $b_1, b_2 \in B$ we have:*

$$\alpha_D \left(\frac{\{a_1, a_2, x_n\}}{\{b_1, b_2\}} \right) + \alpha_D \left(\frac{\{a_1, a_2\}}{\{b_1, b_2, x_n\}} \right) \le \hat{\alpha}_D \left(\frac{\{a_1, a_2\}}{\{b_1, b_2\}} \right). \tag{5.27}$$

Proof   For ease of notation, we use xy to denote the distance $d(x, y)$ between any two taxa x and y. Let $\frac{A}{B}$ be a split on $\{x_1, \ldots, x_{n-1}\}$. Assume, for the sake of contradiction, that the claimed inequality fails for some choice of $a_1, a_2 \in A$ and $b_1, b_2 \in B$, that is, assume

$$\alpha_D \left(\frac{\{a_1, a_2, x_n\}}{\{b_1, b_2\}} \right) + \alpha_D \left(\frac{\{a_1, a_2\}}{\{b_1, b_2, x_n\}} \right) > \hat{\alpha}_D \left(\frac{\{a_1, a_2\}}{\{b_1, b_2\}} \right). \tag{5.28}$$

Note that all three quantities are strictly positive, because neither term on the left can alone be larger than the right-hand term, by definition of the isolation index. Consider the isolation index $\hat{\alpha}_D \left(\frac{\{a_1, x_n\}}{\{b_1, b_2\}} \right)$. Since this quantity is strictly positive, we have that

$$\frac{1}{2} \left(\max\{a_1 b_1 + x_n b_2, a_1 b_2 + x_n b_1\} - a_1 x_n - b_1 b_2 \right) > 0. \tag{5.29}$$

Assume, without loss of generality, that $a_1 b_2 + x_n b_1 \geq a_1 b_1 + x_n b_2$. Then

$$\hat{\alpha}_D \left(\frac{\{a_1, x_n\}}{\{b_1, b_2\}} \right) = \frac{1}{2} (a_1 b_2 + x_n b_1 - a_1 x_n - b_1 b_2). \tag{5.30}$$

This implies

$$\begin{aligned}
& \tfrac{1}{2}(a_1 b_2 + x_n b_1 - a_1 x_n - b_1 b_2 + \max\{a_1 b_1 + a_2 x_n, a_1 x_n + a_2 b_1\} \\
& \qquad - a_1 a_2 - b_1 x_n) \\
& = \hat{\alpha}_D \left(\frac{\{a_1, x_n\}}{\{b_1, b_2\}} \right) + \hat{\alpha}_D \left(\frac{\{a_1, a_2\}}{\{b_1, x_n\}} \right) \\
& \geq \alpha_D \left(\frac{\{a_1, a_2, x_n\}}{\{b_1, b_2\}} \right) + \alpha_D \left(\frac{\{a_1, a_2\}}{\{b_1, b_2, x_n\}} \right) \\
& > \hat{\alpha}_D \left(\frac{\{a_1, a_2\}}{\{b_1, b_2\}} \right) \\
& = \tfrac{1}{2} (\max\{a_1 b_1 + a_2 b_2, a_1 b_2 + a_2 b_1\} - a_1 a_2 - b_1 b_2). \tag{5.31}
\end{aligned}$$

Comparing the first and last of these expressions we get

$$\begin{aligned}
& a_1 b_2 - a_1 x_n + \max\{a_1 b_1 + a_2 x_n, a_1 x_n + a_2 b_1\} \\
& > \max\{a_1 b_1 + a_2 b_2, a_1 b_2 + a_2 b_1\}. \tag{5.32}
\end{aligned}$$

For this inequality to hold, we must have $\max\{a_1 b_1 + a_2 x_n, a_1 x_n + a_2 b_1\} = a_1 b_1 + a_2 x_n$, because otherwise the left-hand side of the inequality would yield

$$\begin{aligned}
a_1 b_2 - a_1 x_n + \max\{a_1 b_1 + a_2 x_n, a_1 x_n + a_2 b_1\} \\
= a_1 b_2 - a_1 x_n + a_1 x_n + a_2 b_1 \\
= a_1 b_2 + a_2 b_1, \tag{5.33}
\end{aligned}$$

which cannot be larger than the right-hand side of the inequality. In consequence, we have:

$$a_1 b_2 - a_1 x_n + a_1 b_1 + a_2 x_n > \max\{a_1 b_1 + a_2 b_2, a_1 b_2 + a_2 b_1\}, \tag{5.34}$$

that is,

$$\begin{aligned}
a_1 b_2 - a_1 x_n + a_1 b_1 + a_2 x_n > a_1 b_2 + a_2 b_1 \quad \text{and} \\
a_1 b_2 - a_1 x_n + a_1 b_1 + a_2 x_n > a_1 b_1 + a_2 b_2,
\end{aligned} \tag{5.35}$$

which implies

$$a_1 b_1 + a_2 x_n > a_2 b_1 + a_1 x_n \quad \text{and} \quad a_1 b_2 + a_2 x_n > a_2 b_2 + a_1 x_n. \tag{5.36}$$

By considering the isolation index $\hat{\alpha}_D \left(\frac{\{a_2, x_n\}}{\{b_1, b_2\}} \right)$ at the beginning of the proof, we can also deduce the reverse of these two inequalities, thus obtaining a contradiction, showing that the lemma must be correct. $\qquad \square$

With this lemma in hand, we can now prove Theorem 5.9.8 by induction:

Proof Assume we are given a distance matrix D on $\mathcal{X} = \{x_1, \ldots, x_n\}$. We use $\mathcal{X}_j = \{x_1, \ldots, x_j\}$ to denote the first j elements of $\mathcal{X}$.

To start the induction, we first show that the result is true for any D-split $S = \frac{A}{B}$ on $\mathcal{X}_{n-1} = \{x_1, \ldots, x_{n-1}\}$. To this end, choose $a_1, a_2 \in A$ and $b_1, b_2 \in B$ such that $\alpha_D\left(\frac{A}{B}\right) = \hat{\alpha}_D\left(\frac{\{a_1, a_2\}}{\{b_1, b_2\}}\right)$ holds. Then we have:

$$
\begin{aligned}
\alpha_D\left(\tfrac{A\cup\{x_n\}}{B}\right) + \alpha_D\left(\tfrac{A}{B\cup\{x_n\}}\right) &\le \alpha_D\left(\tfrac{\{a_1, a_2, x_n\}}{\{b_1, b_2\}}\right) + \alpha_D\left(\tfrac{\{a_1, a_2\}}{\{b_1, b_2, x_n\}}\right) \\
&\le \hat{\alpha}_D\left(\tfrac{\{a_1, a_2\}}{\{b_1, b_2\}}\right) \\
&= \alpha_D\left(\tfrac{A}{B}\right),
\end{aligned}
\tag{5.37}
$$

where the first inequality follows by definition of the isolation index and the second one by Lemma 5.9.9. Now, assume by induction that the claim is true for all D-splits on $\mathcal{X}_j$, with $j \le n - 1$. By backward induction, we can show that this implies that the claim holds for any D-split $S = \frac{A}{B}$ on $\mathcal{X}_{j-1}$, too:

$$
\begin{aligned}
&\sum \left\{ \alpha_D(S') \mid S' \text{ is a } D\text{-split on } \mathcal{X} \text{ that extends } \tfrac{A}{B} \text{ on } \mathcal{X}_{j-1} \right\} \\
&= \sum \left\{ \alpha_D(S') \mid S' \text{ is a } D\text{-split on } \mathcal{X} \text{ that extends } \tfrac{A\cup\{x_j\}}{B} \text{ on } \mathcal{X}_j \right\} \\
&\quad + \sum \left\{ \alpha_D(S') \mid S' \text{ is a } D\text{-split on } \mathcal{X} \text{ that extends } \tfrac{A}{B\cup\{x_j\}} \text{ on } \mathcal{X}_j \right\} \\
&\le \alpha_D\left(\tfrac{A\cup\{x_j\}}{B}\right) + \alpha_D\left(\tfrac{A}{B\cup\{x_j\}}\right) \le \alpha_D\left(\tfrac{A}{B}\right),
\end{aligned}
\tag{5.38}
$$

where the first inequality follows by induction and the second one by Lemma 5.9.9. This completes the proof. $\qquad\square$

One consequence of this result is that a set of D-splits never provides an over-estimation of the distance between two taxa x and y, that is, we have

$$
\sum_{S \in \mathcal{S}(x, y)} \alpha_D(S) \le d(x, y),
\tag{5.39}
$$

summing over the set $\mathcal{S}(x, y)$ of all D-splits that separate x and y. To see this, apply the theorem to the partial D-split $\{x\}|\{y\}$.

5.9.4 Computing the split decomposition

Suppose we are given a distance matrix D on $\mathcal{X}$. As we have established in the previous section, the set of all D-splits on $\mathcal{X}$ is weakly compatible and for any pair of taxa x, y we have that $d(x, y)$ is greater or equal to the sum of weights of all D-splits that separate x and y. The set of all D-splits on $\mathcal{X}$ is called the *split decomposition* of D. The following algorithm computes the split decomposition:

Algorithm 5.9.10 (Split decomposition) *Given a distance matrix D on $\mathcal{X} = \{x_1, \ldots, x_n\}$, compute the set of all weighted D-splits on $\mathcal{X}$ as follows:*

	a	b	c	d	e
a	0	2	3	2	2
b	2	0	2	3	3
c	3	2	0	2	2
d	2	3	2	0	2
e	2	3	2	3	0

(a) Distance matrix D

$\{a\} \mid \{b,c,d,e\}$
$\{b\} \mid \{a,c,d,e\}$
$\{c\} \mid \{a,b,d,e\}$
$\{d\} \mid \{a,b,c,e\}$
$\{e\} \mid \{a,b,c,d\}$

(b) All D-splits

	a	b	c	d	e
a	0	1	2	1	1
b	1	0	1	2	2
c	2	1	0	1	1
d	1	2	1	0	1
e	1	2	1	2	0

(c) Split prime residue D_0

Figure 5.15 The distance matrix D on $\mathcal{X} = \{a, b, c, d, e\}$ shown in (a) admits only five D-splits, namely the trivial splits listed in (b), each with an isolation index of 0.5. The remaining split prime residue D_0 on $\mathcal{X}$ is shown in (c).

Initially, set $\mathcal{X}_1 = \{x_1\}$ and $\mathcal{S}_1 = \emptyset$. Now, assume that we have computed the set of all D-splits $\mathcal{S}_i$ on the first i taxa $\mathcal{X}_i = \{x_1, \ldots, x_i\}$. To obtain $\mathcal{S}_{i+1}$ on $\mathcal{X}_{i+1} = \{x_1, \ldots, x_{i+1}\}$, for each split $\frac{A}{B} \in \mathcal{S}_i$ do:

- If $\alpha_D\left(\frac{A \cup \{x_{i+1}\}}{B}\right) > 0$, then add $\frac{A \cup \{x_{i+1}\}}{B}$ to $\mathcal{S}_{i+1}$.
- If $\alpha_D\left(\frac{A}{B \cup \{x_{i+1}\}}\right) > 0$, then add $\frac{A}{B \cup \{x_{i+1}\}}$ to $\mathcal{S}_{i+1}$.
- If $\alpha_D\left(\frac{A}{\{x_{i+1}\}}\right) > 0$, then add $\frac{A}{\{x_{i+1}\}}$ to $\mathcal{S}_{i+1}$.

The result is given by $\mathcal{S}_n$.

This algorithm works because extending a partial split to more taxa can only maintain or decrease its isolation index. The worst-case time requirement of this algorithm is $O(n^6)$, because the number of iterations of the main loop is n, the number of splits present in the i-th iteration is $O(n^2)$ (as we show below), and the time required to compute the isolation index of a split in the main loop is $O(n^3)$. In practice, the run-time of the algorithm is usually much better than this analysis suggests, since the number of non-trivial D-splits tends to drop quite rapidly as the number of taxa increases.

We can also use this algorithm to efficiently compute all splits of the Buneman tree, by simply replacing the isolation index α_D by the Buneman index β_D.

Let D be a distance matrix on $\mathcal{X}$ and let $\mathcal{S}$ be the set of all D-splits on $\mathcal{X}$. We define the *residue* distance matrix D_0 as

$$d_0(x, y) = d(x, y) - \sum_{S \in \mathcal{S}(x,y)} \alpha_D(S) \geq 0, \qquad (5.40)$$

where $\mathcal{S}(x, y)$ denotes the set of all splits that separate x and y, as usual.

Exercise 5.9.11 (Residue is a metric) *Assuming that D is a metric, show that the residue matrix D_0 is a metric.*

Consider the distance matrix D on $\mathcal{X} = \{a, \ldots, e\}$ shown in Figure 5.15a. For this example, the only splits on $\mathcal{X}$ that have a positive isolation index are the five trivial splits shown in Figure 5.15b. In each case, the isolation index is 0.5. Subtracting the contribution of each of these D-splits, we obtain the split prime distance matrix D_0 shown in Figure 5.15c.

Theorem 5.9.12 (Residue is split prime)   *If D_0 is the residue of some distance matrix D, then D_0 does not admit any D_0-split. In this case D_0 is called split prime.*

In the following, for any split S on $\mathcal{X}$, define the *split metric* δ_S on $\mathcal{X}$ as $\delta_S(x, y) = 1$, if S separates taxa x and y, and 0, else. The proof of the theorem is based on the following technical result:

Lemma 5.9.13 (Elimination lemma)   *Let D be a distance matrix on $\mathcal{X}$, let $S' = \frac{A'}{B'}$ be a fixed D-split on $\mathcal{X}$ and let λ be any number with $\alpha_D(S') \geq \lambda \geq 0$. If D' is a distance matrix on $\mathcal{X}$ defined as*

$$d'(x, y) = d(x, y) - \lambda\delta_{S'}(x, y) \geq 0 \tag{5.41}$$

for all taxa x and y in $\mathcal{X}$, then for any split $\frac{A}{B}$ on $\mathcal{X}$ we have

$$\alpha_{D'}\left(\frac{A}{B}\right) = \begin{cases} \alpha_D\left(\frac{A'}{B'}\right) - \lambda, & \text{if } \frac{A}{B} = S', \quad \text{and} \\ \alpha_D\left(\frac{A}{B}\right), & \text{else.} \end{cases} \tag{5.42}$$

Proof Before proving the lemma, note that, since S' is a D-split on $\mathcal{X}$, we have that $\alpha_D(S')$ is strictly positive. This implies that $\hat{\alpha}_D\left(\frac{\{w,x\}}{\{y,z\}}\right) \geq \alpha_D\left(\frac{A'}{B'}\right) > 0$ for all $w, x \in A'$ and for all $y, z \in B'$. It follows that, for all $w, x \in A'$ and for all $y, z \in B'$, we have:

$$\begin{aligned} \hat{\alpha}_D&\left(\frac{\{w,x\}}{\{y,z\}}\right) \\ &= \tfrac{1}{2}(\max\{d(w, x) + d(y, z), d(w, y) + d(x, z), d(w, z) + d(x, y)\} \\ &\qquad\qquad - d(w, x) - d(y, z)) \\ &= \tfrac{1}{2}(\max\{d(w, y) + d(x, z), d(w, z) + d(x, y)\} - d(w, x) - d(y, z)). \end{aligned} \tag{5.43}$$

Moreover, we have that:

$$\begin{aligned} \hat{\alpha}_D&\left(\frac{\{w,x\}}{\{y,z\}}\right) \\ &= \tfrac{1}{2}(\max\{d(w, x) + d(y, z), d(w, y) + d(x, z), d(w, z) + d(x, y)\} \\ &\qquad\qquad - d(w, x) - d(y, z)) \\ &\geq \alpha_D(\tfrac{A'}{B'}) \geq \lambda. \end{aligned} \tag{5.44}$$

It follows that, for all $w, x \in A'$, and for all $y, z \in B'$, we have:

$$\max\{d(w, z) + d(x, y) - 2\lambda, d(w, y) + d(x, z) - 2\lambda\} \geq d(w, x) + d(x, z). \tag{5.45}$$

To prove the lemma, we only need to show that, for any quadruple $Q = \{w, x, y, z\}$ on $\mathcal{X}$, we have:

$$\hat{\alpha}_{D'}\left(\frac{\{w, x\}}{\{y, z\}}\right) = \begin{cases} \hat{\alpha}_D\left(\frac{\{w,x\}}{\{y,z\}}\right) - \lambda, & \text{if } S'|_{\{w,x,y,z\}} = \frac{\{w,x\}}{\{y,z\}}, \\ \hat{\alpha}_D\left(\frac{\{w,x\}}{\{y,z\}}\right), & \text{else.} \end{cases} \tag{5.46}$$

To this end, we need to consider four different cases, based on how the quadruple $Q = \{w, x, y, z\}$ is partitioned by the split S'.

First, consider the case that Q is completely contained in A' or B'. In this case $\delta_{S'}(x, y) = 0$ and $d'(x, y) = d(x, y)$ for all taxa x and y in Q. Then clearly $\hat{\alpha}_{D'}(\hat{Q}) = \hat{\alpha}_D(\hat{Q})$ for all choices of quartet topologies $\hat{Q}$ on Q.

Second, it may be that three taxa are contained in one part of S', and the fourth is contained in the other; without loss of generality, $\{w, x, y\} \subseteq A'$ and $\{z\} \subseteq B'$, say. Then we have:

$$\begin{aligned}
\hat{\alpha}_{D'}&\left(\frac{\{w,x\}}{\{y,z\}}\right) \\
&= \tfrac{1}{2}(\max\{d'(w, x) + d'(y, z), d'(w, y) + d'(x, z), d'(w, z) + d'(x, y)\} \\
&\qquad\qquad\qquad\qquad\qquad\qquad\qquad - d'(w, x) - d'(y, z)) \\
&= \tfrac{1}{2}(\max\{d(w, x) + d(y, z) - \lambda, d(w, y) + d(x, z) - \lambda, d(w, z) \\
&\qquad\qquad\qquad + d(x, y) - \lambda\} - d(w, x) - d(y, z) + \lambda) \\
&= \tfrac{1}{2}(\max\{d(w, x) + d(y, z), d(w, y) + d(x, z), d(w, z) + d(x, y)\} \\
&\qquad\qquad\qquad\qquad\qquad\qquad\qquad - d(w, x) - d(y, z)) \\
&= \hat{\alpha}_D\left(\frac{\{w,x\}}{\{y,z\}}\right).
\end{aligned} \tag{5.47}$$

Third, if Q is partitioned equally by S', and if $S'|_Q = \hat{Q}$, then we can assume, without loss of generality, that $\{x, w\} \subseteq A'$ and $\{y, z\} \subseteq B'$. Then:

$$\begin{aligned}
\hat{\alpha}_{D'}&\left(\frac{\{w,x\}}{\{y,z\}}\right) \\
&= \tfrac{1}{2}(\max\{d'(w, x) + d'(y, z), d'(w, y) + d'(x, z), d'(w, z) + d'(x, y)\} \\
&\qquad\qquad\qquad\qquad\qquad\qquad\qquad - d'(w, x) - d'(y, z)) \\
&= \tfrac{1}{2}(\max\{d(w, x) + d(y, z), d(w, y) + d(x, z) - 2\lambda, d(w, z) + d(x, y) \\
&\qquad\qquad\qquad\qquad\qquad - 2\lambda\} - d(w, x) - d(y, z)) \\
&= \tfrac{1}{2}(\max\{d(w, y) + d(x, z), d(w, z) + d(x, y)\} - d(w, x) - d(y, z)) - \lambda \\
&= \hat{\alpha}_D\left(\frac{\{w,x\}}{\{y,z\}}\right) - \lambda.
\end{aligned} \tag{5.48}$$

The third equivalence of this result follows from the fact that:

$$\max\{d(w, z) + d(x, y) - 2\lambda, d(w, y) + d(x, z) - 2\lambda\} \geq d(w, x) + d(y, z), \tag{5.49}$$

as demonstrated at the beginning of the proof.

Fourth, if Q is partitioned equally by S', and if $S'|_Q \neq \hat{Q}$, then we can assume without loss of generality that $\{x, y\} \subseteq A'$ and $\{w, z\} \subseteq B'$. Then:

$$
\begin{aligned}
\hat{\alpha}_{D'} &\left(\tfrac{\{w,x\}}{\{y,z\}} \right) \\
&= \tfrac{1}{2}(\max\{d'(w, x) + d'(y, z), d'(w, y) + d'(x, z), d'(w, z) + d'(x, y)\} \\
&\qquad\qquad - d'(w, x) - d'(y, z)) \\
&= \tfrac{1}{2}(\max\{d(w, x) + d(y, z) - 2\lambda, d(w, y) + d(x, z) - 2\lambda, d(w, z) \\
&\qquad\qquad + d(x, y)\} - d(w, x) - d(y, z) + 2\lambda) \\
&= \tfrac{1}{2}(\max\{d(w, x) + d(y, z), d(w, y) + d(x, z), d(w, z) + d(x, y)\} \\
&\qquad\qquad - d(w, x) - d(y, z)) \\
&= \hat{\alpha}_D \left(\tfrac{\{w,x\}}{\{y,z\}} \right).
\end{aligned}
\tag{5.50}
$$

The third equivalence of this result follows from the fact that:

$$
\begin{aligned}
\max\{d(w, x) + d(y, z) - 2\lambda, d(w, y) + d(x, z) - 2\lambda\} &\geq d(w, z) \\
+ d(x, y) &\geq d(w, z) + d(x, y) - 2\lambda,
\end{aligned}
\tag{5.51}
$$

using Eq. (5.45).   □

Using this lemma, we can now prove Theorem 5.9.12:

Proof Let D be a distance matrix on $\mathcal{X}$ and let $\mathcal{S} = \{S_1, \ldots, S_k\}$ be the set of all D-splits on $\mathcal{X}$. Let D_0 be the residue of D, given by

$$
\begin{aligned}
d_0(x, y) &= d(x, y) - \sum_{S \in \mathcal{S}(x,y)} \alpha_D(S) \\
&= d(x, y) - \alpha_D(S_1)\delta_{S_1}(x, y) \cdots - \alpha_D(S_k)\delta_{S_k}(x, y)
\end{aligned}
\tag{5.52}
$$

for all taxa x and y.

By repeated application of Lemma 5.9.13, we can sequentially eliminate each split $S_k, S_{k-1}, \ldots, S_1$ from the distance matrix D, while not changing any of the other D-splits, until finally we are left with D_0, which admits no split and is thus split prime.   □

Exercise 5.9.14 (Split prime) *Verify that the distance matrix D_0 shown in Figure 5.15(c) is split-prime.*

The relationship between the different concepts can be summarized as follows:

Lemma 5.9.15 (Split decomposition) *Let D be a distance matrix on $\mathcal{X}$ and let $\mathcal{S}$ be the set of all D-splits on $\mathcal{X}$. The* split decomposition *of D is given by the unique decomposition*

$$
d(x, y) = \sum_{S \in \mathcal{S}(x,y)} \alpha_D(S) + d_0(x, y)
\tag{5.53}
$$

for all $x, y \in \mathcal{X}$, where D_0 is the split prime residue of D.

In applications, a distance matrix D on $\mathcal{X}$ is represented by the split network N of all its D-splits, computed as described in Chapter 7. If the residue D_0 of D is zero for all $x, y \in \mathcal{X}$, then D is *totally decomposable*. In this case the split network N provides an exact representation of D, as the length of a shortest path between any two taxa x and y in N equals $d(x, y)$.

To measure how well the distances in N (or, more precisely, in the split decomposition of D) represent the distances in D, we define the *percent fit* as

$$\text{fit}(N, D) = 100 \times \frac{\sum_{x,y \in \mathcal{X}} \sum_{S \in \mathcal{S}(x,y)} \alpha_D(S)}{\sum_{x,y \in \mathcal{X}} d(x, y)} \geq 0, \tag{5.54}$$

making use of the fact that distances based on the D-splits never exceed the distances given in D. In practice, a fit of 80% or more is usually required for a split network N based on split decomposition to be considered a reliable representation of a distance matrix D.

For example, the split network N associated with the distance matrix D on $\mathcal{X} = \{a, \ldots, e\}$ shown in Figure 5.15a is a star tree consisting of five edges, each of length 0.5, representing the five trivial splits listed in Figure 5.15b. This is a very poor representation of the distance matrix D, as the distances contained in the split prime residue D_0 shown in Figure 5.15c are quite large. In this case, the fit is only $\approx 41.7\%$.

A main attraction of the split decomposition method is that the number of splits that it can produce is larger than the number of splits produced by a tree-building method (such as the Buneman tree), while not being too large:

Lemma 5.9.16 (Size of the split decomposition) *Let D be a distance matrix on $\mathcal{X} = \{x_1, \ldots, x_n\}$. The number of D-splits on $\mathcal{X}$ is at most $\frac{n(n-1)}{2}$.*

Proof Let D be a distance matrix on $\mathcal{X}$ and let $\mathcal{S}$ be the set of D-splits on $\mathcal{X}$. To start the proof, assume that $\sum_{S \in \mathcal{S}} \lambda_S \delta_S(x, y) = 0$ for some choice of real numbers λ_S and for all taxa $x, y \in \mathcal{X}$. Partition the set of splits $\mathcal{S}$ into three classes, $\mathcal{S}^+$, $\mathcal{S}^-$ and $\mathcal{S}^0$, consisting of those splits S whose assigned number λ_S is positive, negative or zero, respectively. Consider the metrics defined by the two sides of the following equation:

$$\sum_{S \in \mathcal{S}^+} \lambda_S \delta_S(x, y) = \sum_{S \in \mathcal{S}^-} -\lambda_S \delta_S(x, y). \tag{5.55}$$

Because the split decomposition of a distance matrix is unique, we must have $\mathcal{S}^+ = \mathcal{S}^-$ and thus $\mathcal{S}^+ = \mathcal{S}^- = \emptyset$. This implies $\mathcal{S}^0 = \mathcal{S}$ and therefore $\lambda_S = 0$ for all $S \in \mathcal{S}$. This shows that the set of all split metrics δ_S ($S \in \mathcal{S}$) is linearly independent. Because the dimension of the set of all $n \times n$ distance matrices (that is, symmetric matrices with zero diagonal) is $\frac{n(n-1)}{2}$, the claim is proven. $\square$

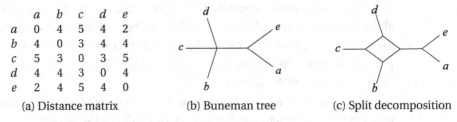

	a	b	c	d	e
a	0	4	5	4	2
b	4	0	3	4	4
c	5	3	0	3	5
d	4	4	3	0	4
e	2	4	5	4	0

(a) Distance matrix   (b) Buneman tree   (c) Split decomposition

Figure 5.16 (a) Distance matrix D on $\mathcal{X} = \{a, \ldots, e\}$. (b) Buneman tree for D. (c) Split decomposition for D.

When presented with an additive distance matrix, the split decomposition method can be used to reconstruct the associated phylogenetic tree:

Lemma 5.9.17 (Split decomposition on additive distances) *Let D be a distance matrix on $\mathcal{X}$. Then D is additive if and only if the set of D-splits on $\mathcal{X}$ is compatible and the residue D_0 of D is zero.*

Exercise 5.9.18 (Proof) *Prove the lemma.*

The split decomposition method has been used in numerous publications to represent distance matrices that show a significant amount of non-tree-likeness, or in situations in which there is reason to believe that the best representation of the data is not a phylogenetic tree. In practice, the method is limited to relatively small datasets of less than 100 taxa, or even much smaller, depending on the actual data. This is because the isolation index of a split S is defined as the minimum quartet isolation index over all quartets that span S and so, for a larger number of taxa, or for more divergent taxa, it is quite likely that one of the quartets will have an isolation index of 0, thus preventing S from being a D-split.

In Section 10.4 we discuss the neighbor-net method, an alternative method for constructing a set of (not necessarily compatible) splits from a distance matrix, which does not suffer from this practical limitation.

Because the isolation index is a relaxation of the Buneman index, it follows that the set of splits computed by the Buneman tree method is always a subset of the set of splits computed by the split decomposition method. In consequence, the Buneman tree can always be obtained from a split network that represents the split decomposition by *contracting* certain splits, as defined in the next section.

In Figure 5.16, we display a distance matrix D on $\mathcal{X} = \{a, \ldots, e\}$ and show both the Buneman tree and split decomposition network associated with D. In this particular example, the distance matrix is not additive. Hence, the splits computed by the Buneman tree method do not provide a total decomposition of D. Indeed, the fit is only $\approx 64\%$. However, the matrix D is totally decomposable by the split decomposition method and so the fit is 100% in this case.

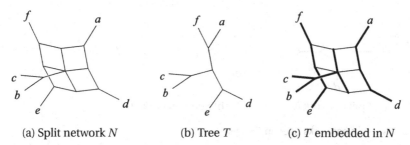

(a) Split network N (b) Tree T (c) T embedded in N

Figure 5.17   (a) A split network N on $\mathcal{X} = \{a, b, c, d, e, f\}$ and (b) a tree T on $\mathcal{X}$. In (c) the edges of N that represent the splits from T are highlighted. One sees clearly that contraction of the other edges gives rise to the tree T and thus that N contains T.

5.10 Representing trees in a split network

Suppose we are given a split network $N = (V, E)$ that represents a set of splits $\mathcal{S}$ on $\mathcal{X}$. For a given split $S \in \mathcal{S}$, we can obtain a new split network N' as follows: Let $E_S \subseteq E$ be the set of edges representing S in N. For each edge $e = (v, w)$ in E_S merge the source node v and the target node w and then delete e. We refer to this operation as the *contraction* of the split S. It follows from the definition of a split network that N' is a split network representing $\mathcal{S} - \{S\}$. This implies the following result:

Lemma 5.10.1 (Tree contained in split network)   *If $\mathcal{S}' \subset \mathcal{S}$ is the split encoding of some tree T on $\mathcal{X}$, then the canonical split network N representing $\mathcal{S}$ contains T in the sense that contracting all splits in $\mathcal{S} - \mathcal{S}'$ will result in a network N' that equals T (in the sense that it has the same labeled topology).*

So, if we are given a set of trees $\mathcal{T} = \{T_1, \ldots, T_k\}$, for example, obtained by phylogenetic analysis of a set of k different genes, we can represent all these trees in one split network N computed on the set of splits $\mathcal{S}$ of all trees in $\mathcal{T}$. To "see" a particular tree T contained in the network, one simply contracts all splits that are not contained in the split encoding of T, as demonstrated in Figure 5.17.

Exercise 5.10.2 (Trees in split network)   *Determine a second tree contained in the split network displayed in Figure 5.17. Construct a split network that contains at least three different bifurcating trees on $\mathcal{X} = \{a, b, c, d, e, f\}$.*

In practice, the idea of using a split network to represent multiple phylogenetic trees simultaneously needs refinement, as the full collection of splits may lead to a network that is too large and complicated to be useful. Also, often the phylogenetic trees are on overlapping, but non-identical, taxon sets. These two issues lead to

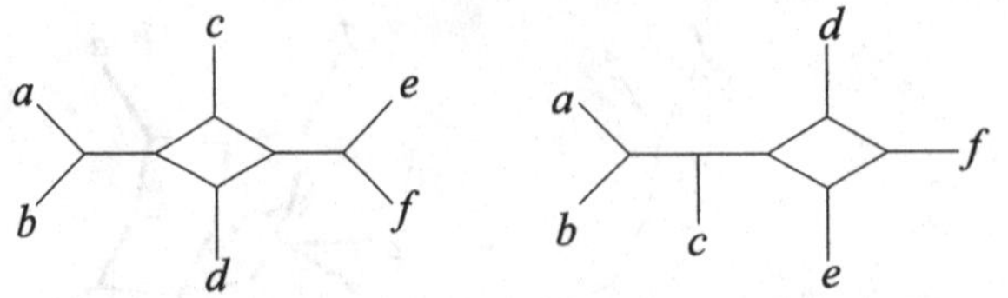

(a) Split network N_1　　　　　(b) Split network N_2

Figure 5.18 Two split networks used in Exercise 5.11.2.

the concepts of consensus split networks and super networks, which we discuss in Chapter 11.

5.11 Comparing split networks

In Section 6.14, we show that comparing two rooted phylogenetic networks is not a straightforward problem and we discuss a number of different distance functions that have been proposed to address this task. The comparison of split networks is much simpler, because there is very close relationship between a set of splits S on $\mathcal{X}$ and the split network N that represents those splits. To compare two split networks N_1 and N_2 on $\mathcal{X}$, we compare the corresponding set of splits S_1 and S_2, respectively, using the Robinson-Foulds distance [204]:

Definition 5.11.1 (Robinson-Foulds distance) *The Robinson-Foulds distance between two sets of splits S_1 and S_2 on $\mathcal{X}$ is based on their symmetric difference:*

$$d_{RF}(S_1, S_2) = \frac{|S_1 \triangle S_2|}{2}. \tag{5.56}$$

Note that the Robinson-Foulds distance is a proper metric for canonical split networks.

Exercise 5.11.2 (Robinson-Foulds distance on split networks) *Compute the Robinson-Foulds distance for the two sets of splits represented by the two split networks N_1 and N_2 depicted in Figure 5.18.*

5.12 T-theory

In this section, we give a very brief introduction to *T-theory*, which is a mathematical theory [63] that extends both the work on split decomposition [9] and earlier work on finite metric spaces [137]. This is provided for the sake of completeness and the material presented here is not used elsewhere in the book. T-theory is not for the mathematically faint-hearted.

Let D be a distance matrix on $\mathcal{X} = \{x_1, \ldots, x_n\}$. Throughout this section, we assume that D is a metric on $\mathcal{X}$, in particular fulfilling the identity requirement of

Definition 3.12.1, and we refer to $(\mathcal{X}, D)$ as a *finite metric space*. As we have seen, if D is additive, then we can represent $(\mathcal{X}, D)$ by an unrooted phylogenetic tree T in such a way that the distance $d(x, y)$ between any two taxa x and y equals their distance $d_T(x, y)$ in the tree.

From a mathematically more abstract point-of-view, we can interpret an unrooted tree as a compact, simply connected, one-dimensional polytope, and this leads to the following question posed in T-theory: If D is *not* additive, can we characterize an appropriate low-dimensional compact polytope into which we can embed $(\mathcal{X}, D)$ *isometrically*, that is, preserving all distances between elements of $\mathcal{X}$? This polytope can then serve as a mathematical generalization of a phylogenetic tree.

Another motivation for the development of T-theory is the desire to give a direct characterization of the phylogenetic tree T associated with an additive distance matrix D that is not based, explicitly or implicitly, on a tree construction algorithm.

The key concept in T-theory is a construction called the *tight span* that provides a (at most $\lfloor \frac{n}{2} \rfloor$-dimensional) compact polytope that contains a given finite metric space. The main aim of T-theory is to investigate and understand the combinatorial structure of the tight span.

Let D be a distance function on $\mathcal{X} = \{x_1, \ldots, x_n\}$. We first define an unbounded n-dimensional polytope associated with $(\mathcal{X}, D)$:

$$P(\mathcal{X}, D) = \left\{ \mathbf{v} = \begin{pmatrix} v_1 \\ \ldots \\ v_n \end{pmatrix} \in \mathbb{R}^n \mid v_i + v_j \geq d(x_i, x_j), \; i, j = 1, \ldots, n \right\}, \tag{5.57}$$

in particular with $v_i \geq 0$ for all i, which follows from $v_i + v_i \geq d(x_i, x_i) = 0$.

For two n-dimensional vectors $\mathbf{u}$ and $\mathbf{v}$ we define $\mathbf{u} \leq \mathbf{v}$ to mean $u_i \leq v_i$ for all $i = 1, \ldots, n$. We now come to the main definition:

Definition 5.12.1 (Tight span) *For a finite metric space $(\mathcal{X}, D)$ we define the tight span $T(\mathcal{X}, D)$ as the (compact) polytope consisting of all minimal points in $P(\mathcal{X}, D)$, given by:*

$$T(\mathcal{X}, D) = \left\{ \mathbf{v} \in P(\mathcal{X}, D) \mid \mathbf{w} \in P(\mathcal{X}, D) \text{ and } \mathbf{w} \leq \mathbf{v} \text{ implies } \mathbf{w} = \mathbf{v} \right\}. \tag{5.58}$$

We shall consider the tight span $T(\mathcal{X}, D)$ together with the distance function

$$||\mathbf{v}, \mathbf{w}||_\infty = \max \{|v_i - w_i| : i = 1, \ldots, n\} \tag{5.59}$$

as a metric space.

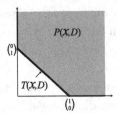

Figure 5.19 For a metric space $(\mathcal{X}, D)$ on two points $\mathcal{X} = \{x_1, x_2\}$ with $d(x_1, x_2) = 1$, we indicate the unbounded polytope $P(\mathcal{X}, D)$ in gray. The tight span $T(\mathcal{X}, D)$ consists of all points along the segment from $\binom{1}{0}$ to $\binom{0}{1}$.

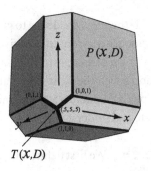

Figure 5.20 The polyhedron $P(\mathcal{X}, D)$ for three taxa with pairwise unit distances is contained in the positive quadrant of $\mathbb{R}^3$, as indicated here in gray. The tight span consists of the three line segments highlighted in bold.

Let us study a simple example given by $\mathcal{X} = \{x_1, x_2\}$ and $d(x_1, x_2) = 1$. In this case we have

$$P(\mathcal{X}, D) = \left\{ \binom{v_1}{v_2} \,\middle|\, v_1 \geq 0, v_2 \geq 0, v_1 + v_2 \geq 1 \right\} \tag{5.60}$$

and

$$T(\mathcal{X}, D) = \left\{ \binom{v_1}{v_2} \,\middle|\, v_1 \geq 0, v_2 \geq 0, v_1 + v_2 = 1 \right\}, \tag{5.61}$$

as illustrated in Figure 5.19.

In Figure 5.20 we show $P(\mathcal{X}, D)$ and $T(\mathcal{X}, D)$ for a set of three taxa $\mathcal{X} = \{x_1, x_2, x_3\}$ with distances $d(x_i, x_j) = 1$, if $i \neq j$, and $= 0$, else.

Exercise 5.12.2 (Tight span for three elements) *Construct $P(\mathcal{X}, D)$ and $T(\mathcal{X}, D)$ for the metric $(\mathcal{X}, D)$ on three elements with $d(x_1, x_2) = 1$, $d(x_1, x_3) = 2$ and $d(x_2, x_3) = 3$.*

In the following we briefly present three fundamental results from T-theory that show the relevance of the tight span construction for phylogenetic trees and networks.

Lemma 5.12.3 (Isometric embedding) *Let $(\mathcal{X}, D)$ be a finite metric space on $\mathcal{X} = \{x_1, \ldots, x_n\}$. The Kuratowski map $\kappa : \mathcal{X} \to \mathbb{R}^n$, defined for all taxa x_t in $\mathcal{X}$ as*

$$x_t \mapsto \kappa(x_t) = \begin{pmatrix} d(x_1, x_t) \\ \ldots \\ d(x_t, x_t) \\ \ldots \\ d(x_n, x_t) \end{pmatrix}, \tag{5.62}$$

maps the metric space $(\mathcal{X}, D)$ isometrically into the tight span $T(\mathcal{X}, D)$.

Proof Consider $\kappa(x_t)$ for an arbitrary taxon x_t. We first claim that $\kappa(x_t)$ lies in $P(\mathcal{X}, D)$. To see this, set

$$\mathbf{v} = \begin{pmatrix} v_1 \\ \ldots \\ v_n \end{pmatrix} = \begin{pmatrix} d(x_1, x_t) \\ \ldots \\ d(x_n, x_t) \end{pmatrix}, \tag{5.63}$$

and then for any pair of indices i and j we have

$$v_i + v_j = d(x_i, x_t) + d(x_j, x_t) \geq d(x_i, x_j), \tag{5.64}$$

by the triangle inequality, showing that the claim is correct.

Secondly, we claim that $\kappa(x_t)$ lies in $T(\mathcal{X}, D)$. To prove this, consider an arbitrary vector $\mathbf{u} = \begin{pmatrix} u_1 \\ \ldots \\ u_n \end{pmatrix}$ with $\mathbf{u} \leq \mathbf{v}$ and $\mathbf{u} \neq \mathbf{v}$. The latter two conditions imply that there exists an index i with $u_i < v_i$. For any other index j, we then have $u_i + u_j < v_i + v_j \leq d(x_i, x_t) + d(x_j, x_t) \leq d(x_i, x_j)$, by the triangle inequality. Thus, $\mathbf{u}$ is not contained in $P(\mathcal{X}, D)$, implying that $\kappa(x_t)$ is a minimal element of $P(\mathcal{X}, D)$ and thus lies in $T(\mathcal{X}, D)$.

Finally, we claim that the Kuratowski map preserves distances. To see this, consider two arbitrary elements x_i and x_j in $\mathcal{X}$. For the distance between their two images under the Kuratowski map we have

$$\|\kappa(x_i) - \kappa(x_j)\|_\infty = \max_i \begin{pmatrix} |d(x_1, x_i) - d(x_1, x_j)| \\ \ldots \\ |d(x_j, x_i) - d(x_j, x_j)| \\ \ldots \\ |d(x_n, x_i) - d(x_n, x_j)| \end{pmatrix} = d(x_i, x_j), \tag{5.65}$$

by application of the triangle equality to every entry of the vector. $\qquad \square$

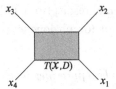

Figure 5.21 The tight span $T(\mathcal{X}, D)$ of a distance matrix D on $\mathcal{X} = \{x_1, \ldots, x_4\}$ that is not additive. The polytope consists of all points contained in the lines and in the gray area. The labels $x_1, \ldots, x_4$ indicate the locations of taxa under the Kuratowski embedding.

Returning to the example illustrated in Figure 5.19, we get

$$\kappa(x_1) = \begin{pmatrix} d(x_1, x_1) \\ d(x_2, x_1) \end{pmatrix} = \begin{pmatrix} 0 \\ 1 \end{pmatrix} \text{ and } \kappa(x_2) = \begin{pmatrix} d(x_1, x_2) \\ d(x_2, x_2) \end{pmatrix} = \begin{pmatrix} 1 \\ 0 \end{pmatrix}. \tag{5.66}$$

For the next result, we first need to establish a link between polytopes and trees. Let P be a compact n-dimensional polytope. We call P an $\mathbb{R}$-*tree*, if it is the set of all points contained in a graph-theoretical tree embedded in n-dimensional space. We state the following without proof.

Lemma 5.12.4 (Additivity and tight span) *Let* $(\mathcal{X}, D)$ *be a finite metric space. The distance matrix* D *is additive if and only if the tight span* $T(\mathcal{X}, D)$ *is an* $\mathbb{R}$-*tree.*

In Figure 5.21 we show the tight-span of a metric on four taxa that is not additive. Here is the third result, again stated without proof:

Lemma 5.12.5 (Split decomposition and tight span) *Let* D *be a distance matrix on* $\mathcal{X}$. *If* D *is totally decomposable then the canonical split network* N *representing the split decomposition of* D *is contained in the* 1-*skeleton (set of 1-dimensional faces) of* $T(\mathcal{X}, D)$.

Chapter notes

The mathematical foundations for the split decomposition and split networks were laid down in [9]. In particular, Sections 5.8 and 5.9 are based on that paper, although we have attempted to simplify the treatment. Our description of the Buneman graph is based on [217], however, the concept of a projection is new. The presented definition of a split network is from [62]. Our brief introduction to T-theory is based on [63].

6

Clusters and rooted phylogenetic networks

The concept of a cluster plays a fundamental role in classification: For a set of taxa $\mathcal{X}$ we may want to group some of the elements together in the form of a cluster, for example, because they share an attribute that distinguishes them from other elements. In the context of phylogenetic analysis, clusters should be clades or monophyletic groups that reflect the evolutionary history of a set of taxa.

Any rooted phylogenetic tree represents a set of *compatible* clusters. In this chapter, we study rooted phylogenetic networks that can be used to represent sets of incompatible clusters, as well. In phylogenetic analysis, incompatible sets of clusters may occur due to reticulate evolutionary events, such as hybridization or horizontal gene transfer, or they might reflect uncertainties due to insufficient data or inadequate analysis methods.

Ideally, a rooted phylogenetic network should explicitly explain the evolutionary history of a set of taxa in terms of evolutionary events such as mutations, speciations and reticulate events. Methods are beginning to emerge to address this challenge, but, as we shall see, practical methods for computing such networks have not yet been established.

In this chapter, we attempt to give a unified approach to rooted phylogenetic networks, bringing together a number of different topics and issues that have so far been treated separately in the literature. This chapter contains some material that is new and has not appeared in print before.

6.1 Overview

Figure 6.1 shows the relationships between some of the main concepts introduced in this chapter. The focus is on *rooted phylogenetic networks*. As a first example, we study *cluster networks* and then point out that any rooted phylogenetic network can be interpreted in two different ways, namely either as a *hardwired* network or as a *softwired* network. Cluster networks and rooted phylogenetic trees are the two

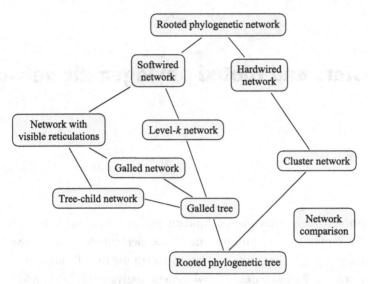

Figure 6.1 Overview of the main concepts introduced in this chapter. From bottom to top, each link represents a "can be interpreted as" relationship.

main examples of hardwired networks, and it is computationally easy to determine these from a given set of clusters.

However, when rooted phylogenetic networks are interpreted in a softwired sense, then many of the basic computational problems, such as determining whether a given network represents some given cluster, become computationally hard to solve, in general. Thus, a number of different definitions have been introduced in the literature to provide topological constraints on the types of networks studied, with the aim of making the computational questions easier to solve.

The concepts of a *galled tree*, *level-k network* and *galled network* all put constraints on how tangled the cycles in a rooted phylogenetic network may be, whereas the *tree-child* and *visible reticulations* properties aim at eliminating complicated patterns of cycles that do not directly influence the data generated on the network. In the last part of this chapter, we discuss the problem of comparing two rooted phylogenetic networks.

6.2 Clusters, compatibility and incompatibility

Recall the definition of a cluster given in Section 1.6:

Definition 6.2.1 (Cluster) *Let X be a set of taxa. A cluster C is any non-empty proper subset C of X. We explicitly exclude both the empty set $\emptyset$ and the full set X from this definition. A cluster C is called* trivial, *if it contains only one element. A weighted cluster is a cluster C that has been assigned a weight $\omega(C) \geq 0$.*

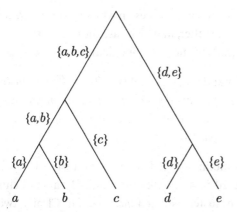

Figure 6.2   A rooted phylogenetic tree whose edges are labeled by the clusters that they represent.

Throughout this book, we usually assume that any set of clusters mentioned contains all trivial clusters, unless otherwise stated.

Let C be a set of clusters on $\mathcal{X}$. We say that C *separates* two taxa x, $y \in \mathcal{X}$ if there exists a cluster $C \in C$ such that either x is contained in C or y is contained in C, but not both. Because we usually assume that all trivial clusters are present, the sets of clusters considered usually separate all pairs of taxa in $\mathcal{X}$.

Let C be a set of clusters on $\mathcal{X}$ and let $\mathcal{X}' \subset \mathcal{X}$ be a subset of taxa. We define the set of clusters *induced* by $\mathcal{X}'$, or the *restriction* of C to the subset $\mathcal{X}'$, as

$$C|_{\mathcal{X}'} = \{C \cap \mathcal{X}' \mid C \in C \text{ and } C \cap \mathcal{X}' \neq \emptyset\}. \tag{6.1}$$

Every edge e of a rooted phylogenetic tree T on $\mathcal{X}$ is said to *represent* the cluster $C \subset \mathcal{X}$ that is defined as the set of all taxa that label nodes that are descendants of the target node of e. In the literature, alternative formulations of this concept are to say that e induces the cluster C or that e *displays* the cluster C. Equivalently, the cluster represented by e can be obtained as follows: Deletion of the edge e from T produces two connected components and the cluster C is then given by the set of all taxa that label leaves that are contained in the connected component that *does not* contain the root ρ. Figure 6.2 shows an example of a rooted phylogenetic tree and the clusters represented by its edges. We use $C(T)$ to denote the set of all clusters represented by edges in T. If the edges of the tree have lengths or weights, then these can be assigned to the corresponding clusters, as well.

Let us take a closer look at the relationships between different clusters in a rooted phylogenetic tree. At the lowest level of the tree, the leaf edges represent trivial clusters. Going up the tree toward the root, larger clusters are formed as the union of smaller clusters. Observe that if a cluster contains some other clusters as subsets, then the edges representing the subsets lie below the edge representing the superset. Also, note that for any two different clusters we have that either the two edges representing them are in different subtrees, in which case the two clusters are disjoint, or one edge lies above the other, in which case the one cluster contains the other. This latter observation leads to the following definition:

Definition 6.2.2 (Compatibility and incompatibility)   *Two different clusters C_1 and C_2 on $\mathcal{X}$ are called* compatible, *if they are disjoint or if one contains the other, that is, if $C_1 \cap C_2 = \emptyset$, $C_1 \subset C_2$ or $C_2 \subset C_1$. A set of clusters C on $\mathcal{X}$ is called* compatible, *if every pair of clusters in C is compatible.*

Compatible clusters are also sometimes called *nested*. Two clusters C_1 and C_2 on $\mathcal{X}$ that are not compatible are called *incompatible* and, similarly, a set of clusters that contains a pair of incompatible clusters is called *incompatible*. Clearly, two clusters C_1 and C_2 on $\mathcal{X}$ are incompatible if and only if we have $C_1 \cap C_2 \neq \emptyset$, $C_1 - C_2 \neq \emptyset$ and $C_2 - C_1 \neq \emptyset$.

As argued above, the set of clusters $\mathcal{C}(T)$ represented by a rooted phylogenetic tree T is always compatible. Vice versa, in Theorem 6.4.4, we will see that any set of compatible clusters C (containing all trivial clusters) on $\mathcal{X}$ can be represented by a rooted phylogenetic tree.

Exercise 6.2.3 (Incompatible clusters not on a tree)   *Show that the set of clusters $C = \{\{a, b\}, \{b, c\}, \{a\}, \{b\}, \{c\}\}$ is incompatible and prove that it cannot be represented by a rooted phylogenetic tree.*

A main topic of this chapter is how to represent a set of incompatible clusters. To this end, in Section 6.3, we first introduce the Hasse diagram as a means of representing set systems and then use it as a basis to develop the concept of a cluster network in Section 6.4.

Definition 6.2.4 (Incompatibility graph)   *The* incompatibility graph $IG(\mathcal{C})$ *of a set of clusters C is the graph (V, E) that has node set $V = C$ and edge set $E = \{\{C_1, C_2\} \mid C_1$ and C_2 are incompatible clusters $\}$.*

A cluster $C \in \mathcal{C}$ is compatible with all other clusters in $\mathcal{C}$ if and only if it is an isolated node in the incompatibility graph, see Figure 6.3. In consequence, a set of clusters $\mathcal{C}$ is compatible if and only if every connected component in the incompatibility graph $IG(\mathcal{C})$ has cardinality one.

As we will see later in the book, the incompatibility graph plays an important role in the computation of rooted phylogenetic networks, because it can be used to decompose the computational problem into subproblems that correspond to the non-trivial connected components of the incompatibility graph.

Exercise 6.2.5 (Incompatibility graph)   *Design a set of clusters C whose incompatibility graph $IG(\mathcal{C})$ is the same as the graph shown in Figure 6.4.*

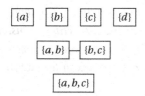

Figure 6.3   For the set of clusters $\mathcal{C} = \{\{a\}, \{b\}, \{c\}, \{d\}, \{a, b\}, \{b, c\}, \{a, b, c\}\}$, here we show the incompatibility graph $IG(\mathcal{C})$. This graph has seven nodes and only one edge, which connects the two nodes labeled $\{a, b\}$ and $\{b, c\}$.

Figure 6.4   Graph used in Exercise 6.2.5.

The following result can be used to decompose a collection of clusters into two (or more) mutually compatible sets of clusters. It is used in Section 8.2:

Lemma 6.2.6 (Decomposition lemma)   *Let $\mathcal{C}$ be a set of two or more clusters on $\mathcal{X}$. Assume that there exists a cluster C on $\mathcal{X}$ such that $\mathcal{C} = \mathcal{C}|_C \cup \mathcal{C}|_{\mathcal{X}-C}$ with $\mathcal{C}|_C \neq \emptyset$ and $\mathcal{C}|_{\mathcal{X}-C} \neq \emptyset$ (if $C \notin \mathcal{C}$), or $\mathcal{C} = \mathcal{C}|_C \cup \mathcal{C}|_{\mathcal{X}-C} \cup \{C\}$ (if $C \in \mathcal{C}$). Then, the incompatibility graph $IG(\mathcal{C})$ for $\mathcal{C}$ has at least two components.*

Exercise 6.2.7 (Proof)   *Prove Lemma 6.2.6.*

6.2.1 Optimal compatible subsets

Let $\mathcal{C}$ be a set of clusters on $\mathcal{X}$. If $\mathcal{C}$ is incompatible, then there are two basic computational problems that have been studied in the literature. Both aim at constructing a set of compatible clusters from the given data and also provide some measure of incompatibility for $\mathcal{C}$. The first problem is to remove a minimum number of clusters so that the remaining set of clusters is compatible:

Problem 6.2.8 (Maximum Compatibility problem)   *Determine a maximum-size subset of clusters $\mathcal{C}' \subset \mathcal{C}$ that is compatible.*

Solving the Maximum Compatibility problem is equivalent to finding a maximum *clique* in a given graph, that is, finding a subgraph of maximum size in which all pairs of nodes are connected. As the latter *Max Clique* problem is known to be NP-hard [146], it follows that the Maximum Compatibility problem is, too.

Exercise 6.2.9 (Proof of equivalence)   *Show that the Maximum Compatibility problem is equal to the Max Clique problem. (Hint: When constructing an instance of the Maximum Compatibility problem from a graph G that is an instance of the Max Clique problem, define one taxon per node of G and one taxon for every pair of two nodes that are not connected by an edge in G.)*

The second problem is to remove a minimum number of taxa so that the set of clusters restricted to the remaining taxa is compatible:

Problem 6.2.10 (Maximum Compatible Subset problem)   *Determine a maximum-size subset of taxa $\mathcal{X}' \subset \mathcal{X}$ such that the set of clusters $C|_{\mathcal{X}'}$ induced on $\mathcal{X}'$ is compatible.*

Unfortunately, this problem is also known to be NP-hard [227], which can be shown by reduction of the three dimensional matching problem [146].

6.3 Hasse diagrams

Hasse diagrams are used in mathematics to visualize partial orders on finite sets.

Definition 6.3.1 (Partial order)   *A partial order is a binary relation "$\preceq$" on a set C that is* reflexive, transitive *and* antisymmetric, *that is, for all $u, v, w \in C$ we have:*

(i) $u \preceq u$ *(reflexivity).*
(ii) *If $u \preceq v$ and $v \preceq w$ then $u \preceq w$ (transitivity).*
(iii) *If $u \preceq v$ and $v \preceq u$ then $u = v$ (antisymmetry).*

An important example of a partial order is set inclusion $\subseteq$. Any set of clusters C on a taxon set $\mathcal{X}$ is partially ordered by set inclusion, and this is the partial order that we are most interested in. A *partially ordered set* is given by a finite set C together with a partial order $\preceq$.

One simple way to represent a partially ordered set $(C, \preceq)$ is to define a directed graph $G = (V, E)$ whose node set consists of all members, $V = C$, and whose edge set represents all relationships between members, $E = \{(v, w) \mid v \preceq w\}$ (see Figure 6.5a). This graph will usually contain many edges that are redundant because they can be inferred from the properties listed in Definition 6.3.1: For each node v there will be an edge (v, v) that can be removed without loss of information, because reflexivity implies $v \preceq v$. Further, for any triangle of edges $\{(v, u), (u, w), (v, w)\} \subseteq E$ one can remove the edge (v, w) from E because transitivity implies that $v \preceq w$ must hold. Finally, note that antisymmetry implies that G contains no directed cycles and is thus a DAG. These considerations lead to the following classic definition:

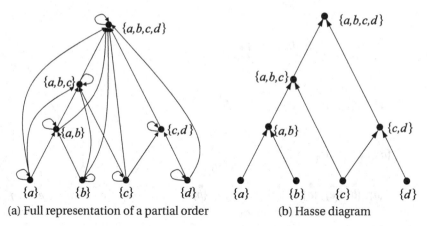

(a) Full representation of a partial order   (b) Hasse diagram

Figure 6.5   For $C = \{\{a, b, c, d\}, \{a, b, c\}, \{a, b\}, \{c, d\}, \{a\}, \{b\}, \{c\}, \{d\}\}$, we show **(a)** the full representation of the partial ordered set $(C, \subseteq)$ and **(b)** the Hasse diagram that represents $(C, \subseteq)$.

Definition 6.3.2 (Hasse diagram)   *Let $(C, \preceq)$ be a partially ordered set. The* Hasse diagram $H = (V, E)$ *that represents $(C, \preceq)$ is a DAG that is defined as follows:*

(i) *Set $V = C$.*
(ii) *For each pair of nodes v, w in V we define an edge $e = (v, w)$ in E if and only if $v \precneqq w$ and there exists no third node u in V with $v \precneqq u \precneqq w$.*

It is customary to draw a Hasse diagram H from bottom to top, for any edge (v, w) in H the node w is placed higher than the node v, and to omit the arrowheads, see Figure 6.6a.

Exercise 6.3.3 (Pseudo-code for Hasse diagram)   *Write an algorithm in pseudocode that takes as input a system of clusters C and produces as output the Hasse diagram that represents $(C, \subseteq)$.*

6.4 Cluster networks

Let C be a set of clusters on $\mathcal{X}$. In the previous section, we saw that C can be represented by a Hasse diagram in which each node is one of the clusters in C. However, this type of representation is rarely used in phylogenetics because it does not provide a direct generalization of phylogenetic trees: in a rooted phylogenetic tree, each cluster is usually associated with an edge, not a node. Also, the edges in a Hasse diagram are directed in the wrong direction, from smaller clusters to larger clusters. These two minor difficulties are overcome in the definition of a cluster network (see Figure 6.6).

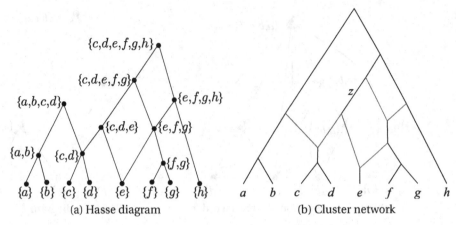

(a) Hasse diagram   (b) Cluster network

Figure 6.6   (a) Hasse diagram H and (b) cluster network N for the same set of clusters C on $\mathcal{X} = \{a, b, \dots, h\}$. Each tree edge in N represents exactly one cluster in C. For example, the edge labeled z represents the cluster $\{c, d, e\}$.

Recall that a rooted DAG is a (weakly) connected, directed acyclic graph that has precisely one node of indegree 0, called the *root* of G. Recall also that a node of outdegree 0 is called a *leaf*. A node of indegree ≤ 1 is called a *tree node*, whereas any node of indegree ≥ 2 is called a *reticulate node* or *reticulation*. An edge $e = (v, w)$ is called a *tree edge*, if its target node w is a tree node and, otherwise, it is called a *reticulate edge*.

Given a rooted DAG $G = (V, E)$, we use $L(G) \subset V$ to denote the set of leaves in G, and for any node v in G we use $L(v)$ to denote the set of leaves that are descendants of v. First we define a cluster network as a graph, without any reference to clusters [121]:

Definition 6.4.1 (Cluster network) *Let $\mathcal{X}$ be a set of taxa. A cluster network $N = (G, \lambda)$ on $\mathcal{X}$ consists of a rooted DAG $G = (V, E)$, together with a bijective leaf-labeling $\lambda : \mathcal{X} \rightarrow L(G)$ (see Figure 6.6b). For any pair of nodes v and w in G we require the following properties:*

  (i) *We have $L(v) \subseteq L(w)$ if and only if v is a descendant of w.*

  (ii) *We have $L(v) = L(w)$ if and only if $v = w$ or v is a reticulate node and parent of w or vice versa.*

  (iii) *If v is a child of w, then there exists no node u with $L(v) \subsetneq L(u) \subsetneq L(w)$.*

Additionally, we require that:

  (iv) *Every reticulate node in G has exactly one out-edge, which is a tree edge.*

We refer to the properties (i), (ii) and (iii) as the *nestedness, uniqueness* and *reducedness* condition, respectively. We refer to (iv) as the *reticulation-separation* property.

Every tree edge $e = (v, w)$ in a cluster network N defines a cluster $\gamma(e)$ on $\mathcal{X}$, namely the set of labels of all nodes in $L(w)$. The set $\mathcal{C}(N)$ of all clusters obtainable in this way is called the set of clusters *represented* by N. Note that we do not associate a cluster with any reticulate edge. Two main properties of a cluster network can also be formulated in terms of the two clusters $\gamma(e)$ and $\gamma(f)$ assigned to any pair of tree edges e and f:

(i') *Nestedness:* We have $\gamma(e) \subseteq \gamma(f)$, if and only if e lies below f, that is, there exists a directed path (possibly of length 0) from the target of f to the source of e.

(ii') *Uniqueness:* If $\gamma(e) = \gamma(f)$, then $e = f$.

Additionally, we have

(iii') *Reducedness:* If an edge e lies below an edge f then there does not exist an edge g that connects the source of f to the target of e.

We now discuss an algorithm, called the *cluster-popping algorithm* [121], which can be used to efficiently compute a cluster network that represents a given set of clusters $\mathcal{C}$ on $\mathcal{X}$. The algorithm proceeds in three steps: it first constructs (lines 3–13) the Hasse diagram for $\mathcal{C} \cup \{\mathcal{X}\}$ (recall that the full set $\mathcal{X}$ is not considered a member of $\mathcal{C}$), then inserts additional tree edges to ensure the reticulation-separation property holds (lines 14–17), and then finally, it sets up the taxon labeling (lines 18–19). By construction, each node v of the original Hasse diagram (except the root) corresponds to a cluster $C \in \mathcal{C}$ and is the target of exactly one tree edge e, and so this edge is used to represent C. The algorithm maintains a mapping v between the set of clusters $\mathcal{C}$ and the set of nodes in the network.

Algorithm 6.4.2 (Cluster-popping algorithm)   *Let $\mathcal{C}$ be a set of clusters on a taxon set $\mathcal{X}$. To construct the cluster network N that represents $\mathcal{C}$ (and only $\mathcal{C}$), do the following:*

1   **for each** *cluster $C \in \mathcal{C}$ create a node u and define $v(C) = u$.*
     Create a root node ρ
3   **for each** $C \in \mathcal{C}$ *in order of non-increasing cardinality* **do**
         Unmark all nodes
         Push ρ onto a stack S and mark ρ
         while *S is not empty* **do**
             Pop v off S
             Set isBelow $= false$

> **for each** *child w of v* **do**
>> **if** $C \subset v^{-1}(w)$ **then**
>>> *Set isBelow* $= true$
>>> **if** *w is unmarked* **then** *Mark w and push w onto S*
>> **if** *isBelow* $= false$ **then** *Create a new edge* $(v, v(C))$

14   **for each** *node v with indegree* ≥ 2 *and outdegree* $\neq 1$ **do**
> *Create a new node* v'
> *Redirect all in-edges of v to* v'
> *Create a new edge* (v', v)

18   **for each** *taxon* $a \in \mathcal{X}$ **do**
> *Set* $\lambda(a) = v^{-1}(\{a\})$

end

Figure 6.7 illustrates the cluster-popping algorithm on a small example. Throughout the execution of the algorithm, the map v provides a bijection of the clusters, together with the full taxon set $\mathcal{X}$, on to the set of nodes of the network. The local variable *isBelow* is set to true in the case that the current node v has a child that will be placed "above" (that is, is an ancestor of) the node $v(C)$. When processing a cluster, all nodes are initially unmarked; only unmarked nodes are pushed onto the stack, and then marked, so as to prevent any node from being pushed onto the stack more than once.

Because computing the Hasse diagram is order independent and the remaining part of the algorithm applies some modifications that are order independent, the final result is independent of the order in which C is processed.

Assume that the number of taxa is n and the number of clusters is m. As each node is incident to some tree edge and there are m tree edges, the number of nodes is at most $2m$ and the number of reticulate edges is at most $\binom{2m}{2}$. An upper bound for the running time of the algorithm is given by $O(nm^3)$.

Exercise 6.4.3 (Cluster network from trees)   *Consider the two rooted phylogenetic trees T_1 and T_2 depicted in Figure 6.8. Write down all the clusters represented in the tree T_1 and all the clusters represented in the tree T_2. Construct a cluster network that represents the total set of clusters.*

Theorem 6.4.4 (Compatible clusters correspond to tree)   *A cluster network N representing a set of clusters C (containing all trivial clusters) on $\mathcal{X}$ is a rooted phylogenetic tree if and only if C is compatible.*

Proof   Assume that C is compatible and that the associated cluster network N contains a reticulate node r. By the properties of a cluster network, r has exactly one child u, which is a tree node and is associated with a cluster $L(u) \in C$. Moreover, r has at least two parent nodes, v and w, say. By the reticulation-separation property,

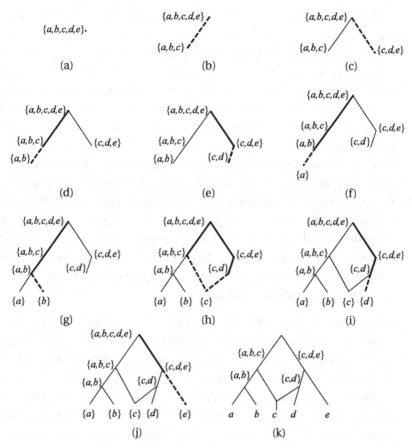

Figure 6.7   For clusters $C = \{\{a, b, c\}, \{c, d, e\}, \{a, b\}, \{c, d\}, \{a\}, \{b\}, \{c\}, \{d\}, \{e\}\}$ on $X = \{a, b, c, d, e\}$, we illustrate how the cluster-popping algorithm constructs the associated cluster network N. (a) We start with a single node representing the set of taxa X. We then process the set of clusters C in order of descending cardinality. In (b–j) we illustrate the result of processing each cluster, respectively, emphasizing in bold the edges that are used by the algorithm. Dashed lines indicate new edges inserted to accommodate a cluster. The graph in (j) is the Hasse diagram for $C \cup \{X\}$. In (k) we show the resulting cluster network N, with each non-trivial cluster labeling the edge that it is represented by.

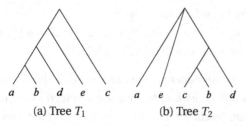

(a) Tree T_1 (b) Tree T_2

Figure 6.8   Two trees used as input for Exercise 6.4.3.

both v and w must be tree nodes, and so they both correspond to clusters $L(v) \in C$ and $L(w) \in C$, respectively. Now, we must have $L(v) \cap L(w) \neq \emptyset$, because both sets must contain $L(u)$. By uniqueness, we cannot have $L(v) = L(w)$. Moreover, $L(v) \subset L(w)$ and $L(w) \subset L(v)$ are both impossible due to the nestedness and reducedness properties of a cluster network. Hence, the two clusters $L(v)$ and $L(w)$ are incompatible, contradicting the assumption that C is compatible. Thus the cluster network N cannot contain a reticulate node and so N must be a tree, by Lemma 1.4.3.

Now consider a set of clusters C on X that contains two incompatible clusters C_1 and C_2, and let x be a taxon that is contained in the non-empty intersection $C_1 \cap C_2$. Any cluster network N representing C contains a leaf u labeled by x and also two distinct tree nodes v and w with $L(v) = C_1$ and $L(w) = C_2$, say. By nestedness, the two nodes v and w cannot be connected by a directed path. However, each is connected to u by a directed path. These two paths must both enter some node r along two different in-edges before they reach u, and thus N contains a reticulate node, namely r. □

Cluster networks provide a nice generalization of rooted phylogenetic trees: in a cluster network, each cluster is represented by precisely one tree edge, whereas the role of the additional reticulate edges is to facilitate the sharing of taxa that is required by incompatible clusters.

Cluster networks can be drawn either as cladograms, in which edge lengths have no direct interpretation, or as phylograms, in which the tree edges are drawn to scale to represent evolutionary distances. Both approaches are discussed in Chapter 13.

6.5 Rooted phylogenetic networks

Cluster networks are a first example of a general class of networks that we call *rooted phylogenetic networks.*

Definition 6.5.1 (Rooted phylogenetic network) *Let X be a set of taxa. A rooted phylogenetic network $N = (V, E, \lambda)$ on X consists of a directed graph $G = (V, E)$ and a node labeling $\lambda : X \to V$ such that:*

  (i) *The graph G is a directed, acyclic graph (DAG).*
  (ii) *The graph G is rooted, that is, it has precisely one node ρ of indegree 0.*
  (iii) *The node labeling λ assigns a taxon to each leaf of G.*

We usually assume that the network does not contain any node that is suppressible, in other words, any unlabeled node with indegree 1 and outdegree 1. In addition we usually assume that the taxon labeling λ is a bijection between the taxon set X and the set of leaves of N.

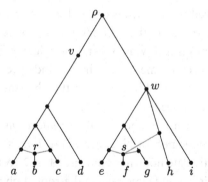

Figure 6.9   In this rooted phylogenetic network N, the leaves are labeled by taxa $a, \ldots, i$, the root node is ρ, there are two reticulate nodes r and s, and all other nodes are tree nodes. The node v is suppressible. All tree nodes except the node w are bifurcations. The reticulate node r is a bicombination, whereas s is not.

Recall that Lemma 1.4.2 states that a rooted DAG is always connected. Here is a summary of all the types of nodes and edges that we would like to distinguish among (see Figure 6.9):

Definition 6.5.2 (Types of nodes and edges)   *Let N be a rooted phylogenetic network on a taxon set $\mathcal{X}$.*

  (i)   *The one node with indegree 0 is called the* root.
 (ii)   *A node with outdegree 0 is called a* leaf *node.*
(iii)   *A node with at most one in-edge is called a* tree *node.*
(iv)   *Any edge leading to a tree node is called a* tree *edge.*
 (v)   *A node with at least two in-edges is called a* reticulate *node or a* reticulation.
(vi)   *Any edge leading to a reticulate node is called a* reticulate *edge.*
(vii)   *An unlabeled node with indegree 1 and outdegree 1 is called a* suppressible *node. As defined in Section 1.2, to* suppress *such a node v with in-edge (u, v) and out-edge (v, w) means to delete v and then to create a new edge (u, w), if such an edge does not already exist.*

Definition 6.5.1 is the most general definition of a rooted phylogenetic network N that is given in this book. In analogy to the term $\mathcal{X}$-tree mentioned in Chapter 3, we could also call N a *rooted $\mathcal{X}$-network*, as here we also allow taxa to appear as labels of internal nodes and also allow nodes to be labeled by multiple taxa. However, we do not use this general term.

The definition of a rooted phylogenetic network can be restricted in several ways. For example in the context of biological processes such as speciation by hybridization, horizontal gene transfer or recombination, it is generally assumed that only two species can be involved in any such event and so sometimes the set of possible indegrees of nodes is restricted to the values 0, 1 and 2, corresponding to the root, tree nodes and reticulations involving only two in-edges, respectively. However, this is not a mandatory requirement, as a higher indegree may also indicate uncertainty in the order of a set of multiple events, rather than the existence of one event involving more than two lineages.

As in the case of phylogenetic trees, a tree node is called a *bifurcation*, if it has outdegree 2, and a rooted phylogenetic network is called *bifurcating*, if every tree node has outdegree 2 and every reticulate node has outdegree 1 or 2. A reticulate node is called a *bicombination*, if it has indegree two, and a rooted phylogenetic network is called *bicombining*, if every reticulate node has indegree two. A node whose indegree is larger than two is called a *multicombining* node and if such nodes are present, then we speak of a *multicombining network*. A rooted phylogenetic network that is both bifurcating and bicombining and in which every reticulate node has outdegree one is called *resolved*.

Exercise 6.5.3 (Rooted network as cluster network) *After suppressing the node v in the rooted phylogenetic network N shown in Figure 6.9, the network can be interpreted as a cluster network. Determine the corresponding set of clusters. The network is not resolved, does there exist a resolved network that represents the same set of clusters?*

Let N be a rooted phylogenetic network on $\mathcal{X}$. We say that a rooted phylogenetic network N' is *embedded* in N, if N' can be obtained from N by performing a series of node deletions, edge deletions and node suppressions, as defined in Section 1.2. This is illustrated in Figure 6.10.

6.6 The lowest stable ancestor

In a rooted tree T, any two nodes v and w have a unique *lowest common ancestor* *(LCA)* that is defined as the lowest node $lca(v, w)$ that is an ancestor of both v and w. In a rooted phylogenetic network N, it can happen that two nodes do not have a *unique* LCA. In this case, the LCA is defined as the set of all lowest nodes that are ancestors of both v and w. For example, in Figure 6.11b, the LCA of nodes b and c consists of the two nodes p and q. Both in trees and networks, the LCA of a set of more than two nodes $\{v_1, v_2, v_3, \dots\}$ is defined as $lca(v_1, lca(v_2, lca(v_3, \dots)))$.

The "lowest stable ancestor" (sometimes also called the "lowest single ancestor") is a related concept that plays a role in the context of rooted phylogenetic networks [121]:

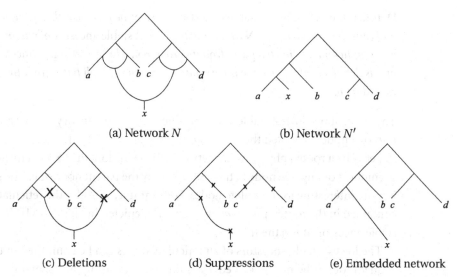

(a) Network N (b) Network N'

(c) Deletions   (d) Suppressions   (e) Embedded network

Figure 6.10   In (a) and (b) we show two rooted phylogenetic networks N and N'. (In this example, the latter is a tree, for simplicity.) To demonstrate that N' is embedded in N, starting with N, we first delete one node and one edge, as indicated by the two crosses in (c). We then suppress five nodes of degree two as indicated by the five crosses in (d). The resulting network shown in (e) is topologically the same as the desired network N'.

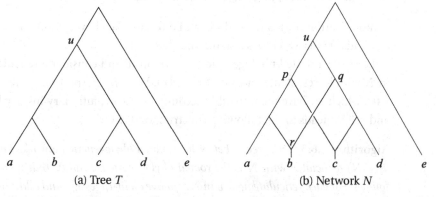

(a) Tree T (b) Network N

Figure 6.11   (a) In this rooted phylogenetic tree T, the lowest common ancestor of the two nodes labeled b and c is the node u. (b) In this rooted phylogenetic network N, the two nodes labeled b and c have two lowest common ancestors, namely the nodes p and q. In contrast, their lowest stable ancestor is uniquely defined; it is the node u. The lowest stable ancestor of the reticulate node r is u.

Definition 6.6.1 (Lowest stable ancestor)　*Let N be a rooted phylogenetic network and consider any node v in N that is not the root. A stable ancestor of v is any ancestor of v that lies on all directed paths from the root to v. The lowest stable ancestor (LSA) of v is defined as the last node $lsa(v)$ that is contained on all paths from the root to v, excluding v.*

The concept of a lowest stable ancestor is not new. In the theory of flow graphs, the LSA of a node v is called the *immediate dominator* of v [164].

Let N be a rooted phylogenetic network. By definition, the LSA of any node in N is unique. For any tree node v, the LSA is simply the parent node of v. The situation is more interesting for reticulate nodes. For example, consider the reticulate node r contained in the rooted phylogenetic network depicted in Figure 6.11b. The lowest stable ancestor of r is the node u.

The lowest stable ancestors of all reticulate nodes can be computed in time that is quadratic in the number of edges in the network using the following approach [121]:

Algorithm 6.6.2 (Computation of all LSAs)　*To compute the LSA of each individual reticulate node of N, one can proceed in a single postorder traversal of the network: For each reticulate node v and every node w in the network, keep track of which (reversed) paths from v toward the root go through w. If, at any point, a node u is encountered such that the set of all paths through u equals the set of all paths currently in existence for v, then the LSA of v has been found; it equals u.*

A modification of this algorithm can be used to label every node u by the set of descendants for which u is a stable ancestor.

Let N be a rooted phylogenetic network on $\mathcal{X}$ and consider a reticulate node r in N with lowest stable ancestor $\ell = lsa(r)$. We can interpret r as a descendant of ℓ for which we have incompatible accounts of the evolutionary history between ℓ and r. This leads to the following construction [121]:

Algorithm 6.6.3 (LSA tree)　*Let N be a rooted phylogenetic network. The LSA tree $T_{\text{LSA}}(N)$ associated with N is the rooted phylogenetic tree on $\mathcal{X}$ that is computed as follows: For each reticulate node v in N, remove all in-edges of v and add a new in-edge $e = (lsa(v), v)$ from the LSA of v to v. Then repeatedly remove all unlabeled leaves and all nodes that have both in- and outdegree one, until no further such removal is possible. The result is a rooted phylogenetic tree $T_{\text{LSA}}(N)$ called the LSA tree of N (see Figure 6.12).*

To see that the resulting graph is indeed a rooted phylogenetic tree, simply note that after application of the algorithm, all nodes (except the root of N) will have indegree one and then apply Lemma 1.4.3. The LSA tree of a rooted phylogenetic network plays an important role in algorithms used for drawing such networks, as

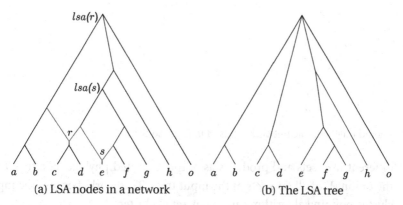

(a) LSA nodes in a network   (b) The LSA tree

Figure 6.12   (a) A rooted phylogenetic network N containing two reticulations, r and s, that have different lowest stable ancestors, $lsa(r)$ and $lsa(s)$. The node labeled e has four stable ancestors, namely the node labeled s, the node labeled $lsa(s)$ and the two nodes above it. (b) The LSA tree $T_{lsa}(N)$ associated with N.

discussed in Chapter 13. In Section 6.7, we look at other ways of obtaining trees from a rooted phylogenetic network.

Let $\mathcal{X}' \subset \mathcal{X}$ be a non-empty subset of taxa and N a rooted phylogenetic network on $\mathcal{X}$. Let $lsa(\mathcal{X}')$ denote the last node contained on all paths from the root of N to any leaf labeled by a taxon in $\mathcal{X}'$. The network *induced* by $\mathcal{X}'$ in N is defined as the unique network embedded in N that contains all nodes and edges that lie on any directed path from the node $lsa(\mathcal{X}')$ to any leaf labeled by a taxon in $\mathcal{X}'$.

6.6.1 LSA consensus tree

Let $\mathcal{T}$ be a collection of rooted phylogenetic trees on $\mathcal{X}$ and let $\mathcal{C}(\mathcal{T})$ be the set of all clusters represented by any tree in $\mathcal{T}$. We obtain the *LSA consensus tree* for $\mathcal{T}$, denoted by $T_{LSA}(\mathcal{T})$, by first constructing the cluster network $N(\mathcal{C}(\mathcal{T}))$ for the set of all clusters $\mathcal{C}(\mathcal{T})$ and then returning the LSA tree $T_{LSA}(N)$ of N. More formally, $T_{LSA}(\mathcal{T}) = T_{LSA}(N(\mathcal{C}(\mathcal{T})))$. The LSA consensus tree $T_{LSA}(\mathcal{T})$ is closely related to the Adams consensus tree, which was introduced in Section 3.17, see Figure 3.27.

Lemma 6.6.4 (LSA and Adams consensus trees are compatible)   *Let $\mathcal{T}$ be a collection of rooted phylogenetic trees on $\mathcal{X}$. The LSA consensus tree $T_{LSA}(\mathcal{T})$ and Adams consensus tree $T_{ADAMS}(\mathcal{T})$ are compatible in the sense that they possess a common refinement, that is, there exists a rooted phylogenetic tree T on $\mathcal{X}$ such that both $T_{LSA}(\mathcal{T})$ and $T_{ADAMS}(\mathcal{T})$ can each be obtained from T by contracting some edges in T.*

We leave the proof as a challenging exercise:

Exercise 6.6.5 (Proof)   *Prove Lemma 6.6.4.*

The LSA tree construction can also be used as a *super-tree method*, called the *LSA super-tree* method, that takes as input a set of rooted phylogenetic trees $\mathcal{T}$ on

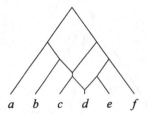

Figure 6.13 Rooted phylogenetic network N_1 used in Exercise 6.7.2.

different taxa sets and produces as output a rooted phylogenetic tree $T_{\text{LSA}}(T)$ on the union $\mathcal{X}$ of all taxa sets of the input trees. This works because the input of the cluster-popping algorithm can be any set of clusters.

6.7 Representing trees in rooted networks

Let $T = \{T_1, T_2, \ldots, T_m\}$ be a collection of rooted phylogenetic trees on $\mathcal{X}$. For example, these might be the result of a multiple gene analysis or they might be different estimations of a single gene-based phylogeny. In practice, such a collection of trees is often combined into a consensus tree $\bar{T}$ that is then interpreted as representing the parts of the evolutionary history upon which the different trees agree.

An alternative idea is to use a rooted phylogenetic network N to represent T in the following sense:

Definition 6.7.1 (Rooted trees represented by a rooted network) *Let N be a rooted phylogenetic network on a taxon set $\mathcal{X}$. We say that a rooted phylogenetic tree T on $\mathcal{X}$ is represented by N, or that T is displayed by N, if T is embedded in N. We use $T(N)$ to denote the set of all rooted phylogenetic trees on $\mathcal{X}$ that are represented by N.*

Exercise 6.7.2 (Networks and trees) *Draw all rooted phylogenetic trees represented by the rooted phylogenetic network N_1 shown in Figure 6.13. Then construct the cluster network N_2 associated with the clusters of the obtained trees. Are the two networks N_1 and N_2 the same or different? Explain.*

Let T be a set of rooted phylogenetic trees on $\mathcal{X}$. How can one compute a rooted phylogenetic network N that represents T, that is, for which $T \subseteq T(N)$ holds? Let $C(T)$ be the set of all clusters contained in any of the rooted trees in T. A first simple idea is to use the cluster-popping algorithm to compute the cluster network N that represents $C(T)$. This actually works quite well because of the following result:

Lemma 6.7.3 (Cluster network contains rooted trees) *Let T be a set of rooted phylogenetic trees on $\mathcal{X}$ and let C be the set of clusters of T. The cluster network N for C represents a refinement of every tree in T.*

The idea of the proof of this result is to show that each edge in a given tree T maps onto a path in the cluster network N and that two such paths can only intersect at their end nodes. As the paths do not necessarily intersect, in general, the embedding is that of a refinement of T, rather than of T itself.

Proof In the following, let $L_N(v)$ denote the set of taxa that label leaves below a node v in a rooted phylogenetic network N. Consider any tree $T \in \mathcal{T}$ and let $e = (v, w)$ be an edge of T. By the nestedness property, there exist two nodes v' and w' in N such that $L_N(v') = L_T(v)$, $L_N(w') = L_T(w)$ and v' is an ancestor of w'. We choose a path $P(v', w')$ from v' to w' to represent T in N. We need to show that different edges in T correspond to disjoint paths in N as then the claim follows from the nestedness property of N. Let $f = (p, q)$ be a second edge in T with path $P(p', q')$ in N. Assume that some node z in $P(p', q')$ is contained in the interior of $P(v', w')$. We cannot have $z = p'$ or $z = q'$ because then we would have $L_T(v) \subset L_T(p) \subset L_T(w)$ or $L_T(v) \subset L_T(q) \subset L_T(w)$, respectively, contradicting the fact that (v, w) is an edge in T. Similarly, we cannot have that v' or w' is contained in the interior of $P(p', q')$. Assume that the two paths intersect on an interior node z. Then $L(w) \cup L(q) \subset L(v) \cap L(p)$ and so, by compatibility, we have $L(p) \subset L(v)$ or $L(v) \subset L(p)$. The former case implies $L(w) \subset L(w) \cup L(q) \subset L(p) \subset L(v)$ and the latter implies $L(q) \subset L(w) \cup L(q) \subset L(v) \subset L(p)$, in both cases contradicting the fact that (v, w) and (p, q) are edges in T. □

The following exercise emphasizes that the word "refinement" cannot be omitted in the formulation of the above lemma:

Exercise 6.7.4 (Trees in cluster network) *Determine two rooted phylogenetic trees T and T' (not necessarily bifurcating) such that the cluster network N for $C(T) \cup C(T')$ represents neither T nor T'.*

So the cluster-popping algorithm provides a fast way of obtaining a rooted phylogenetic network that represents a set of rooted phylogenetic trees $\mathcal{T}$, see Figure 6.14a–c. However, in practice, the resulting network may sometimes be too large and messy to be of real use. One way to address this problem, discussed in Section 11.4, is to only represent those clusters that occur in at least p percent of the input trees, where p is a user-defined parameter. The resulting network will then no longer represent all trees in their full resolution, as some of them will occur only in a contracted form.

A second idea is to search for a rooted phylogenetic network N that represents $\mathcal{T}$ using a minimum number of reticulations, nodes and edges. This is discussed in Section 11.5 for the case of two trees.

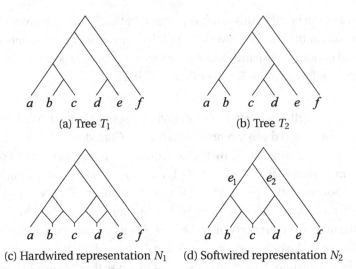

(a) Tree T_1 (b) Tree T_2

(c) Hardwired representation N_1 (d) Softwired representation N_2

Figure 6.14 In (a) and (b) we show two rooted phylogenetic trees T_1 and T_2. (c) A cluster network N_1 that represents all clusters contained in T_1 and T_2 (in the hardwired sense). (d) A second rooted phylogenetic network N_2 representing the same set of clusters in the softwired sense. In this particular example, both input trees are also represented by both networks.

6.8 Hardwired and softwired clusters

In this section, we discuss the problem of defining the set of clusters $C(N)$ associated with a given rooted phylogenetic network N. As we will see, this can be done in two different ways and we propose to use the terms "hardwired" and "softwired" to distinguish between the two different approaches [131].

Let N be a rooted phylogenetic network on $\mathcal{X}$. If N is a cluster network as defined in Section 6.4, then each tree edge e in N represents precisely one cluster $\gamma(e)$, which is given by the set of all taxa that appear as labels of nodes below e, that is, all labels of nodes that are descendants of the target node of e, see Definition 6.4.1. We use the term *hardwired clusters* to refer to the set of all clusters $C_{hard}(N)$ that are obtainable from a rooted phylogenetic network N is this way. An example is shown in Figure 6.14c.

An alternative way to define the set of clusters represented by N is to use the set of all clusters obtainable from the set of trees $\mathcal{T}(N)$ represented by N. We refer to this as the set $C_{soft}(N)$ of clusters represented by N in the *softwired* sense, for reasons that should now become apparent. To obtain these clusters directly from the network N, one must treat the in-edges leading to each reticulation r as a set of alternatives, one of which is "on" if and only if all others are "off". A softwired cluster C is then obtained directly from the network by first deciding for each reticulation which reticulation edge is on and which is off and then collecting all taxa that are reachable below some fixed tree edge e without using any reticulation edge that is off. An example is shown in Figure 6.14d.

It is important to understand and to distinguish between these two different ways of interpreting a rooted phylogenetic network and so we summarize them here:

Definition 6.8.1 (Hardwired and softwired intepretation)   *Let N be a rooted phylogenetic network on $\mathcal{X}$. Under a hardwired interpretation, any tree edge e in N represents precisely one cluster $\gamma(e)$, namely the set of all taxa that label nodes that are descendants of the target node of e. Under a softwired interpretation, any tree edge e in N represents a set $C(e)$ of one or more clusters. Each member of this set is determined as follows: First, for each reticulation r, turn one in-edge on and all others off. Then, for each such set of choices, a cluster belonging to $C(e)$ is given by the set of all taxa that label nodes that lie below e and can be reached without using any reticulate edge that is off.*

To understand the relationship between $\mathcal{C}_{hard}(N)$ and $\mathcal{C}_{soft}(N)$, consider Figure 6.14. In (a) and (b) we show two different rooted phylogenetic trees T_1 and T_2 on the taxon set $\mathcal{X} = \{a, b, c, d, e, f\}$. In (c) and (d) we show two different rooted phylogenetic networks N_1 and N_2 that both represent or *display* the two trees T_1 and T_2. Now, note that all clusters that are represented by (or *contained in*) T_1 and T_2 are also represented by (or *contained in*) the cluster network N_1 in the hardwired sense. However, network N_2 does not contain all such clusters in the hardwired sense, for example, the cluster $\{a, b\}$ is missing, because in this network, any edge that lies above a and b also lies above c.

In the rooted phylogenetic network N_2, the two clusters $\{a, b\}$ and $\{a, b, c\}$ are both represented by the same edge e_1 and the two clusters $\{d, e\}$ and $\{c, d, e\}$ are both represented by the same edge e_2. This illustrates that a rooted phylogenetic network that is interpreted in the softwired sense will usually require fewer edges to represent a set of clusters than a hardwired one, because individual tree edges can represent more than one cluster.

Exercise 6.8.2 (Hardwired and softwired clusters)   *Let N be a rooted phylogenetic network on $\mathcal{X}$. Show that $\mathcal{C}_{hard}(N) \subseteq \mathcal{C}_{soft}(N)$ holds.*

Assume we are given a set of rooted phylogenetic trees $\mathcal{T}$ on $\mathcal{X}$. If N is a rooted phylogenetic network that represents all trees in $\mathcal{T}$, then N will also represent all clusters $\mathcal{C}(\mathcal{T})$ of the trees in $\mathcal{T}$ in the softwired sense. However, N will not necessarily represent those clusters $\mathcal{C}(\mathcal{T})$ in the hardwired sense, as the above example shows.

Exercise 6.8.3 (Clusters in networks)   *Consider Figure 6.14. Write down all hardwired clusters contained in the rooted phylogenetic network N_1 and all softwired*

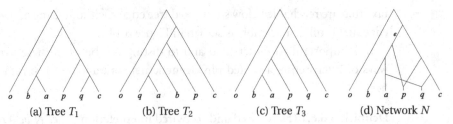

(a) Tree T_1 (b) Tree T_2 (c) Tree T_3 (d) Network N

Figure 6.15 All clusters represented in the three rooted phylogenetic trees T_1–T_3 shown in (a–c) are all represented in the rooted phylogenetic network N shown in (d) in the softwired sense. However, N does not contain all three trees, as T_3 is missing.

clusters contained in N_2. Verify that each of the two networks contains all clusters represented by the two rooted phylogenetic trees T_1 and T_2.

When a given rooted phylogenetic network N is to be explicitly interpreted in the hardwired or softwired sense, then we sometimes refer to it as a *hardwired network* or *softwired network*, respectively. Cluster networks are defined as hardwired networks. However, all other rooted phylogenetic networks are usually interpreted in the softwired sense.

If we are given a set of trees T on X, then it is natural to ask whether a given rooted phylogenetic network that represents all the clusters of the trees will also automatically represent the trees themselves. We have already seen that this is true for cluster networks, up to refinement. However, it is not true in general. In Figure 6.15 we show three trees T_1, T_2 and T_3, and a rooted phylogenetic network N that represents all clusters of the three trees in the softwired sense. This network represents trees T_1 and T_2, but not T_3.

Lemma 6.8.4 (Representing clusters and trees in softwired network) *Let T be a set of rooted phylogenetic trees on X and let $C(T)$ be the set of all clusters contained in these trees. A rooted phylogenetic network N that represents $C(T)$ in the softwired sense does not necessarily contain all trees in T.*

Proof Figure 6.15 shows an example of three rooted phylogenetic trees T_1, T_2 and T_3 and a rooted phylogenetic network N such that N contains all the clusters present in the three trees, but not all three trees. (Example by Magnus Bordewich.)   □

How many clusters does a rooted phylogenetic network represent? A hardwired network represents one cluster per tree edge. The number of softwired clusters represented by a rooted phylogenetic network N is exponential in the number of reticulations contained in N, in the worst case. In slightly more detail, the number is bounded from above by

$$\sum_{e \text{ a tree edge}} 2^{\text{sp}(e)}, \tag{6.2}$$

where $\text{sp}(e)$ is the number of reticulations that *span* e, that is, the number of reticulate nodes that lie below e but can also be reached from the root of N by an alternative path that does not contain e. In other words, this is the number of reticulate nodes for which the target node of e is an ancestor, but not a stable ancestor.

For example, the tree edge labeled e in Figure 6.15 is spanned by both reticulations of the network N and so we have $2^{\text{sp}(e)} = 2^2 = 4$. This value is indeed attained, as the edge e represents the four softwired clusters $\{a, b\}$, $\{a, b, p\}$, $\{a, b, q\}$ and $\{a, b, p, q\}$. Two of these clusters, the first and third, are also represented by another edge in the network.

Definition 6.8.5 (Representation of clusters) *Assume that we are given a set of clusters C and a rooted phylogenetic network N, on X. We say that N represents C in the* hardwired *sense, or softwired sense, if $C \subseteq C_{hard}(N)$, or $C \subseteq C_{soft}(N)$, respectively.*

6.9 Minimum rooted phylogenetic networks

Let C be a set of clusters on X and let N be some rooted phylogenetic network that represents C in the softwired sense. In a biological interpretation of N, the reticulate nodes will ideally correspond to reticulate evolutionary events such as speciation by hybridization, horizontal gene transfer, or recombination of closely related sequences, depending on the context. If we assume that such evolutionary events occur only rarely, then this justifies that we require that a rooted phylogenetic network N representing C should contain as few reticulations as possible. More formally, we pose the following computational problem:

Problem 6.9.1 (Minimum Rooted Phylogenetic Network problem) *For a given set of clusters C on X, compute a bicombining rooted phylogenetic network N that represents C in the softwired sense and contains a minimum number of reticulate nodes and then a minimum number of edges.*

There are two ways to extend this to multicombining networks. On the one hand, we can simply count all reticulate nodes, whether bicombining or multicombining. This is done, for example, in the computation of galled networks, as described in Section 8.4. On the other hand, we can modify the optimization goal so as to minimize the value

$$r(N) = \sum_v (\text{indeg}(v) - 1), \tag{6.3}$$

where the sum is taken over all reticulate nodes v in the given rooted phylogenetic network N and $\text{indeg}(v)$ is the indegree of v. If all reticulations are bicombining then $r(N)$ is simply the number of reticulations in N. Otherwise, $r(N)$ counts the number of reticulations that any bicombining refinement of N would have. This generalization of the problem is more widely used. Problem 6.9.1 is hard:

Theorem 6.9.2 (Minimum network is hard) *Let C be a set of clusters on $\mathcal{X}$. The problem of computing a bicombining rooted phylogenetic network N that represents C in the softwired sense and contains a minimum number of reticulate nodes is NP-hard.*

Proof The following result is shown in [135]: Let T_1 and T_2 be two bifurcating, rooted phylogenetic trees on $\mathcal{X}$. The minimum number of reticulations used in any rooted phylogenetic network that represents both T_1 and T_2 is equal to the minimum number of reticulations used by any softwired network that represents all clusters of the two trees. The following result is shown in [27, 241]: Let T_1 and T_2 be two bifurcating, rooted phylogenetic trees on $\mathcal{X}$. The problem of determining the minimum number of reticulations contained in any rooted phylogenetic network that represents both T_1 and T_2 is NP-hard, see also Section 11.5. Together, these two results imply the theorem. □

In Section 6.10 we argue that an unconstrained minimization of the number of reticulations in a network does not make sense, as this may lead to completely unrelated parts of the network being brought together so as to save reticulations. This leads to the concept of decomposability. Then, in Section 6.11 we discuss one approach to making the problem computationally simpler that operates by imposing topological constraints on the networks that are taken under consideration.

Exercise 6.9.3 (Cluster network is minimum?) *Is it true that a cluster network N minimizes the number of reticulate nodes used to represent a given set of clusters C on $\mathcal{X}$ in the hardwired sense? Does it also minimize the number of edges used?*

6.10 Decomposability

Let C be a set of clusters on $\mathcal{X}$ and let N be a rooted phylogenetic network that represents C, either in the hardwired or softwired sense. There is a close relationship between the non-trivial connected components of the incompatibility graph $IG(C)$ and the non-trivial biconnected components of N. For example, in Figure 6.16 we list a set C of 13 clusters on $\mathcal{X} = \{a, \ldots, f\}$ whose incompatibility graph has two non-trivial connected components. This is accompanied by a hardwired network N

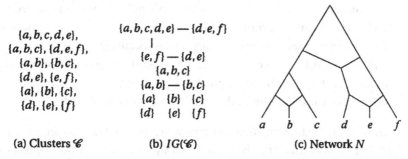

$\{a, b, c, d, e\},$
$\{a, b, c\}, \{d, e, f\},$
$\{a, b\}, \{b, c\},$
$\{d, e\}, \{e, f\},$
$\{a\}, \{b\}, \{c\},$
$\{d\}, \{e\}, \{f\}$

(a) Clusters $\mathscr{C}$

$\{a, b, c, d, e\} - \{d, e, f\}$
|
$\{e, f\} - \{d, e\}$
$\{a, b, c\}$
$\{a, b\} - \{b, c\}$
$\{a\} \quad \{b\} \quad \{c\}$
$\{d\} \quad \{e\} \quad \{f\}$

(b) $IG(\mathscr{C})$

(c) Network N

Figure 6.16 For the set of clusters C listed in (a) we show the incompatibility graph $IG(C)$ in (b). This graph has two non-trivial connected components, one containing four clusters $C' = \{\{a, b, c, d, e\}, \{d, e, f\}, \{e, f\}, \{d, e\}\}$ and the other containing two clusters $C'' = \{\{a, b\}, \{b, c\}\}$. In (c) we show a rooted phylogenetic network N that represents the set of clusters C in the hardwired sense. This network has two non-trivial biconnected components, the larger containing those edges that represent the three clusters in C' and the smaller one containing those edges that represent the two clusters in C''.

for C that has two corresponding non-trivial biconnected components. As we will see, it is desirable to enforce a one-to-one correspondence between the two types of components and this can be achieved using an appropriate divide-and-conquer strategy.

Let N be a softwired network on $\mathcal{X}$ and assume that C is a cluster that is represented by some tree edge e in N. There may be more than one such edge that represents C. In particular, if the target node of e has exactly two out-edges f and g, where f is a tree edge and g is a reticulate edge, then any cluster that is represented by e, with the reticulate edge g turned off, can also be represented by f.

For simplicity, throughout this book we assume that for any rooted phylogenetic network N representing a set of clusters C on $\mathcal{X}$ there is an *edge assignment map* α that chooses one suitable tree edge $\alpha(C)$ for each cluster in C. However, this map is usually not mentioned explicitly. The main requirement that we place on this map is that $\alpha(C)$ is contained in as small a number of reticulation cycles as possible. As an example, consider the network depicted in Figure 6.16c. There the edge assignment map chooses the leaf edge leading to the leaf labeled a to represent the trivial cluster $\{a\}$ rather than the predecessor of that edge.

Definition 6.10.1 (Decomposable network) *Let C be a set of clusters on $\mathcal{X}$. A phylogenetic network N that represents C is said to have the* decomposition property, *or to be* decomposable *with respect to C, if there exists an edge assignment α such that for any two clusters $C_1, C_2 \in C$ the following holds: The two tree edges $\alpha(C_1)$ and $\alpha(C_2)$ that represent C_1 and C_2 are contained in the same biconnected component of N if and only if C_1 and C_2 lie in the same connected component of the incompatibility graph $IG(C)$.*

The concept of a decomposable phylogenetic network was introduced in [96] and [127], where similar proprieties are defined for a phylogenetic network that is obtained from a set of sequences or splits, respectively.

In Figure 6.17 of Section 6.10.2 we present an example of a decomposable network and of a non-decomposable one. The following result states that one direction of the condition in this definition is always fulfilled:

Lemma 6.10.2 (Incompatible clusters in biconnected components) *Let C be a set of clusters on $\mathcal{X}$ and let N be a rooted phylogenetic network that represents C (either in the hardwired or softwired sense). If C_1 and C_2 are two different clusters contained in the same connected component of the incompatibility graph $IG(C)$ then the two tree edges e_1 and e_2 that represent C_1 and C_2 are contained in the same biconnected component of N.*

Proof Let N be a rooted phylogenetic network representing a set of clusters C on $\mathcal{X}$. Let e_1 and e_2 be two tree edges in N that represent two clusters C_1 and C_2. By transitivity, it suffices to show that e_1 and e_2 must be contained in the same biconnected component of N if C_1 and C_2 are incompatible. For the sake of contradiction, let us assume that C_1 and C_2 are incompatible and that e_1 and e_2 are not contained in the same biconnected component of N. Then there must exist either a cut edge h or at least a cut node v that separates e_1 and e_2. In the latter case the cut node v must be an unresolved node of degree ≥ 4 and in this case we can introduce a new edge h that is a cut edge between e_1 and e_2. In either case, the edge $h = (v, w)$ separates all leaves below w from all other leaves. This implies that any cluster that is represented by an edge that does not lie below w must be compatible with every cluster that is represented by an edge below w. Therefore C_1 and C_2 must be compatible, a contradiction.   □

6.10.1 Decomposability of hardwired networks

In this section we show that cluster networks are decomposable. However, if we more generally consider the class of *all* hardwired networks, then decomposability does not hold because one can add superfluous edges to such a network so as to merge some of its non-trivial biconnected components.

Lemma 6.10.3 (Cluster networks are decomposable) *Let C be a set of clusters on $\mathcal{X}$ and let N be a cluster network that represents C in the hardwired sense. Then N is decomposable.*

Proof Consider two clusters C_1 and C_2 that are contained in the same connected component of the incompatibility graph $IG(C)$. Then the two edges e_1 and e_2 representing them are contained in the same biconnected component of N by Lemma 6.10.2.

Now, consider two clusters C_1 and C_2 that are represented by two edges e_1 and e_2 that are contained in the same biconnected component of N. Then, by the properties of a biconnected component, there exists an undirected cycle in N that contains both e_1 and e_2. We must show that C_1 and C_2 are contained in the same connected component of $IG(\mathcal{C})$.

For the purpose of this proof, we define a *reticulation cycle* Z to consist of two disjoint directed paths P_1 and P_2, called the *arms* of the cycle, from one node p (the apex of the cycle) to a reticulate node $r \neq p$ (the reticulation of the cycle). At least one of the arms, say P_1, must contain more than one edge, because otherwise r and p would be directly connected by two edges, which we do not allow. Moreover, P_2 must contain more than one edge, too. Indeed, by the reticulation-separation property p and v cannot be reticulate nodes, where v is the parent node of r belonging to P_1. This implies that, for each node v in P_1, we have $L(r) \subsetneq L(v) \subsetneq L(p)$. But this contradicts the reducedness property. So P_2 must contain more than one edge. Finally, the reticulation-separation property also implies that both arms must contain at least one tree edge.

First, let us assume that e_1 and e_2 are incomparable, that is, that e_1 is not below e_2 and e_2 is not below e_1. Then e_1 and e_2 are contained in different arms of a reticulation cycle Z, say e_1 in P_1 and e_2 in P_2. Then the two clusters C_1 and C_2 must have some taxa in common, namely those that lie below node r. By the nestedness property, C_1 must contain additional taxa not contained in C_2, and vice versa. But then C_1 and C_2 are incompatible and are thus contained in the same connected component of $IG(\mathcal{C})$.

Second, assume that e_1 and e_2 are comparable and contained in the same arm of a reticulation cycle Z. Adapting the previous argument, C_1 and C_2 are incompatible with every cluster represented by a tree edge in the other arm of Z. This implies that C_1 and C_2 are contained in the same connected component of $IG(\mathcal{C})$.

Third, assume that e_1 and e_2 are comparable, with $C_1 \subset C_2$, say, and are not contained in the same arm of Z. Let e'_1 be the lowest tree edge in the arm P_1 containing e_1 such that the associated cluster C'_1 is not comparable with C_2, which always exists. Then C'_1 and C_2 are incompatible by the first case and C'_1 and C_1 are incompatible by the second case.

Finally, assume that e_1 and e_2 are contained in some undirected cycle in N, but in two different reticulation cycles. A simple induction shows that there must exist a series of reticulation cycles $Z_1, \ldots, Z_t$ in N such that e_1 is contained in Z_1, e_2 is contained in Z_t and for any $i = 1, \ldots, t-1$ we have that the two reticulation cycles Z_i and Z_{i+1} share some tree edge f_i. As we have seen, any pair of tree edges f_i and f_{i+1} contained in the same reticulation cycle represent two incompatible clusters, which are contained in the same connected component of $IG(\mathcal{C})$. Hence, the two clusters C_1 and C_2 represented by e_1 and e_2 must be contained in the same connected component of $IG(\mathcal{C})$, too. $\square$

6.10.2 Decomposability of softwired networks

By definition, the softwired interpretation of a rooted phylogenetic network does not forbid the presence of superfluous reticulate edges and this fact is easily used to construct an example of a softwired network that does not have the decomposition property.

However, even if we restrict our attention to "minimum" softwired networks, then the decomposition property still does not necessarily hold [97, 131]. Here, by minimum we mean that each tree edge represents at least one cluster in C that is not represented by any other tree edge, and the network is bicombining and contains a minimum number of reticulate nodes over all such networks.

To see this, consider the softwired network N depicted in Figure 6.17a. It has one reticulate node r and represents the set of clusters C listed in Figure 6.17b. This network is minimum in the sense mentioned above. Now add the cluster $A = \{a, b, c, d\}$ listed in Figure 6.17c to C to obtain a new set of clusters C_1. Does N represent C_1 as well? The only edge in N that could possibly represent A is the edge e, as it is the only edge that lies above all the nodes labeled a, b, c and d. But e does not represent A since every cluster represented by e contains h. A simple modification N_1 of N to accommodate the cluster A can be obtained by adding a new reticulation edge u from the source of e to a new node placed in the middle of the reticulation edge q, as shown in Figure 6.17d. Then, to represent the cluster A by the edge e on N_1, we turn on the edges u and q to *avoid* the taxon h, that is, to set things up such that every directed path from the edge e to the leaf with label h contains some reticulate edge that is off. This modified network represents $C_1 = C \cup \{\{a, b, c, d\}\}$ in the softwired sense and contains two bicombining reticulations.

Exercise 6.10.4 *Verify that it is not possible to represent C_1 using only one bicombining reticulation.*

We now make a new network consisting of two copies of N_1, called N_1 and N_2, with taxon sets $\mathcal{X}_1 = \{a, b, c, d, o, h\}$ and $\mathcal{X}_2 = \{a', b', c', d', o', h'\}$, and cluster sets C_1 and C_2 (analogous to C_1), respectively. This new network N^* on $\mathcal{X}^* = \mathcal{X}_1 \cup \mathcal{X}_2$ will represent the set of clusters $C^* = C_1 \cup \{\{a, b, c, d, o, h\}\} \cup C_2 \cup \{\{a', b', c', d', o', h'\}\}$. The network is depicted in Figure 6.17e. It has four bicombining reticulations and is decomposable.

The edges u and u' are used (only) to avoid the taxa h and h', respectively. But we can also achieve this goal using one less reticulation, as shown in Figure 6.17f. Here, we turn on edges u' and q to avoid h when representing $\{a, b, c, d\}$ on the edge e and we turn on edges u and q' when representing $\{a', b', c', d'\}$ on the edge e'.

The network N^* does not have the decomposition property because it has only one non-trivial biconnected component, while the incompatibility graph for C^*

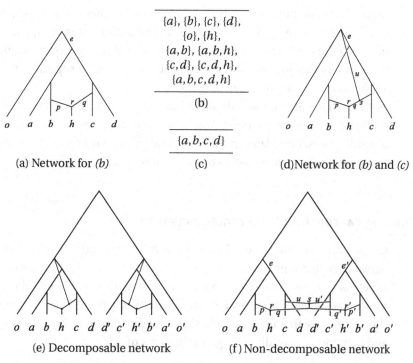

$\{a\}, \{b\}, \{c\}, \{d\},$
$\{o\}, \{h\},$
$\{a,b\}, \{a,b,h\},$
$\{c,d\}, \{c,d,h\},$
$\{a,b,c,d,h\}$

(b)

$\{a,b,c,d\}$

(c)

(a) Network for *(b)*   (d)Network for *(b)* and *(c)*

(e) Decomposable network   (f) Non-decomposable network

Figure 6.17   (a) A rooted phylogenetic network that represents all clusters listed in (b) in the softwired
sense. Note that this network does not represent the cluster $\{a, b, c, d\}$ listed in (c). (d) A
minimum rooted phylogenetic network that represents all clusters listed in (b) and also the
one listed in (c). The edge u is required to avoid h when representing the cluster $\{a, b, c, d\}$
on the edge e. The network shown in (e) contains two copies of the one shown in (a),
using a total of four reticulate nodes. It represents all clusters contained in (b) and (c), and
a second set of corresponding ones on the taxon set $\{o', a', b', c', d', h'\}$. It is decomposable
on the union of these two sets of clusters. The network shown in (f) also represents all
described clusters, while using only three reticulations. However, this improvement is gained
at an undesirable price: decomposability is abandoned and two completely unrelated parts
of the phylogeny are linked together via reticulation edges.

has two connected components coming from C_1 and C_2. This example implies the
following result:

Theorem 6.10.5 (Minimum softwired networks not decomposable) *Softwired
networks are not necessarily decomposable, even when restricted to networks containing
a minimum number of reticulations and containing no superfluous edges.*

Decomposing the problem of computing a minimum network is an attractive
approach. However, this theorem states that solutions based on a decomposition of

the problem may not necessarily be optimal. On the other hand, non-decomposable solutions can be very misleading: in the example above, two completely unrelated parts of a dataset (these could be, for example, alpha-proteobacteria and bats) are brought together just to save one reticulation. Hence, we suggest that decomposability should always be required of a solution.

As we have just argued, decomposability is a desirable property to have. In Section 8.2 we present a *divide-and-conquer* strategy that is used by some of the algorithms described later in the book, and show that networks computed using this strategy always have the decomposition property.

6.11 Topological constraints on rooted networks

Given a set of clusters C on X, the problem of computing a rooted phylogenetic network N that represents C in the softwired sense and has a minimum number of reticulations is a computationally hard problem, as mentioned above. One way to address this issue is to consider topologically constrained classes of networks. In this section, we first introduce the simplest class of non-tree networks, *galled trees* and then look at two different generalizations of this class.

6.11.1 Galled trees

A rooted phylogenetic network N that does not have any non-trivial biconnected components cannot contain any undirected cycles and must therefore be a rooted phylogenetic tree. Now, assume that N has one or more non-trivial biconnected components. Because N is a DAG, each such biconnected component must contain at least one reticulation. We say that a reticulation r is *properly contained* in a non-trivial biconnected component H of N, if all reticulation edges leading to r are contained in H.

The topologically simplest type of rooted phylogenetic network that is not a tree, is given by the following definition (see Figure 6.18):

Definition 6.11.1 (Galled tree) *A rooted phylogenetic network N is called a* galled tree, *if every non-trivial biconnected component of N properly contains exactly one reticulation. Galled trees are sometimes assumed to be bicombining.*

In the original definition of a galled tree [98, 241], undirected cycles were additionally required to be node-disjoint and a gall was defined as a biconnected component that contains exactly one reticulation [98, 99, 241]. However, none of the existing algorithms for computing galled trees use this additional assumption and so requiring only edge-disjointness seems more natural. We define a *gall* as a biconnected

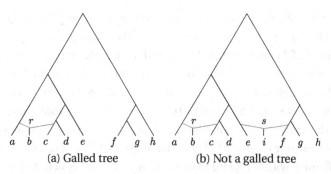

(a) Galled tree   (b) Not a galled tree

Figure 6.18 (a) A galled tree on the taxa $\mathcal{X} = \{a, \ldots, h\}$. (b) Addition of a new taxon i destroys the galled tree property, as the two reticulations r and s are not independent.

component of a rooted phylogenetic network that properly contains exactly one reticulation.

While the condition that a galled tree must be bicombining is appropriate in the context of recombination networks, see Section 9.9, it can be dropped when constructing rooted phylogenetic networks from clusters, as discussed in Section 8.3.

Exercise 6.11.2 (Galled tree for clusters)   *Given a set of clusters C, does there always exist a galled tree N that represents C in the softwired sense? To answer this, consider the following set of clusters*

$$C = \big\{\{a, b\}, \{b, c\}, \{c, d\}, \{a, d\}\big\} \tag{6.4}$$

on $\mathcal{X} = \{a, b, c, d\}$. By inspection of all possible galled trees on $\mathcal{X}$, show that the set of clusters C cannot be represented by a galled tree.

Let C be a set of clusters on $\mathcal{X}$ and let N be a galled tree that represents C. We say that N is a *minimum galled tree*, if it minimizes the following three quantities (in this order of priority):

(i)   the number of reticulate nodes,
(ii) the number of clusters represented by edges contained in reticulate cycles, and
(iii) the number of edges in the network.

We have the following result:

Lemma 6.11.3 (Minimum galled trees are decomposable)   *Let C be a set of clusters on $\mathcal{X}$ and let N be a minimum galled tree that represents C in the softwired sense (if one exists), then N is decomposable.*

Proof Let C be a set of clusters on $\mathcal{X}$ and let N be a minimum galled tree that represents C. Let r be any reticulate node in N and let Z denote a reticulation cycle associated with r. We show that any two clusters represented by tree edges in Z are contained in the same connected component of the incompatibility graph $IG(\mathcal{C})$. As in the proof of Lemma 6.10.3, let P_1 and P_2 denote the two arms of Z. Let $h_1 = (p, r)$ and $h_2 = (q, r)$ denote the two reticulate edges at the bottom of the two arms. By the definition of a galled tree, all edges in the two arms, except the two reticulate edges h_1 and h_2, must be tree edges. As in that proof, we argue that at least one of the arms, say P_1, must contain more than one edge and let $e = (a, p)$ denote the lowest tree edge in P_1. Let A denote the set of all taxa whose labels can be reached from p without going through r. Similarly, let R denote the set of taxa whose labels can be reached from r.

If P_2 consists only of h_2, then the node q lies above a in the path P_1. Note that q is connected to a by a path of tree edges in P_1, and let f be the first tree edge on that path. We can assume that f represents some cluster $A \cup B$, where B is disjoint from R, because otherwise the edge h_2 could attach below f, by minimality. Similarly, we can assume that the edge e must represent the cluster $A \cup R$, because otherwise the edge h_1 could attach above e. Now, every cluster represented by an edge in P_1 below f and above e will either contain A, a subset of B and nothing from R, and thus will be incompatible with $A \cup R$, or will contain A and R and a subset of B, and then will be incompatible with $A \cup B$. Because $A \cup B$ and $A \cup R$ are incompatible, it follows that all the clusters represented by tree edges in Z lie in the same component of the incompatibility graph.

Now, consider the case in which P_2 contains at least one tree edge and let $f = (b, q)$ denote the lowest tree edge in P_2. Let B denote the set of taxa that can be reached from q without going through r.

Every cluster represented by a tree edge in P_1 must be of the form $A \cup R$, $A \cup R \cup D$, $A \cup D$ or A. Here, D is a set of additional taxa whose labels can be reached from a node in P_1 without going through r. The first and last clusters come from the edge e, while the others come from tree edges higher up in P_1.

In the first case and second case, the clusters are incompatible with the cluster $B \cup R$. In the third case, the cluster is incompatible with $A \cup R$. The fourth case cannot occur due to minimality, because the cluster A can be represented outside of the cycle by a tree edge located above the subtree containing the nodes with labels in A.

Applying the same argument to the arm P_2, we obtain that every cluster represented by a tree edge in Z is either incompatible with $A \cup R$ or with $B \cup R$. Finally, note that both $A \cup R$ and $B \cup R$ must be present in $\mathcal{C}$ (or the corresponding edges would be superfluous, contradicting minimality). As $A \cup R$ and $B \cup R$ are incompatible with each other, we have that any two clusters represented by tree edges in Z are contained in the same connected component of the incompatibility graph $IG(\mathcal{C})$. $\square$

In Section 8.3 we show how to determine whether a given set of clusters $\mathcal{C}$ on $\mathcal{X}$ can be represented by a galled tree and we describe an algorithm for constructing a galled tree for a given set of clusters that takes advantage of the decomposition property. In Section 11.6, the same algorithm is used to compute a galled tree that represents two given rooted phylogenetic tree, T_1 and T_2 on $\mathcal{X}$, if one exists, based on the following result:

Lemma 6.11.4 (Galled tree contains both trees) *Let T_1 and T_2 be two rooted (not necessarily bifurcating) phylogenetic trees on $\mathcal{X}$ and let $\mathcal{C} = \mathcal{C}(T_1) \cup \mathcal{C}(T_2)$ be the set of all clusters represented by the two trees. If there exists a rooted phylogenetic network N on $\mathcal{X}$ that is a galled tree that represents $\mathcal{C}$ and has the reticulation-separation property, then N contains both trees T_1 and T_2, if the two trees are bifurcating, or refinements of them, otherwise.*

Proof Let us assume that neither T_1 nor a refinement of T_1 is contained in a galled tree N computed for $\mathcal{C}$. This implies that there exist two disjoint (and non-empty) sets of taxa C and D such that the clusters C and $C \cup D$ are represented in T_1, but the edge $e = (v_e, w_e)$ representing C in N does not lie below the edge $f = (v_f, w_f)$ representing $C \cup D$ in N. Let $L(C)$ and $L(D)$ denote the set of all leaves in N that are labeled by taxa in C and D, respectively. We show that, if the edge representing C does not lie below the edge representing $C \cup D$ in N, then this can easily be fixed by adjusting the (implicitly given) edge assignment map α (introduced in Section 6.10). There are two possible cases.

First, let us assume that e lies above f, as illustrated in Figure 6.19a. If w_e is connected to v_f by a directed path of tree edges, then there must be a path of tree edges from w_f to a reticulate node r that is a stable ancestor for $L(D)$. This is because we need to be able to disrupt all paths from w_e to elements of $L(D)$ to obtain the cluster C on e. We cannot use more than one reticulate node to do so because each such reticulate node would be contained in a different cycle that contains e, which is not possible in a galled tree. To be able to obtain the cluster C on e, the lowest stable ancestor of r must lie above v_e. In this configuration, the clusters C and $C \cup D$ can be obtained on both e and f, and so we can switch the edge assignments for C and $C \cup D$ on e and f, as indicated by the arrow in Figure 6.19a. What if there is a reticulate node r' on the path from w_e to v_f? Then the set of leaves $L(C)$ can only be reached if the edge to r' is turned on. We can now argue as before to establish the existence of the reticulate node r, but then the edges e and f will lie in a cycle that goes through two different reticulate nodes, r and r', which is not possible in a galled tree.

The other case is that e and f are incomparable in N, as illustrated in Figure 6.19b. Because w_e and w_f must both lie above $L(C)$, the two edges are contained in a reticulation cycle with reticulate node r, which is a stable ancestor

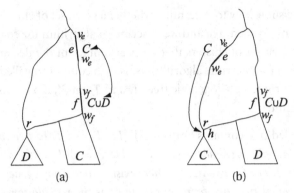

(a)   (b)

Figure 6.19 The two possible configurations for the two edges e and f in N, as discussed in the proof
of Lemma 6.11.4. Either f lies below e, as shown in (a). Or e and f are incomparable, as
shown in (b). In either case, the assignment of clusters to edges in N can be modified so
as to ensure that the cluster C appears below $C \cup D$ in N, as indicated by the arrows.

of $L(C)$. By the reticulation-separation property, there is a tree edge $h = (r, u)$
such that u is the lowest stable ancestor of $L(C)$. Thus, we can change the edge
assignment so as to use the edge h to represent C, as indicated by the arrow in
Figure 6.19b.   □

Galled trees were initially introduced in the context of *recombination networks*
and we discuss this in Section 9.9.

6.11.2 Level-k networks

Let N be a rooted phylogenetic network. If N is a galled tree, then by definition any
biconnected component of N contains precisely one reticulation. In the literature,
the maximum number of reticulations in any biconnected component of N is
sometimes called the *level* of N and we have the following definition [51]:

Definition 6.11.5 (Level-k network)   *Let N be a bicombining, rooted phylogenetic
network on $\mathcal{X}$. If the maximum number of reticulations properly contained in any
biconnected component of N is k, then N is called a* level-k *network.*

The definition of a multicombining level-k network is slightly different: in this case,
we require that any biconnected component of N contains at most $k + 1$ reticulate
edges, as discussed in Section 6.9.

Note that, according to this definition, level-0 networks are trees and level-1
networks are galled trees. The name "level" is perhaps slightly misleading, as it
suggests a quantity that changes gradually as a network changes, for example, when
introducing additional reticulations. However, for any number k one can easily

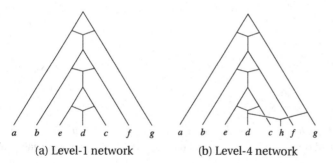

(a) Level-1 network   (b) Level-4 network

Figure 6.20   (a) A level-1 network with three reticulations. (b) A level-4 network obtained by introducing one additional reticulation associated with h.

construct a level-1 network that immediately jumps to level-k after addition of one new reticulation, as indicated for $k = 4$ in Figure 6.20.

If a set of clusters C on X can be represented by some level-k network N, then does there exist such a level-k network N' that is decomposable? The answer is given by the following result [135]:

Theorem 6.11.6 (Level-k networks and decomposability)   *Let C be a set of clusters on X. If there exists a level-k network N that represents C in the softwired sense for some number k, then there also exists such a network N' that is decomposable with respect to C.*

Proof   To prove the result it suffices to specify how to obtain a suitable level-k network N_H from the given network N for an arbitrary non-trivial component H of the incompatibility graph $IG(C)$, as we can then apply the divide-and-conquer strategy described in Section 8.2 to obtain the desired decomposable network N'. The basic idea is to define N_H as the rooted phylogenetic network that is induced in N by the set of taxa mentioned in H. The crucial point is that the level of the induced network will not exceed that of the original network N.

More formally, assume that we are given a level-k network N that represents a set of clusters C on X for some number k. Let C' be the set of clusters contained in some non-trivial component H of the incompatibility graph $IG(C)$ and let X' be the set of taxa that are present in C'. Consider the equivalence relation on X' in which two taxa are *equivalent* if and only if they are not separated by C'. Let $\bar{X}'$ denote a subset of X' obtained by choosing one taxon for each equivalence class. Then it holds that the network $\bar{N}$ induced in N by $\bar{X}'$ represents $C'|_{\bar{X}'}$, as required by the divide-and-conquer strategy. Lemma 6.10.2 states that all edges in N that represent clusters from C' must be contained in the same biconnected component Q of N. Hence, it follows that all edges in $\bar{N}$ come from the same biconnected component, Q. As the computation of an induced subnetwork does not involve creating new

(a) Galled tree   (b) Galled network   (c) Not a galled network

Figure 6.21 (a) A galled tree. (b) A galled network. (c) A rooted phylogenetic network that is neither a galled tree nor a galled network. The first network is a level-1 network, whereas the other two are both level-2 networks.

reticulate nodes, the level of $\bar{N}$ will be at most the number of reticulations in Q, which is $\leq k$.   □

In Section 8.5 we describe an algorithm that takes as input a set of clusters C on X and computes as output a level-k network that represents C, with the aim of minimizing k.

6.11.3 Galled networks

Let N be a rooted phylogenetic network N. A *tree cycle* in N is an undirected cycle Z that consists of two disjoint paths, each consisting of a chain of tree edges followed by exactly one reticulate edge, and both joining a fixed tree node a, called the *apex*, to the same reticulate node r.

Galled trees are rooted phylogenetic networks that contain isolated reticulations and so one can view them as a first generalization of rooted phylogenetic trees. However, they have the draw-back that a given set of clusters might not possess a representation in terms of a galled tree. This is also true for level-k networks, for any fixed k. We now introduce a third type of topological constraint that does not suffer from this problem [129, 131], see Figure 6.21b for an example:

Definition 6.11.7 (Galled network) *A bicombining rooted phylogenetic network N is called a* galled network, *if every reticulation in N has a tree cycle.*

This definition can easily be extended to multicombining networks. In this case, we require that for each reticulate node r and every pair of in-edges p and q of r there exists a tree cycle that contains r and goes through p and q.

Equivalently, a rooted (possibly multi-combining) phylogenetic network is a galled network if and only if it holds that no two reticulate nodes can be connected by a directed path that is contained in a single biconnected component.

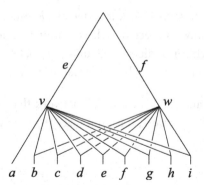

Figure 6.22 Any set of clusters on $\mathcal{X}$ can be represented by a galled network using a construction that is illustrated here for $\mathcal{X} = \{a, \ldots, i\}$. For all leaves except the one labeled by the first taxon (in this example, a), the source of each leaf edge is a reticulate node that is connected to two parent nodes v and w. Any cluster C of $\mathcal{X}$ that contains the first taxon is represented by the edge e, whereas any cluster that does not contain the first taxon is represented by edge f, both in the softwired sense.

For any set of clusters $\mathcal{C}$ on $\mathcal{X}$ there exists a galled network N that represents $\mathcal{C}$. We illustrate how to construct such a network in Figure 6.22. Hence, an algorithm that aims at computing a galled network for a given input dataset never has to fail.

The most important mathematical property of a galled network N is that every reticulation r separates the network into two sets of nodes, namely (a) the set of all nodes that can be reached from the root ρ by a directed path that avoids r and (b) the set of all nodes that can only be reached by a path that passes through r. We call this property the *separation property* [129]:

Lemma 6.11.8 (Separation Lemma)   *Let $N = (V, E, \lambda)$ be a (not necessarily bicombining) galled network and r a reticulation in N. Let $R \subseteq V$ be the set of all nodes that are descendants of r. Then every (directed or undirected) path from any node v in $V - R$ to any node w in R must pass through r.*

Proof We first show that any edge e from a node in $V - R$ to a node in R must be a reticulate edge. Let $e = (v, w)$ be any edge with $v \in V - R$ and $w \in R$. If $w = r$, then e is a reticulate edge, because r is a reticulate node, by assumption. Otherwise, by definition of R, there exists a path P from r to w. By definition of $V - R$, the path P does not contain e. Hence, the indegree of w must be at least two and so e is a reticulate edge.

Now, let $e = (v, w)$ be any edge from a node in $V - R$ to a node in R and assume that $w \neq r$. Let $f = (u, w)$ be an edge from some node u in R to w, which must exist, by definition of R. Any undirected cycle containing e and f must contain another edge connecting a node in $V - R$ to a node in R. Since this edge is a reticulate edge, it follows that any undirected cycle containing e and f must contain a third reticulate edge, which contradicts the assumption that N is a galled network. Hence, we must have $w = r$ and so N has the separation property. $\square$

Let C be a set of clusters on $\mathcal{X}$ and let N be a galled network that represents C. We say that N is a *minimum galled network*, if it minimizes the following three quantities (in this order of priority):

(i) the number of reticulate nodes,
(ii) the number of clusters represented by edges contained in reticulate cycles, and
(iii) the number of edges in the network.

The following result justifies the fact that we restrict our attention to decomposable galled networks:

Lemma 6.11.9 (Minimum galled networks are decomposable) *Let C be a set of clusters on $\mathcal{X}$ and let N be a minimum galled network that represents C in the softwired sense (if one exists), then N is decomposable.*

A proof of this result for the restricted case of galled networks that fulfill the tree-child property can be found in [127]. For the general case, see [152].

In Section 8.4 we discuss how to compute a minimum galled network N for a given set of clusters C in $\mathcal{X}$.

6.11.4 Tree-child networks and visible reticulations

One attractive property to require of a rooted phylogenetic network is the following [41]:

Definition 6.11.10 (Tree-child property) *Let N be a rooted phylogenetic network on $\mathcal{X}$. A node v of N is said to have the* tree-child *property, if it has at least one child that is a tree node. We say that N has the* tree-child *property, if all internal nodes of N have the property.*

An easy consequence of this is that every internal node v in such a network N is the ancestor of at least one leaf via a path of tree edges. All three networks shown in Figure 6.21 have the tree-child property, while the one displayed in Figure 6.22 does not because there all children of the node w are reticulate nodes. The following result is not immediately obvious:

Lemma 6.11.11 (Galled trees are tree-child) *If a rooted phylogenetic network N is a galled tree, then it has the tree-child property.*

Proof To see that the claim is true, consider an internal node v and assume that v does not have the tree-child property, in other words, that all children of v are reticulate nodes. Then v cannot be a reticulate node, because v and any child of v are contained in the same biconnected component of N, and no such component may contain more than one reticulate node, by definition of a galled tree. Then either v is the root node, or v can only have one in-edge, which is a tree edge. In the former case, any reticulate child of v must either be connected to v by two reticulate edges, which is not allowed, or is otherwise the descendant of one of the other reticulate children of v, in contradiction to the definition of a galled tree. In the latter case, v must have at least two children, because otherwise v would be a suppressible node (that has both indegree one and outdegree one), which is not allowed in N. In summary, v has at least two (or more) children that are reticulate nodes and their undirected cycles both contain the in-edge of v, in contradiction to the assumption that N is a galled tree. □

However, this result does not hold for galled networks:

Exercise 6.11.12 (Galled networks are not tree-child) *Construct an example of a galled network that does not have the tree-child property.*

We now introduce a property that is weaker than the tree-child property, but that suffices to ensure that the cluster containment problem, as discussed below, can be efficiently solved for such networks, although the problem is computationally hard in general.

Definition 6.11.13 (Visible reticulation) *Let N be a rooted phylogenetic network. A reticulate node r in N is called* visible, *if it is a stable ancestor of some leaf v.*

We say that a rooted phylogenetic network N has the *visibility property*, if all its *reticulate* nodes are visible. When sampling a cluster C from N we can "see" whether a reticulation edge leading to reticulation r is turned on or off, by simply checking whether the taxon x associated with a leaf v that fulfills the requirements is present in C, see Figure 6.23. By definition, every tree-child network has the visibility property, moreover:

Lemma 6.11.14 (Reticulations in a galled network are visible) *If a rooted phylogenetic network N is a galled network, then all its reticulations are visible.*

Proof Let r be a reticulate node in N. If r has outdegree 0, then r is a leaf and we are done. Now, assume that the outdegree of r is ≥ 1. Let H denote the set of all nodes that can be reached from r by some directed path of tree edges, including r. If H contains a leaf, then r is visible and we are done. Otherwise, let r' be a reticulate node that can be reached from some node in H by crossing a single reticulate edge.

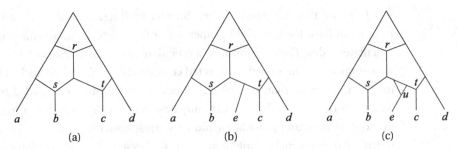

Figure 6.23   In (a), the two reticulations *s* and *t* are visible, but *r* is not. In (b), *r* is visible, because there exists a path of tree edges from *r* to the leaf labeled *e*. However, in (c) there is no path of tree edges from *r* to the leaf labeled *e* because of the reticulate node *u*. Nevertheless, *r* is visible because *r* is a stable ancestor of *e*.

Let z be some leaf that is a descendant of r'. By the Separation Lemma 6.11.8, all paths from the root of N to z must go through r', and so r' is a stable ancestor of z. This implies that r is visible in this case, too.   □

6.11.5 Tree-sibling networks

In a network, two nodes u and v are called *siblings*, if they share a parent. If a rooted phylogenetic network N does not satisfy the tree-child property, then it might still satisfy the following requirement [38]:

Definition 6.11.15 (Tree-sibling property)   *Let N be a rooted phylogenetic network on X. A node v of N is said to have the* tree-sibling *property, if it has at least one sibling that is a tree node. We say that N has the* tree-sibling *property, if all reticulate nodes of N have the property.*

Since the tree-child property is clearly more restrictive than the tree-sibling one, we have the following result:

Lemma 6.11.16 (Galled trees are tree-sibling)   *If a rooted phylogenetic network N is a galled tree, then it has the tree-sibling property.*

Exercise 6.11.17 (Proof)   *Prove Lemma 6.11.16.*

This result does not hold for galled networks:

Exercise 6.11.18 (Galled networks are not tree-sibling)   *Construct an example of a galled network that does not have the tree-sibling property.*

6.11.6 Time-consistent rooted networks

Consider a rooted phylogenetic network N that represents the evolution of some set of taxa $\mathcal{X}$ and assume that the reticulate nodes represent putative hybridization

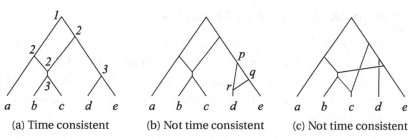

(a) Time consistent	(b) Not time consistent	(c) Not time consistent

Figure 6.24 (a) A rooted phylogenetic network whose inner nodes are consistently labeled by time stamps. (b) A rooted phylogenetic network for which no consistent labeling by time stamps exists. This is because the parent p of the reticulate node r is also an ancestor of q, the other parent of r. (c) Another network for which no consistent labeling exists, even though no reticulation is formed between a node and one of its ancestors.

events, say. For a hybrid species to form, the two parent species must exist at the same time. This requirement implies a topological constraint on the network N [16]:

Definition 6.11.19 (Time-consistent network)   *A rooted phylogenetic network N on $\mathcal{X}$ is called* time consistent, *if there exists a consistent labeling of its nodes by time stamps such that:*

(i) *The time stamp of a tree node is larger than the time stamp of its parent node.*
(ii) *The parents of a reticulate node, and the reticulate node itself, all have the same time stamp.*

The labeling describes a possible ordering of speciation and reticulation events, but not necessarily actual times, of course. In Figure 6.24 we show an example of a time-consistent network and two examples of networks that do not have this property. Time consistency is used later in Section 6.14.

Exercise 6.11.20 (Prove not time consistent)   *Verify that the rooted phylogenetic network shown in Figure 6.24c is not time consistent.*

Note that the network shown in Figure 6.24b is a galled tree and is not time consistent. This implies that galled networks, galled trees, tree-child and tree-sibling networks may all fail to be time consistent. Although time consistency seems to be quite a natural requirement, not even cluster networks always fulfill it.

Exercise 6.11.21 (Cluster networks are not time consistent)   *Construct a cluster network on $\mathcal{X} = \{a, \ldots, d\}$ that is not time consistent, for some suitable set of clusters $\mathcal{C}$.*

6.12 Cluster containment in rooted networks

One of the main goals of phylogenetic analysis is to determine clusters that represent clades of closely related species and one of the main uses of a rooted phylogenetic tree or network is to provide a representation of such clusters. Hence, an important question is: *How easy is it to determine whether a particular cluster is contained in a given tree or network?*

Problem 6.12.1 (Cluster containment problem) *For a given rooted phylogenetic network N and a given cluster C, on $\mathcal{X}$, the* cluster containment problem *is to determine whether the cluster C is represented by N.*

6.12.1 Cluster containment in hardwired networks

First, consider the case of a rooted phylogenetic tree T on $\mathcal{X} = \{x_1, \ldots, x_n\}$ (viewed as a special case of a network). We can determine, in linear time in n, whether a given cluster C is represented in T.

Exercise 6.12.2 (Linear time cluster containment for trees) *Design an algorithm that determines whether a given cluster C is represented by a given rooted phylogenetic tree T in linear time. (This is a difficult exercise that requires additional results from the literature).*

Next, consider the case of a cluster network N on $\mathcal{X} = \{x_1, \ldots, x_n\}$. We denote the number of clusters represented by N as k. To naively determine whether some given cluster C on $\mathcal{X}$ is represented in N, search for an appropriate edge e in a post-order traversal. As the number of edges in N is $O(k^2)$, determining containment requires at most $O(n \times k^2)$ steps.

Now consider any rooted phylogenetic network N on $\mathcal{X} = \{x_1, \ldots, x_n\}$, not necessarily a cluster network. Again, we can easily check whether a given cluster C is represented in N in the *hardwired* sense using a postorder traversal of N. In this more general case, the time required is proportional to the number of edges times n.

6.12.2 Cluster containment in softwired networks

Consider a rooted phylogenetic network N on $\mathcal{X}$ that we interpret in the *softwired* sense. In this case a simple traversal of the network does not suffice to determine whether a given cluster C on $\mathcal{X}$ is represented in N. This is because, in this case, for each reticulate node in N, we must decide which incoming reticulate edge to turn on and which ones to turn off and the choices made at different reticulate nodes are interdependent in a non-trivial way. In fact, it turns out that the problem is difficult [145]:

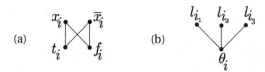

(a)

(b)

Figure 6.25 (a) Each variable x_i is represented in N by this "variable gadget" configuration. (b) Each clause $\theta_i = \ell_{i_1} \vee \ell_{i_2} \vee \ell_{i_3}$ is represented in N by this "clause gadget" configuration. Copies of these two types of gadgets are used to construct the network shown in Figure 6.26.

Theorem 6.12.3 (Softwired cluster containment is hard) *Let N be a rooted phylogenetic network and let C be a cluster on $\mathcal{X}$. The problem of deciding whether N represents the cluster in the softwired sense (that is, "contains" the cluster C) is NP-complete.*

To show that a decision problem is NP-complete, one must first show that the problem is contained in the class *NP*, that is, that a proposed solution can be verified in polynomial time. We must then show that some other problem that is known to be *NP-complete* reduces to the new problem in polynomial time. In other words, the aim is to show that any algorithm that can solve the new problem in polynomial time would also solve all other problems that are NP-complete in polynomial time.

If we are given a concrete choice of reticulate edge to turn on for each reticulation, then to verify whether the corresponding rooted phylogenetic tree contains the given cluster C can be solved in polynomial time, as discussed above. Hence, the problem of cluster containment in the softwired case is contained in NP.

In the following, we show that the problem is NP-complete by constructing a reduction from the *3-SAT* problem, or, to be precise, the *1-in-3-SAT* problem [83]. In the 3-SAT problem, one is given a Boolean expression $\phi = \theta_1 \wedge \theta_2 \wedge \cdots \wedge \theta_h$ in which each clause θ_i is the disjunction of exactly three literals, $\theta_i = \ell_{i_1} \vee \ell_{i_2} \vee \ell_{i_3}$, for $i = 1, \ldots, h$. Recall that a literal is an expression of the form x or $\bar{x}$ ("not x") for some Boolean variable x. Let $x_1, \ldots, x_k$ be the set of k different Boolean variables that occur in the expression ϕ. The 3-SAT problem is to determine whether there exists an assignment of Boolean values to the variables $x_1, \ldots, x_k$ such that the expression ϕ evaluates to true.

To transform an instance of the 3-SAT problem into an instance of the cluster containment problem, we construct a rooted phylogenetic network in a number of steps. For each variable x_i we define a small "variable gadget" consisting of four nodes and edges: we create two nodes x_i and $\bar{x}_i$ and two leaves labeled t_i and f_i, representing truth values true and false, respectively. These four nodes are then joined by edges in a "butterfly" configuration as shown in Figure 6.25a.

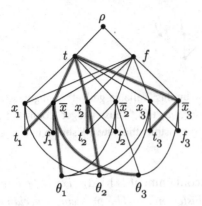

Figure 6.26 The rooted phylogenetic network N constructed for the Boolean expression $\phi = (\bar{x}_1 \vee \bar{x}_2 \vee x_3) \wedge (x_1 \vee x_2 \vee x_3) \wedge (\bar{x}_1 \vee \bar{x}_2 \vee \bar{x}_3)$. Variable gadgets in the middle of the network represent the different variables $x_1, \ldots, x_n$. Clause gadgets at the bottom of the network represent the clauses $\theta_1, \ldots, \theta_k$. Turning on all reticulate edges highlighted in gray demonstrates that this particular expression is satisfiable.

For each clause $\theta_i = \ell_{i_1} \vee \ell_{i_2} \vee \ell_{i_3}$ we build a "clause gadget" by connecting the three literal nodes ℓ_{i_1}, ℓ_{i_2} and ℓ_{i_3} to a new leaf labeled θ_i, see Figure 6.25b.

A rooted phylogenetic network N is then obtained by creating a root node ρ with two children f and t which are both connected to all literal nodes x_i and $\bar{x}_i$, see the example shown in Figure 6.26. Note that the set of taxa is given by the set of labels of all leaves $\mathcal{X} = \{\theta_1, \ldots, \theta_h, t_1, \ldots, t_k, f_1, \ldots, f_k\}$.

Now, to determine whether there exists an assignment of truth values to all variables $x_1, \ldots, x_k$ such that the given expression ϕ evaluates to true, we need to determine whether there exists a choice of reticulation edges to switch on, one for each reticulate node, such that the subtree below node t contains all leaves labeled $\theta_1, \ldots, \theta_h$ and $t_1, \ldots, t_k$, and that the subtree below node f contains all leaves labeled $f_1, \ldots, f_k$.

By construction of this network N, the cluster $\{\theta_1, \ldots, \theta_h, t_1, \ldots, t_k\}$ is represented in N in the softwired sense if and only if the given expression ϕ can be satisfied. Hence, any algorithm that could solve the cluster containment problem in polynomial time would also solve 3-SAT in polynomial time and so the problem is NP-complete.

Exercise 6.12.4 (Another solution) *Find another choice of edges to turn on in Figure 6.26 so that the expression is demonstrated to be satisfiable.*

6.12.3 Cluster containment for visible reticulations

We now turn to rooted phylogenetic networks that have the property that all reticulations are visible. We have the following new result:

Lemma 6.12.5 (Cluster containment for visible reticulations is easy) *Let N be a rooted phylogenetic network that has the visibility property. Let e be some edge of N and let $C \subsetneq X$ be a non-empty cluster. We can determine whether C is represented by e in polynomial time.*

Proof Let C be a cluster on X and let N be a rooted phylogenetic network on X that has the visibility property. Consider an edge $e = (v, w)$ of N. We can efficiently determine whether C is represented by e in a traversal of the part of the rooted phylogenetic network N that lies below the target node w of e, as follows.

First, let H be the set of all nodes below w that can be reached by only using tree edges. If H contains any leaf whose label is not contained in C, then the cluster C cannot be represented by the edge e.

Otherwise, while there is a node u in H that is the source of some reticulate edge f, repeat the following:

Let r be the reticulate node at the other end of f. By visibility, r is a stable ancestor of some leaf z. If the label of z is contained in C, then turn the edge f on, and otherwise turn f off. If f was turned on, extend the set H by adding the set of all nodes for which r is a stable ancestor.

Once the repetitions of this step have been completed, to determine whether the cluster C is represented by e we check that the set of all labels of nodes in H equals C and that no reticulate node r is encountered that has all its input edges turned off (which can only happen if its lowest stable ancestor $lsa(r)$ lies below w). $\square$

Because all tree-child networks, galled trees and galled networks have the required visibility property, we have:

Corollary 6.12.6 (Easy cases of cluster containment) *The cluster containment problem can be solved efficiently for tree-child networks, galled trees and galled networks.*

6.13 Tree containment

The problem of determining whether a given rooted phylogenetic network N contains a given rooted phylogenetic tree T, on X, is NP-complete [145].

6.14 Comparing rooted networks

Suppose we are given two different phylogenetic networks N_1 and N_2 on X, as shown in Figure 6.27, say. How to measure their similarity? For phylogenetic trees, this question can be answered in a number of ways. For example, we could use the Robinson-Foulds distance to compare two unrooted phylogenetic trees, or the cluster distance to compare two rooted phylogenetic trees (Section 3.16). Because

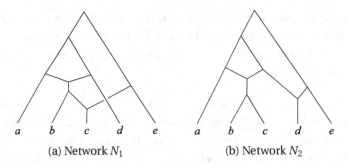

(a) Network N_1 (b) Network N_2

Figure 6.27 Two distinct rooted phylogenetic networks N_1 and N_2 that represent different sets of hard-wired clusters, softwired clusters, and rooted phylogenetic trees.

splits and clusters correspond one-to-one to the edges of an associated unrooted or rooted phylogenetic tree, respectively, comparing the splits or clusters represented by two trees is equivalent to comparing the trees themselves.

For rooted phylogenetic networks, the situation is different. The association between a set of clusters or trees and the phylogenetic network used to represent the data is quite loose. For example, consider the two networks N_3 and N_4 in Figure 6.28. These two networks are topologically distinct, but, in a sense, *indistinguishable*, because they represent the same set of hardwired clusters, the same set of softwired clusters and the same set of rooted phylogenetic trees. So, comparing the clusters or trees represented by two rooted phylogenetic networks is not the same as comparing the networks themselves.

While biologists are not very interested in distinguishing between different networks that are indistinguishable in the above sense, this question has nevertheless obtained some attention in the computational literature and we report on some of the results in this section.

Below, we first discuss a number of distance functions that aim at comparing the represented data and then look at some measures that are based on topological invariants of the networks. With the exception of the subnetwork distance, none of them provide a proper metric for the set of rooted phylogenetic networks on $\mathcal{X}$, as they all return a distance of zero for the two distinct networks N_3 and N_4 displayed in Figure 6.28, and thus fail to fulfill the identity property of Definition 3.12.1.

Exercise 6.14.1 (Two indistinguishable networks) *Show that the two rooted phylogenetic networks N_3 and N_4 displayed in Figure 6.28 represent the same set of hardwired clusters, the same set of softwired clusters and the same set of rooted phylogenetic trees.*

There is another issue to consider. Rooted phylogenetic networks are usually computed from an input set of incompatible clusters or an input set of incongruent

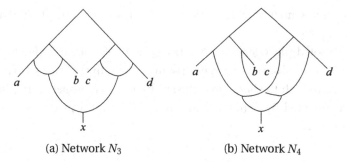

(a) Network N_3 (b) Network N_4

Figure 6.28 Two topologically distinct rooted phylogenetic networks N_3 and N_4 that represent the same
hardwired clusters, softwired clusters and rooted phylogenetic trees.

rooted phylogenetic trees, on $\mathcal{X}$. The size of such an input dataset is usually polynomial in the number of taxa. However, a rooted phylogenetic network N that
is computed to represent such a dataset will, in general, represent an exponential
number of softwired clusters or rooted phylogenetic trees. This is due to the fact
that each reticulation added to the network can potentially double the number of
represented softwired clusters or trees. So, a direct comparison of the softwired
clusters or rooted phylogenetic trees represented by two different networks will
sometimes be infeasible, due to the large number of objects, and might be misleading, if most of the clusters or trees compared are not objects that the network is
intended to represent.

6.14.1 Comparing hardwired networks

Let N_1 and N_2 be two rooted phylogenetic networks on $\mathcal{X}$, interpreted in the
hardwired sense. A natural way to compare two such networks is to compute their
hardwired cluster distance, a simple extension of the cluster distance for rooted
phylogenetic trees (see Section 3.16), defined as follows:

Definition 6.14.2 (Hardwired cluster distance) *Let N_1 and N_2 be two rooted phylogenetic networks on $\mathcal{X}$. We define the* hardwired cluster distance d_{hard} *between N_1
and N_2 as the cardinality of the symmetric difference of the two sets of hardwired
clusters represented by the two networks, divided by two:*

$$d_{hard}(N_1, N_2) = \frac{|\mathcal{C}_{hard}(N_1) \triangle \mathcal{C}_{hard}(N_2)|}{2}. \tag{6.5}$$

The hardwired distance between any two rooted phylogenetic networks on $\mathcal{X}$ is
easy to compute, because the number of hardwired clusters represented by a rooted
phylogenetic network equals the number of tree edges in the network. For example,
the hardwired distance between the two rooted phylogenetic networks N_1 and N_2
displayed in Figure 6.27 equals one, because there is exactly one cluster, $\{c, e\}$,

that is only represented by N_1, and there is exactly one cluster, $\{d, e\}$, that is only represented by N_2.

The two rooted phylogenetic networks N_3 and N_4 shown in Figure 6.28 are topologically distinct, however they both represent exactly the same set of hardwired clusters. So, although the hardwired cluster distance is a proper metric for sets of clusters (represented by hardwired networks), it is not a proper metric on the networks themselves.

6.14.2 Comparing softwired networks

Let N_1 and N_2 be two rooted phylogenetic networks on $\mathcal{X}$, interpreted in the softwired sense. A natural way to compare two such networks is to compute their *softwired cluster distance*, which is another simple extension of the cluster distance for rooted phylogenetic trees (see Section 3.16), defined as follows:

Definition 6.14.3 (Softwired cluster distance)   *Let N_1 and N_2 be two rooted phylogenetic networks on $\mathcal{X}$. We define the* softwired cluster distance d_{soft} *between N_1 and N_2 as the cardinality of the symmetric difference of the two sets of softwired clusters represented by the two networks, divided by two:*

$$d_{soft}(N_1, N_2) = \frac{|\mathcal{C}_{soft}(N_1) \triangle \mathcal{C}_{soft}(N_2)|}{2}. \tag{6.6}$$

For example, the softwired cluster distance between the two rooted phylogenetic networks N_1 and N_2 displayed in Figure 6.27 is $\frac{5}{2}$, because there are exactly four softwired clusters that are only contained in N_1, namely $\{a, b\}$, $\{b, d\}$, $\{c, e\}$, and $\{a, b, d\}$, and there is exactly one that is only contained in N_2, namely $\{d, e\}$.

A rooted phylogenetic network N on $\mathcal{X}$ can represent an exponential number of different softwired clusters, in the worst case, making this distance impractical to use in general. However, in Section 6.8 we saw that the number of softwired clusters depends exponentially on the maximal number of reticulations that span any given tree edge in a rooted phylogenetic network. While this number might be quite large in the context of the population genetics of sexually reproducing organisms, say, in phylogenetic applications this number should often be fairly small.

Moreover, for some topologically restricted classes of rooted phylogenetic networks, such as galled trees and level-k networks for a fixed level k, the number of softwired clusters is polynomial and so, for such networks, the softwired cluster distance can always be computed efficiently.

The two rooted phylogenetic networks N_3 and N_4 shown in Figure 6.28 both represent the same set of softwired clusters, but are topologically distinct. Thus, while the softwired cluster distance is a proper metric on sets of clusters (represented by softwired networks), it is not a proper metric for the networks themselves.

One way to avoid the computational difficulties of the softwired cluster distance is to use the hardwired cluster distance, even when comparing rooted phylogenetic networks that are being interpreted in the softwired sense. This seems to be a reasonable heuristic, because every cluster that is represented by a rooted phylogenetic network in the hardwired sense is also represented by that network in the softwired sense. However, it is unclear how well the two different distance measures are correlated.

The tripartition distance

The *tripartition distance* was introduced in an attempt to provide a metric for softwired networks that is easy to compute [179].

To describe this approach, we need to introduce some additional simple definitions. Let N be a rooted phylogenetic network on $\mathcal{X}$. Recall that an ancestor u of a node v in N is called a stable ancestor, if every path from the root to v passes through u. If this is not the case, then we refer to u as an *unstable ancestor*. By symmetry, we call v a *stable descendant* of u, if u is a stable ancestor of v, and we call v a *unstable descendant* of u, if u is an unstable ancestor of v.

Definition 6.14.4 (Tripartitions of a network) *Let N be a rooted phylogenetic network on $\mathcal{X}$. The tripartition associated with a tree edge $e = (v, w)$ in N is defined as*

$$\theta(e) = (A(e), B(e), C(e)), \tag{6.7}$$

where

(i) $A(e) = \{x \in \mathcal{X} \mid x$ labels a stable descendant of $w\}$,
(ii) $B(e) = \{x \in \mathcal{X} \mid x$ labels an unstable descendant of $w\}$, and
(iii) $C(e) = \{x \in \mathcal{X} \mid x$ labels a node that is not a descendant of $w\}$.

We use $\Theta(N)$ to denote the (multi-)set of tripartitions obtained from all internal tree edges in N.

The set of all tripartitions of a rooted network is easy to compute, and so the following distance based on comparing the tripartitions can be easily calculated:

Definition 6.14.5 (Tripartition distance) *The tripartition distance between two rooted phylogenetic networks N_1 and N_2 on $\mathcal{X}$ is defined as:*

$$d_{tripart}(N_1, N_2) = \frac{|\Theta(N_1) \triangle \Theta(N_2)|}{2}. \tag{6.8}$$

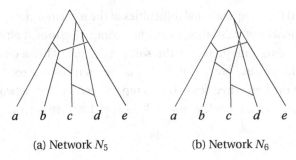

(a) Network N_5 (b) Network N_6

Figure 6.29 Two distinct rooted phylogenetic networks N_5 and N_6 that represent different softwired clusters but have tripartition distance zero.

As an example for this distance function, consider the two rooted phylogenetic networks N_1 and N_2 shown in Figure 6.27. The five internal tree edges of N_1 give rise to the following five tripartitions:

$$(\{a, b, d\}, \{c\}, \{e\}),\ (\{a\}, \{b, c\}, \{d, e\}),\ (\{d\}, \{b, c\}, \{a, e\}),$$
$$(\{e\}, \{c\}, \{a, b, d\}),\ (\{b\}, \{c\}, \{a, d, e\}).$$

The five internal tree edges of N_2 give rise to the following five tripartitions:

$$(\{a, b, c\}, \{d\}, \{e\}),\ (\{a\}, \{b, c\}, \{d, e\}),\ (\emptyset, \{b, c, d\}, \{a, e\}),$$
$$(\{e\}, \{d\}, \{a, b, c\}),\ (\{b, c\}, \emptyset, \{a, b, d, e\}).$$

As the two networks have only one tripartition in common, the tripartition distance is $d_{tripart}(N_1, N_2) = \frac{8}{2} = 4$.

On rooted phylogenetic trees, the tripartition distance equals the cluster distance. In [179], it was claimed that the tripartition distance is a proper metric on rooted phylogenetic networks. However, this assertion is now known to be false. For example, the networks N_5 and N_6 displayed in Figure 6.29 are distinct, and represent distinct sets of softwired clusters, and yet their tripartition distance is zero.

Exercise 6.14.6 (Tripartition distance) *Show that the two networks N_5 and N_6 displayed in Figure 6.29 represent different sets of softwired clusters, and have a tripartition distance of zero.*

The tripartition distance is a proper metric on the set of all rooted phylogenetic networks that have the tree-child property and are time consistent [39]:

Lemma 6.14.7 (Metric on tree-child time-consistent networks) *The tripartition distance is a proper metric on the set of all rooted phylogenetic networks that have the tree-child property and are time consistent.*

6.14.3 Comparing networks representing trees

Let N_1 and N_2 be two rooted phylogenetic networks on $\mathcal{X}$. If the networks are being used to represent a collection of rooted phylogenetic trees on $\mathcal{X}$, then it seems reasonable to base their comparison on the (multi-)sets of trees that they represent, using the *displayed trees distance*:

Definition 6.14.8 (Displayed trees distance) *Let N_1 and N_2 be two rooted phylogenetic networks on $\mathcal{X}$. We define the displayed tree distance d_{trees} between N_1 and N_2 as the cardinality of the symmetric difference of the two sets of rooted phylogenetic trees represented by the two networks, divided by two:*

$$d_{trees}(N_1, N_2) = \frac{|\mathcal{T}(N_1) \bigtriangleup \mathcal{T}(N_2)|}{2}. \tag{6.9}$$

The two rooted phylogenetic networks N_1 and N_2 displayed in Figure 6.27 both represent four rooted phylogenetic trees each (if we distinguish between multiple occurrences of the same tree), but they have only two trees in common. Thus, the displayed tree distance between N_1 and N_2 is two.

As in the case of the softwired clusters distance, the use of this distance function is limited by the fact that the number of objects to be compared is exponential in the worst case.

The two rooted phylogenetic networks N_3 and N_4 displayed in Figure 6.28 represent the same set of trees, so their displayed trees distance is zero, although the two networks are distinct. Hence, the displayed trees distance is not a proper metric for rooted phylogenetic networks.

6.14.4 Comparing the topology of networks

Two rooted phylogenetic networks N_1 and N_2 on $\mathcal{X}$ are called *isomorphic*, if there exists a directed-graph isomorphism between them that preserves the labeling of the leaves. The *isomorphism distance function* assigns a distance of 0 between any two isomorphic networks and 1 between any two non-isomorphic ones. The complexity of determining whether two rooted phylogenetic networks are isomorphic is unknown. However, if both the indegrees and outdegrees of all nodes are bounded, then isomorphism can be checked in polynomial time [168].

The isomorphism distance does not tell us how similar two non-isomorphic networks are. To obtain a more useful distance function, we proceed as follows. For

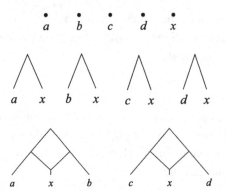

Figure 6.30 The rooted phylogenetic networks N_3 and N_4 in Figure 6.28 share exactly the same set of proper subnetworks list here. The two networks differ only as complete networks.

a rooted phylogenetic network $N = (V, E)$ on $\mathcal{X}$ and any node v of N, we define the *subnetwork* $N(v)$ associated with v to be the rooted phylogenetic network $N|_{V'}$ that is embedded in N and is induced by the set V' of all nodes that are descendants of v, with root v. For an example, see Figure 6.30. We use $\Omega(N)$ to denote the multi-set of all rooted phylogenetic networks that are subnetworks of N and we define [40]:

Definition 6.14.9 (Subnetwork distance) *Let N_1 and N_2 be two rooted phylogenetic networks on $\mathcal{X}$. The* subnetwork isomorphism distance *between N_1 and N_2 is defined as*

$$d_{sub-net}(N_1, N_2) = \frac{|\Omega(N_1) \, \triangle \, \Omega(N_2)|}{2}. \tag{6.10}$$

As an example, consider the two rooted phylogenetic networks N_3 and N_4 displayed in Figure 6.28. They both have the same proper subnetworks, listed in Figure 6.30. But as the full networks themselves are also considered subnetworks, the symmetric difference of the two sets has cardinality two and so the subnetwork distance is $d_{sub-net}(N_1, N_2) = 1$.

In addition to this approach, a number of simpler distance functions have been suggested that try and capture how topologically similar two rooted phylogenetic networks are. They are based on comparing basic topological invariants of the networks that can be computed efficiently. Also, they are easier to use than the isomorphism-based approach when comparing two networks "by hand". In the following, we describe the two that are perhaps most useful.

The path-multiplicity distance

Let N be a rooted phylogenetic network on $\mathcal{X} = \{x_1, \dots, x_n\}$. The *path-multiplicity distance* is based on the idea of counting the number of different paths from any given node to each of the leaves in N. The *path-multiplicity vector* of a node v in N is defined as

$$\mu(v) = (m_1(v), \ldots, m_n(v)), \tag{6.11}$$

where $m_i(v)$ is the number of different directed paths from v to the leaf with label x_i (for $i = 1, \ldots, n$). Let $\mu(N)$ denote the (multi-)set of all path-multiplicity vectors obtained for all internal nodes of N. The set of all path-multiplicity vectors is easy to compute and we define a distance by comparing the sets of path-multiplicity vectors [41]:

Definition 6.14.10 (Path-multiplicity distance) *The path-multiplicity distance (also called the μ-distance) between two rooted phylogenetic networks N_1 and N_2 on $\mathcal{X}$ is defined as*

$$d_{path}(N_1, N_2) = \frac{|\mu(N_1) \triangle \mu(N_2)|}{2}. \tag{6.12}$$

As an example, consider the two rooted phylogenetic networks N_1 and N_2 displayed in Figure 6.27. Both have eight internal nodes. Processing the nodes from top-left to bottom-right, the path vectors associated with network N_1 are:

$(1, 2, 3, 1, 1)$, $(1, 2, 2, 1, 0)$, $(1, 1, 1, 0, 0)$, $(0, 1, 1, 1, 0)$,
$(0, 0, 1, 0, 1)$, $(0, 1, 1, 0, 0)$, $(0, 1, 1, 0, 0)$, $(0, 0, 1, 0, 0)$.

Similarly, for N_2 we obtain:

$(1, 2, 2, 2, 1)$, $(1, 2, 2, 1, 0)$, $(1, 1, 1, 0, 0)$, $(0, 1, 1, 1, 0)$,
$(0, 0, 0, 1, 1)$, $(0, 1, 1, 0, 0)$, $(0, 1, 1, 0, 0)$, $(0, 0, 0, 1, 0)$.

The symmetric difference between the two listed sets contains six elements and so $d_{path}(N_1, N_2) = 3$.

Exercise 6.14.11 (Indistinguishability) *Show that the path-multiplicity distance between the two distinct rooted phylogenetic networks N_3 and N_4 displayed in Figure 6.28 is zero.*

The two networks depicted in Figure 6.28 show that the path multiplicity distance is not a metric on the set of all rooted phylogenetic networks. However, the path-multiplicity distance is a proper metric on the set of all root phylogenetic networks that have the tree-child property [41]:

Lemma 6.14.12 (Metric on tree-child networks) *The path-multiplicity distance is a proper metric on the set of all rooted phylogenetic networks that have the tree-child property.*

Moreover, the path-multiplicity distance is a also proper metric on the following set of rooted phylogenetic networks [38]:

Lemma 6.14.13 (Metric on a subset of tree-sibling networks)   *The path-multiplicity distance is a proper metric on the set of all bicombining rooted phylogenetic networks that have the tree-sibling property and are time consistent.*

The nested-labels distance

The final distance function that we consider is the nested-labels distance. In order to compute this distance function, we assign *nested labels* to all nodes of a rooted phylogenetic network N on $\mathcal{X}$ recursively as follows:

(i) The nested label of a leaf is the taxon that labels the leaf.
(ii) The nested label of an internal node is defined as the multi-set of all nested labels of its children.

Let $\Upsilon(N)$ denote the multi-set of all nested labels assigned to nodes of N. The nested-labels distance can be defined as follows [40, 183]:

Definition 6.14.14 (Nested-labels distance)   *The nested-labels distance between two rooted phylogenetic networks N_1 and N_2 on $\mathcal{X}$ is defined as*

$$d_{nested}(N_1, N_2) = \frac{|\Upsilon(N_1) \bigtriangleup \Upsilon(N_2)|}{2}. \tag{6.13}$$

As an example, consider the two rooted phylogenetic networks N_1 and N_2 shown in Figure 6.27. The leaf nodes give rise to the nested labels $a, b, \ldots, e$. Although the labels are computed bottom-up, we report them here from top-left to bottom-right. The nested labels of network N_1 are:

$$\{\{\{a, \{\{b, \{c\}\}\}\}, \{\{\{b, \{c\}\}\}, d\}\}, \{\{c\}, e\}\},$$
$$\{\{a, \{\{b, \{c\}\}\}\}, \{\{\{b, \{c\}\}\}, d\}\},$$
$$\{a, \{\{b, \{c\}\}\}\},$$
$$\{\{\{b, \{c\}\}\}, d\},$$
$$\{\{b, \{c\}\}\},$$
$$\{b, \{c\}\},$$
$$\{\{c\}, e\},$$
$$\{c\},$$
$$a, \ b, \ c, \ d, \ e.$$

The nodes of the network N_2 give rise to the following nested labels:

$$\{\{\{a, \{\{b, c\}\}\}, \{\{\{b, c\}\}, \{d\}\}\}, \{\{d\}, e\}\},$$
$$\{\{a, \{\{b, c\}\}\}, \{\{\{b, c\}\}, \{d\}\}\},$$
$$\{a, \{\{b, c\}\}\},$$
$$\{\{\{b, c\}\}, \{d\}\},$$
$$\{\{d\}, e\},$$
$$\{\{b, c\}\},$$
$$\{b, c\},$$
$$\{d\},$$
$$a, \ b, \ c, \ d, \ e.$$

The symmetric difference of these two sets contains 16 labels and so we have $d_{nested}(N_1, N_2) = \frac{16}{2} = 8$.

Exercise 6.14.15 (Indistinguishability)   *Show that the nested label distance between the two distinct rooted phylogenetic networks N_3 and N_4 displayed in Figure 6.28 is zero.*

In [40] it is shown that this distance is a proper metric on a number of different classes of networks. Here we report one of the results:

Lemma 6.14.16 (Metric on tree-child networks)   *The nested label distance is a proper metric on the set of all rooted phylogenetic networks that have the tree-child property.*

6.14.5 Alignment of rooted phylogenetic networks

To aid a visual comparison of two rooted phylogenetic networks N_1 and N_2 on $\mathcal{X}$, it would be helpful to algorithmically match nodes that are similarly situated in the two networks. In general terms, an *alignment* of two rooted phylogenetic networks N_1 and N_2 on $\mathcal{X}$ is given by a one-to-one mapping between a subset of the nodes of N_1 and a subset of the nodes of N_2, which defines a set of *aligned pairs* of nodes.

In the computation of the path-multiplicity distance in Section 6.14.4, each internal node in both rooted phylogenetic networks under comparison is assigned a path-multiplicity vector. These vectors can be used to compute an alignment between the two networks [41].

In more detail, consider the *network alignment graph* $A(N_1, N_2) = (V_1 \cup V_2, E)$, whose node set is the union of the set V_1 of internal nodes of network N_1 and the set V_2 of internal nodes of network N_2, and whose edge set E consists of all possible edges between a node in V_1 and a node in V_2.

In this bipartite graph, we assign a weight $\omega(e)$ to each edge $e = (v_1, v_2)$, with $v_1 \in V_1$ and $v_2 \in V_2$, that is base on the negative sum of absolute differences of the

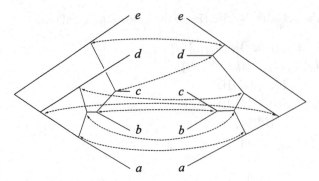

Figure 6.31 An alignment of the two rooted phylogenetic networks N_1 and N_2. Dashed lines are used to connect aligned pairs of nodes. For simplicity, we have omitted the lines between the two root nodes and between the paired leaves.

components of the path-multiplicity vectors $\mu(v_1)$ and $\mu(v_2)$, namely

$$\omega((v_1, v_2)) = W - \sum_{i=1}^{n} |m_i(v_1) - m_i(v_2)|, \qquad (6.14)$$

where W is a large constant that ensures that all weights will be non-negative. Recall from graph theory that a *matching* M is a set of edges in a graph such that no two nodes are incident to more than one edge in M. It is called a *maximum weighted matching*, if, in addition, it maximizes the sum of edge weights $\sum_{e \in M} \omega(e)$. Determining a maximum weighted matching for a bipartite graph is computationally easy [156].

So, let M be a maximum weighted matching in the alignment graph $A(N_1, N_2)$. An alignment of N_1 and N_2 is then given as follows: Any two nodes that are connected by an edge that is contained in the matching M form an aligned pair. In addition, each pair of leaves that are labeled by the same taxon form an aligned pair.

Exercise 6.14.17 (Special case) *In a rooted phylogenetic network N_1, assume that r is a reticulate node that has exactly one child, v. Compare the path vectors $\mu(r)$ and $\mu(v)$. What problem can arise for r and v in a matching-based alignment to some other network N_2? How can the problem be solved?*

To illustrate the alignment of networks, consider the two rooted phylogenetic networks N_1 and N_2 displayed in Figure 6.27. A maximum weighted matching M in $A(N_1, N_2)$ is achieved by pairing the internal nodes in the two networks in the order that they occur when listed from top-left to bottom-right, as illustrated in Figure 6.31.

Chapter notes

This chapter contains many new results and a fair amount of unpublished material. To the best of our knowledge, this is the first attempt to develop a theory of clusters and rooted phylogenetic networks. We became aware of the importance of distinguishing between "softwired" and "hardwired" rooted phylogenetic networks while writing this chapter. The hardness of the cluster containment problem (and also of the tree containment problem) for softwired networks suggests that simply reporting a rooted phylogenetic network without further annotations may be too uninformative and thus additional information indicating how clusters or trees are embedded in the network may be useful (for example, see Figures 4.5 and 4.6).

Part III

Algorithms and applications

In this third part of the book we present a wide range of algorithms that have been developed for computing phylogenetic networks from biological data. For many of the methods, we briefly describe an application and provide references to papers that contain more details. The material is organized by type of input, which is either splits (Chapter 7), clusters (Chapter 8), sequences (Chapter 9), distances (Chapter 10), trees (Chapter 11), triples or quartets (Chapter 12). Additionally, we address the problem of drawing phylogenetic networks (Chapter 13) and give an overview of some of the existing software for computing them (Chapter 14).

7

Phylogenetic networks from splits

In Chapter 5, we introduced splits and split networks and presented the Buneman graph as the canonical split network associated with a given set of splits. Splits can be computed from many different types of data, as we discuss in some of the following chapters. The focus of this chapter is on the problem of computing a split network N for a given set of splits $\mathcal{S}$ on $\mathcal{X}$. We present two different approaches. The first is the *convex hull algorithm* that computes the Buneman graph and can be applied to any set of splits, using an exponential number of nodes and edges in the worst case [11]. It is also used in Section 9.4 to compute median networks. The second is the *circular network algorithm*, which can be applied to any set of circular splits and produces an *outer-labeled planar network* with only a quadratic number of nodes and edges [62].

Both algorithms proceed in two steps. In the first step, all trivial splits in $\mathcal{S} = \{\mathcal{S}_1, \ldots, \mathcal{S}_m\}$ are processed to obtain a *star network* consisting of a central node and one leaf per taxon. Then, in the second step, the remaining splits are inserted one by one so as to obtain the final network.

To avoid some technical complexities, in the following we will assume that $\mathcal{S}$ contains all n trivial splits on $\mathcal{X} = \{x_1, \ldots, x_n\}$. We will use $\mathcal{S}^O = \{S_1^O, \ldots, S_n^O\}$ and $\mathcal{S}^I = \{S_1^I, \ldots, S_r^I\}$ to denote the set of trivial ("outer") and non-trivial ("inner") splits in $\mathcal{S}$, with $n + r = m$.

7.1 The convex hull algorithm

In this section we present the *convex hull algorithm* for constructing a split network. First we introduce the concept of the convex hull of a set of taxa in a split network:

Definition 7.1.1 (Convex hull)   *Let $\mathcal{S}$ be a set of splits on $\mathcal{X}$ and let $N = (V, E, \sigma, \lambda)$ be a split network that represents $\mathcal{S}$. Consider a set of taxa $C \subset \mathcal{X}$. The* convex hull *$H(C)$ of C in N is defined as the set of all nodes in N that lie on a shortest path*

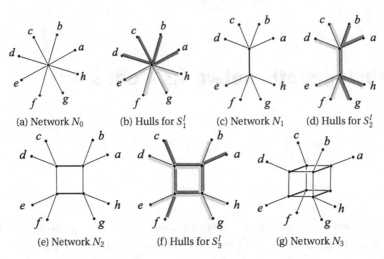

(a) Network N_0　　(b) Hulls for S_1^I　　(c) Network N_1　　(d) Hulls for S_2^I

(e) Network N_2　　(f) Hulls for S_3^I　　(g) Network N_3

Figure 7.1　Construction of the canonical split network for three non-trivial splits S_1^I, S_2^I and S_3^I on $\mathcal{X} = \{a, \ldots, h\}$. (a) The split network N_0 representing all trivial splits on $\mathcal{X}$. (b) The two convex hulls $H(A_1)$ and $H(B_1)$ for the split $S_1^I = \frac{A_1}{B_1} = \frac{\{a,b,c,d\}}{\{e,f,g,h\}}$ (highlighted in dark gray and light gray, respectively). The intersection $H(A_1) \cap H(B_1)$ consists of the central node. (c) The resulting network N_1 obtained using the convex hull algorithm. (d) The two convex hulls for $S_2^I = \frac{A_2}{B_2} = \frac{\{a,b,g,h\}}{\{c,d,e,f\}}$. (e) The resulting network N_2. (f) The two convex hulls for $S_3^I = \frac{A_3}{B_3} = \frac{\{a,c,f\}}{\{b,d,e,g,h\}}$. (g) The resulting network N_3.

between any two nodes v and w that are contained in $\lambda(C)$, the set of all nodes labeled by taxa in C.

Examples can be found in Figure 7.1.

Exercise 7.1.2 (Alternative definition)　*Show that the above definition can be rephrased as follows: the convex hull $H(C)$ consists of all nodes in N that can be reached from some node $v \in \lambda(C)$ using only such edges that are labeled by splits contained in the set $\{\frac{A}{B} \in S \mid A \cap C \neq \emptyset, B \cap C \neq \emptyset\}$.*

Our aim is to construct a split network $N = (V, E, \sigma, \lambda)$ that represents the set of splits $S = S^O \cup S^I$. To accomplish this, we need to define the set of nodes V, set of edges E, labeling of nodes by taxa λ and the labeling of edges by splits σ. First, we process all the trivial splits so as to obtain a *star network* with one central node and one edge for each trivial split, as shown in Figure 7.1a:

Algorithm 7.1.3 (Star network for trivial splits)　*Let $\mathcal{X} = \{x_1, \ldots, x_n\}$ be a set of taxa and let $S^O = \{S_1^O, \ldots, S_n^O\}$ be the set of all trivial splits on $\mathcal{X}$, with $S_i^O = \frac{\{x_i\}}{\mathcal{X} - \{x_i\}}$ for $i = 1, \ldots, n$. We obtain a split network N_0 representing S^O as follows: Create a new node v_0. For $i = 1, \ldots, n$ do:*

(i) *Create a new node v_i and new edge $e_i = \{v_0, v_i\}$.*
(ii) *Set $\lambda(x_i) = v_i$ and $\sigma(e_i) = S_i^O$.*

After constructing the initial star network, we add the non-trivial splits to the network one by one, as follows. For each non-trivial split, compute the convex hull of each of the two split parts, determine and duplicate the intersection of the two hulls, one copy for each hull, and then insert new edges to connect the two copies.

More formally, we process the set of non-trivial splits S^I by applying the following algorithm to each split in $S^I = \{S_1^I, S_2^I, \ldots, S_r^I\}$ in turn (see Figure 7.1b–g):

Algorithm 7.1.4 (Convex hull algorithm) *Let S^O be the set of all trivial splits on X and let $S_t^I = \{S_1^I, S_2^I, \ldots, S_t^I\}$ be a set of $t \geq 0$ non-trivial splits on X. Assume that we have already computed a split network N_t for $S^O \cup S_t^I$. To obtain a split network for $S^O \cup S_t^I \cup \{S_{t+1}^I\}$, where $S_{t+1}^I = \frac{A}{B}$ is a new non-trivial split, modify N_t as follows:*

(i) *Compute the two convex hulls $H(A)$ and $H(B)$ in N_t and let M be the split graph induced in N_t by the set of nodes $H(A) \cap H(B) \neq \emptyset$.*
(ii) *Create a copy M' of M and for each node v and edge e in M let v' and e' denote their copies in M', respectively.*
(iii) *For every edge f that leads from some node u in $H(B) - H(A) \neq \emptyset$ to a node v in M, redirect the edge f so that it leads from u to v'.*
(iv) *Connect each pair of nodes v in M and v' in M' by a new edge $f = (v, v')$ and set $\sigma(f) = S_{t+1}$.*

The algorithm makes two assumptions that we should briefly discuss. Let N be a split network for a set of splits S on X and let $S = \frac{A}{B}$ be a split of X that is not contained in S. Then $H(A) \cap H(B) \neq \emptyset$ must hold, as otherwise there would already exist a set of edges in N that separates $\lambda(A)$ from $\lambda(B)$. This is used in step (i) of the algorithm. Further, the assumption that S contains all trivial splits of X implies $H(A) - H(B) \neq \emptyset$ and $H(B) - H(A) \neq \emptyset$. This is used in step (iii) of the algorithm.

In the worst case, $H(A) \cap H(B)$ will consist of all non-leaf nodes of N_t and then N_{t+1} will have twice as many internal nodes as N_t. As this can happen for every split that is processed by the algorithm, the number of nodes and edges in the final split network can be exponential in the number of splits.

Note that the split network N constructed by this algorithm is uniquely defined and is the same as the one specified by the Buneman graph construction in Section 5.6. Although this is not easy to prove, we leave the proof as an exercise:

Exercise 7.1.5 (Convex hull algorithm computes Buneman graph) *Prove that the convex hull algorithm computes the Buneman graph.*

7.2 The circular network algorithm

A network N is called *planar*, if it can be drawn in the euclidean plane in such a way that no two edges intersect. If a concrete drawing of N is given, then we refer to this as a *plane network* and, for brevity, also denote it by N. A *topological embedding* of N is given by a mapping τ that assigns to each node $v \in V$ an ordering $\tau(v) = (v_1, \ldots, v_k)$ of the set of all its adjacent nodes $\{v_1, \ldots, v_k\}$.

As above, assume that we are given a set of splits $\mathcal{S} = \mathcal{S}^O \cup \mathcal{S}^I$, where $\mathcal{S}^O = \{S_1^O, \ldots, S_n^O\}$ is the set of all trivial splits on $\mathcal{X}$ and $\mathcal{S}^I = \{S_1^I, \ldots, S_r^I\}$ is a set of non-trivial splits. Additionally, we assume that $\mathcal{S}$ is *circular* with respect to the ordering $(x_1, \ldots, x_n)$, as defined in Section 5.7.

Our aim is to construct a planar split network N that represents $\mathcal{S}$. This can be done using the *circular network algorithm*, which operates in a similar fashion to the convex hull algorithm. Just as in that algorithm, the first step is to obtain a star network with one central node and one edge for each trivial split using Algorithm 7.1.3 (see Figure 7.1a).

The main difference between the circular network algorithm and the convex hull algorithm is in the choice of the subnetwork M to be duplicated when processing a non-trivial split: in the circular network algorithm, the subnetwork M will be a single path along the outside of the existing network.

Let $\mathcal{S}$ be a set of splits on $\mathcal{X}$ and suppose we are given an outer-labeled planar split network $= (V, E, \sigma, \lambda)$ for $\mathcal{S}$, in which the taxa appear in the circular order $(x_1, \ldots, x_n)$ as one circles around the network in mathematically positive orientation. For a new non-trivial split $S = \frac{\{x_p, \ldots, x_q\}}{\mathcal{X} - \{x_p, \ldots, x_q\}} \notin \mathcal{S}$, with $1 < p < q \leq n$, we define $M(x_p, x_q)$ to be the unique shortest path from $\lambda(x_p)$ to $\lambda(x_q)$ that runs around the outside of the embedded network N and is incident to every leaf edge that is connected to a node labeled by one of the taxa in $\{x_p, \ldots, x_q\}$.

The second step is to process the non-trivial splits one after the other (see Figure 7.2). To ensure that the resulting network has a minimal number of nodes and edges, and possesses an embedding in the plane in which all edges corresponding to the same split can be drawn as parallel lines of the same length, the algorithm must process the non-trivial splits in order of non-increasing cardinality of the split part that does not contain taxon x_1.

Algorithm 7.2.1 (Circular network algorithm)   *Let $\mathcal{S}^O$ be the set of all trivial splits on $\mathcal{X}$ and let S_t^I be a set of $t \geq 0$ non-trivial splits that is circular on $(x_1, \ldots, x_n)$. Assume that we have already computed a split network N_t for $\mathcal{S}^O \cup S_t^I$ using this algorithm. To obtain a split network for $\mathcal{S}^O \cup S_t^I \cup \{S_{t+1}^I\}$, where $S_{t+1}^I = \frac{\{x_p, \ldots, x_q\}}{\mathcal{X} - \{x_p, \ldots, x_q\}}$ is a new non-trivial split (obeying the ordering requirement formulated above), modify N_t as follows:*

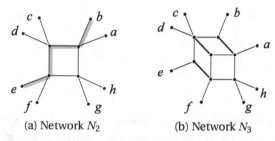

(a) Network N_2 (b) Network N_3

Figure 7.2 To add the split $\frac{A}{B} = \frac{\{a,f,g,h\}}{\{b,c,d,e\}}$ to the split network N_2 shown in (a), with $x_1 = a$, $x_2 = b$, etc., the circular network algorithm duplicates the internal edges of the path $M(b, e)$ leading from the b to e (highlighted in gray) to obtain the network N_3 shown in (b), in which the three edges representing $\frac{A}{B}$ are highlighted in bold.

(i) *Determine the path $M(x_p, x_q)$ and let $\dot{M}$ denote the path obtained by removing the first and last (leaf) edges from $M(x_p, x_q)$.*

(ii) *Create a copy $\dot{M}'$ of $\dot{M}$ and for each node v and edge e in $\dot{M}$ let v' and e' denote their copies, respectively.*

(iii) *Redirect every edge f that leads from some node $u = \lambda(x_i)$ to some node v in $\dot{M}$ so that it leads from u to v', for all $i = p, \ldots, q$.*

(iv) *Connect each pair of nodes v in $\dot{M}$ and v' in $\dot{M}'$ by a new edge $f = (v, v')$ and set $\sigma(f) = S_{t+1}$.*

The number of nodes and edges produced by this algorithm is at most quadratic in the number of splits m. To see this, note that the path $M(x_p, x_q)$ in the algorithm contains at most one edge per split and so processing of a single non-trivial split produces a linear number of edges.

Indeed, $M(x_p, x_q)$ cannot contain two edges corresponding to the same split $S' \in \mathcal{S}_t^I$ because this would imply that the split part of S' that does not contain x_1 is smaller than the split part of S_{t+1}^I that does not contain x_1, which is impossible due to the ordering in which the algorithm processes the splits. The requirement that the non-trivial splits must be processed in order of non-increasing size of the split parts not containing x_1 is not only important for minimizing the number of edges in the network. If we disregard this requirement, then the resulting split network may fail to have the property that edges corresponding to the same split can be drawn as parallel line segments of the same length:

Exercise 7.2.2 (Correct order is required) *Consider the split network N_3 shown in Figure 7.2. Using the circular network algorithm, attempt to modify N_3 so as to obtain a new split network N_4 that also represents the split $S_4^I = \frac{\{a,h\}}{\{b,c,d,e,f,g\}}$ as well. What goes wrong? Rerun the algorithm on all four splits in the correct order.*

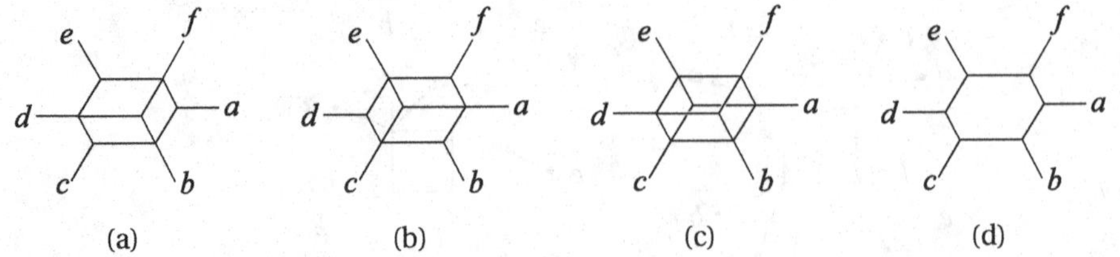

Figure 7.3   All four different split networks shown here represent the same set of splits. The split networks shown in (a) and (b) were computed using the circular network algorithm processing the splits and taxa in two different orders. The one shown in (c) was constructed using the convex hull algorithm. The split network shown in (d) can be obtained by deleting superfluous edges in any of the first three.

Unlike in the case of the convex hull algorithm, the split network constructed by this algorithm is not uniquely defined, as the resulting network depends on the order in which the splits are processed. In Figure 7.3, we show four different representations of the split set S consisting of all six trivial splits on $X = \{a, \ldots, f\}$ and the three non-trivial splits $S_1 = \frac{\{a,b,c\}}{\{d,e,f\}}$, $S_2 = \frac{\{b,c,d\}}{\{e,f,a\}}$ and $S_3 = \frac{\{c,d,e\}}{\{f,a,b\}}$. The first two networks were produced by the circular network algorithm and the third by the convex hull algorithm. The fourth network can be obtained by the deletion of superfluous edges in any of the first three. This example also illustrates that the convex hull algorithm does not necessarily produce a planar network, even when given a circular set of splits.

Chapter notes

In many cases, direct application of the convex hull algorithm leads to an over-complicated network, as indicated in Figure 7.3. In practice, a useful heuristic is to first choose an order of the taxa such that a large subset of the given set of splits is circular. This subset of splits is then processed using the circular network algorithm to obtain an outer-labeled planar network. The remaining splits are then processed using the convex hull algorithm, which will add some non-planar parts to the network. This is the default approach used in SplitsTree [125].

In the following chapters we describe a number of different methods for computing a set of splits from sequences, distances, trees or quartets. All these methods are used in conjunction with the convex hull and circular network algorithms to obtain a split network from biological data.

8

Phylogenetic networks from clusters

In this chapter, we present some methods for computing phylogenetic networks from clusters. Clusters, as opposed to splits, are inherently rooted and this explains why all methods presented in this chapter produce *rooted* phylogenetic networks. We first briefly mention the cluster-popping algorithm for computing a cluster network, and then present a divide-and-conquer approach for computing rooted phylogenetic networks. Based on this, we discuss how to compute minimum galled trees, galled networks and level-k networks from clusters.

8.1 Cluster networks

In Chapter 6, we introduced the concept of a cluster network, which is a rooted phylogenetic network that represents a set of clusters in the hardwired sense. In such a network, every tree edge represents exactly one cluster, which is defined as the set of taxa that label nodes below that edge. Given a set of clusters C on $\mathcal{X}$, the cluster network N that represents C can be efficiently computed using the cluster-popping algorithm presented in Section 6.4. This algorithm is implemented in the Dendroscope program and there it can be used to compute a cluster network from the set of clusters associated with a collection of rooted phylogenetic trees [121].

8.1.1 Application

A typical application of a cluster network is that one is given multiple rooted gene trees for a set of species and one would like to produce a network to illustrate the parts of the phylogeny upon which the trees agree and which parts are resolved in different ways. For example, in [195] the evolutionary relationship between human, chimpanzee and gorilla, together with orangutan and rhesus monkey is investigated using genome-wide data, and strong support for three alternative rooted phylogenetic trees is reported in [65], see Figure 8.1a–c. The cluster network N that represents all clusters contained in the three trees is shown in Figure 8.1d.

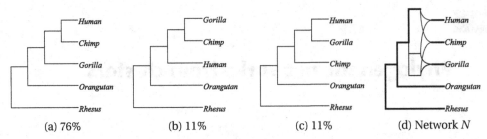

(a) 76%   (b) 11%   (c) 11%   (d) Network N

Figure 8.1   Three rooted phylogenetic trees shown in (a), (b) and (c) supported by 76%, 11% and 11%, respectively, of all genes studied in [65]. In (d) we show the cluster network N computed on the three trees. The line width of each edge is proportional to the number of trees that contain it. Adapted from [121].

8.2 Divide-and-conquer using decomposition

As we argue in Section 6.10, decomposability is a desirable property to have and thus it is sensible to restrict ones attention to such networks that exhibit the decomposition property when computing a rooted phylogenetic network N to represent a given set of clusters C on $\mathcal{X}$. We now describe a *divide-and-conquer* strategy that is based on the idea of decomposition and always produces a decomposable network. We use this approach later in the chapter to compute galled trees, galled networks and level-k networks from clusters.

Given a set of clusters C on $\mathcal{X}$ and a *base method M* that computes a rooted phylogenetic network for a set of *tangled* clusters (as defined below), the divide-and-conquer approach has three steps:

(i) Determine the incompatibility graph $IG(C)$.
(ii) For each non-trivial connected component C' of $IG(C)$, merge all taxa that are not separated by C' and then compute a "local" phylogenetic network N' for C' using the given base method M.
(iii) Merge all the local phylogenetic networks to obtain a "global" network N that represents the complete set of clusters C.

Let us look at this approach in more detail. The process is guided by a *backbone tree* that is defined as follows:

Definition 8.2.1 (Backbone tree)   *Let C be a set of clusters on $\mathcal{X}$ with incompatibility graph $IG(C)$. The set of* backbone clusters *associated with C is defined as*

$$B(C) = \{\mathcal{X}(C') \mid C' \text{ is a connected component of } IG(C)\}, \tag{8.1}$$

where $\mathcal{X}(C') = \cup_{C \in C'} C$ denotes the set of all taxa mentioned in C'. The backbone tree *for C is given by the unique rooted phylogenetic tree $T(B(C))$ that represents $B(C)$.*

Note that this definition requires that the set of backbone clusters is compatible.

Exercise 8.2.2 (Backbone clusters are compatible)   *Let C be a set of clusters on X. Prove that the set of backbone clusters $B(C)$ is compatible.*

Let C be a set of clusters on X. The set of backbone clusters $B(C)$ contains every cluster C in C that is compatible with all other clusters in C. Moreover, each non-trivial connected component Z of $IG(C)$ gives rise to a backbone cluster consisting of all taxa that occur in the clusters contained in Z. A member of $B(C)$ can also arise in both ways, see, for example, the cluster $\{a, b, c, d\}$ obtained from the incompatibility graph shown in Figure 8.2b.

We refer to the set of clusters C' of a non-trivial connected component of the incompatibility graph $IG(C)$ (together with the set of contained taxa) as a *subproblem* of C, as it corresponds to a dataset that is dealt with in step (ii) of the divide-and-conquer approach.

Before applying the given base method M to a subproblem C' to obtain a rooted phylogenetic network N' for the subproblem, we simplify the subproblem further so as to fulfill the requirements of the following definition:

Definition 8.2.3 (Tangled clusters)   *A set of clusters C on X is called* tangled, *if it satisfies two conditions:*

(C1) The incompatibility graph $IG(C)$ is connected and has more than one node.
(C2) Every pair of taxa x and y in X is separated by some cluster C in C.

A subproblem always fulfills property (C1) by definition. To ensure that (C2) holds, we *collapse* a given subproblem C on X by merging any two taxa in X that are not separated by a cluster in C. In more detail, each maximal set A of taxa in X that is not separated by C is treated as a single *composite* taxon $\overline{A}$. Let $\overline{X}$ denote the set of all composite taxa obtained in this way. Clearly, each taxon in X is contained in some composite taxon in $\overline{X}$. Each cluster C in C gives rise to a new cluster $\overline{C}$ on $\overline{X}$, which is defined as $\overline{C} = \{\overline{A} \in \overline{X} \mid A \subseteq C\}$. In other words, $\overline{C}$ contains a composite taxon $\overline{A}$ if and only if the original cluster C contains the elements of the original set A. We use $\overline{C}$ to denote the set of all such *collapsed clusters*.

Here is a detailed description of the divide-and-conquer algorithm. The whole process is illustrated in Figure 8.2.

Algorithm 8.2.4 (Divide-and-conquer)   *Assume that we are given a set of clusters C on X and a base method M for computing a rooted phylogenetic for any tangled set of clusters C'. The decomposition-based divide-and-conquer approach operates as follows.*

(i) *Compute the incompatibility graph $IG(C)$.*
(ii) *Compute the set of backbone clusters $B(C)$ and the associated rooted backbone tree $T(B(C))$, and initially set $N = T(B(C))$.*

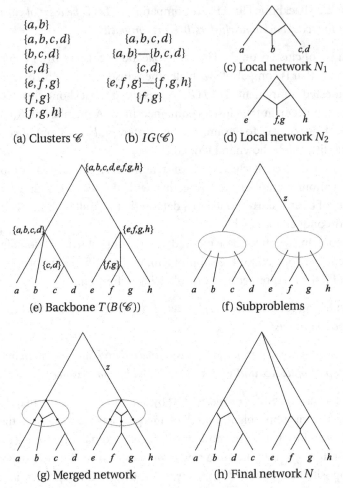

(a) Clusters $\mathscr{C}$

(b) $IG(\mathscr{C})$

(c) Local network N_1

(d) Local network N_2

(e) Backbone $T(B(\mathscr{C}))$

(f) Subproblems

(g) Merged network

(h) Final network N

Figure 8.2   Details of the divide-and-conquer strategy. For the set of clusters C shown in (a) we show the corresponding incompatibility graph $IG(C)$ in (b) (ignoring all trivial clusters). For the two non-trivial connected components in C, namely $C_1 = \{\{a, b\}, \{b, c, d\}\}$ and $C_2 = \{\{e, f, g\}, \{f, g, h\}\}$, we show two corresponding rooted phylogenetic networks N_1 and N_2 in (c) and (d), respectively. In (e), we illustrate the backbone tree $T(B(C))$ for C with all internal nodes labeled by the corresponding taxon sets. The two subproblems for this tree are indicated as ovals in (f). In (g) we show how the two networks N_1 and N_2, obtained using the cluster-popping algorithm, are inserted into the backbone tree. We show the final rooted phylogenetic network N in (h). Note that the edge z in (g) is contracted in (h) because the cluster $\{e, f, g, h\}$ represented by z is not a member of the original input set C.

(iii) *For each edge $e = (u, v)$ in N that corresponds to a subproblem C', with composite taxon set X', do the following:*

    (a) *Collapse C' to $\overline{C'}$ and then compute a corresponding local rooted phylogenetic network N' for $\overline{C'}$ on $\overline{X'}$ using the base method M.*

    (b) *Insert the local network N' into N. This is done by detaching all children of v in N, identifying v with the root of N' and then reattaching each former child w of v to that leaf ℓ of N' that is labeled by a composite taxon $\overline{A} \in X'$ such that $L(w) \subseteq A$ holds.*

(iv) *Contract any tree edge in N that does not represent a cluster in C.*

(v) *Suppress all suppressible nodes.*

A crucial property of the algorithm is that no two biconnected components are merged when inserting a local network N' into the network N. This follows from the fact that, for each node w, there exists exactly one leaf ℓ that fulfills the requirement posed in step (iiib) of the algorithm, which in turn is a direct consequence of the following result:

Lemma 8.2.5 (Insertion lemma)   *Let C be a set of clusters on X and let $T(B(C))$ be the corresponding backbone tree. Consider two successive edges $e_1 = (u, v)$ and $e_2 = (v, w)$ in the backbone tree $T(B(C))$, and let C_1 and C_2 be the two subproblems associated with these two edges, on taxon sets X_1 and X_2, respectively. (Here we consider C_2 to be a subproblem, even if it consists of only one cluster). If C_1 is a cluster in C_1 such that $C_1 \cap X_2 \neq \emptyset$ holds, then $X_2 \subseteq C_1$.*

Proof   For the sake of contradiction, let us assume that the claim is false. Then either there exists a cluster C in C_1 such that $C \subset X_2 \subset X_1$ holds, or there exists a cluster C_1 in C_1 that is incompatible with X_2. In the former case we can apply Lemma 6.2.6, using X_2 as the separating cluster, to show that there must exist some cluster C_1 in C_1 that is incompatible with X_2, because otherwise the incompatibility graph for C_1 cannot be connected. To finish the first case and to take care of the second case, suppose that C_1 and X_2 are incompatible. There are three possibilities to consider: (i) The intersection $C_1 \cap X_2$ is a cluster in C_2. (ii) There exists a cluster C_2 in C_2 that is properly contained in $C_1 \cap X_2$. (iii) There exists a cluster $C_2 \in C_2$ that is incompatible with $C_1 \cap X_2$. In the first two cases, Lemma 6.2.6, using $C1 \cap X_2$ as the separating set, implies that the incompatibility graph $IG(C_2)$ cannot be connected, a contradiction. In the third case, it follows that C_1 and C_2 are incompatible, again a contradiction.   □

Let C be a set of clusters on X. Any rooted phylogenetic network N computed for C using the divide-and-conquer strategy has the decomposition property:

Theorem 8.2.6 (Decomposability) *Let C be a set of clusters on $\mathcal{X}$ and let M be a method for calculating a softwired network N' for any set of tangled clusters C'. If N is a rooted phylogenetic network that was computed for C using Algorithm 8.2.4 with M as base method, then N has the decomposition property with respect to C and represents C in the softwired sense.*

Proof Let C be a set of clusters on $\mathcal{X}$ and let M be a method for calculating a rooted phylogenetic network to represent any given set of tangled clusters. Let N be a rooted phylogenetic network computed using the divide-and-conquer algorithm. By Lemma 6.10.2, we have that any two clusters that are contained in the same connected component of $IG(C)$ are represented by edges that are contained in the same biconnected component of N. By definition, only clusters contained in the same component of $IG(C)$ are considered together in a subproblem. On the other hand, Lemma 8.2.5 ensures that the insertion step of the algorithm never merges two distinct biconnected components into a larger biconnected component. Also, the final step of contracting superfluous tree edges cannot merge any biconnected components. Hence, the resulting network N has the decomposition property. Lemma 8.2.5 also implies the following: For any tree edge e in N that is present in the original backbone tree T we have that the set of leaves reachable from the target node of e in N is the same as in T. This implies that N represents C.   □

If the base method M can fail, such as in the case of the galled tree algorithm for clusters described in Section 8.3, then the divide-and-conquer algorithm should be modified to take such failures into account, for example by aborting the computation and returning *fail*. If the base method M can return more than one possible network for a given subproblem C', then the divide-and-conquer algorithm should be modified so as to store the set of all solutions obtained for each subproblem. Moreover, the user should be provided with the possibility of inspecting the set of solutions for any given subproblem.

Exercise 8.2.7 (Divide-and-conquer) *Using the cluster-popping algorithm as the base method M, apply the divide-and-conquer-approach to the following set of clusters:*

$$\{\{a, b, c, d, e\}, \{b, c, d\}, \{b, c, d, e, f, g\}, \{c, d\}, \{c, d, e\}, \{f, g\}\}.$$

8.3 Galled trees

In Chapter 6, we saw that unfortunately, in general, rooted phylogenetic networks interpreted in the softwired sense are computationally hard to work with. Indeed, even just the Cluster Containment problem of determining whether a given rooted phylogenetic network N contains some given cluster C on $\mathcal{X}$ (in the softwired sense) is NP-complete. To avoid these computational problems, in this chapter we restrict our attention to three topologically constrained classes of networks. In this section we look at the computation of galled trees, as defined in Section 6.11.1 and then, in the next section, we turn to galled networks, as introduced in Section 6.11.3. Finally, we consider the computation of level-k networks, as defined in Section 6.11.2.

In the following we present an algorithm for computing a galled tree N for a given set of clusters C that is tangled (if it exists). If the input cluster set C is not tangled, then one can first use the divide-and-conquer approach described in the previous section to decompose the problem of computing a galled tree for C into a number of subproblems, each consisting of a set of tangled clusters C'. Each subproblem is then solved by the algorithm described below and the resulting networks are merged to obtain a galled tree N for the original input C (if it exists).

Assume that we are given a set of tangled clusters C on $\mathcal{X}$. Our goal is to compute an unrooted phylogenetic network N that represents C and is topologically a galled tree, if such a network exists. The following algorithm determines all possible galled trees on C:

Algorithm 8.3.1 (Galled trees for tangled clusters) *Let C be a set of tangled clusters on $\mathcal{X}$. For each taxon $r \in \mathcal{X}$ such that $C' = C|_{\mathcal{X}-\{r\}}$ is compatible, construct a galled tree N (that is not necessarily bicombining) in the following steps:*

  (i) *Determine the rooted phylogenetic tree T that represents $C' \cup (\mathcal{X} - \{r\})$ and let λ denote the mapping of taxa in $\mathcal{X} - \{r\}$ onto the leaves of T.*
 (ii) *Set $N = T$.*
(iii) *Create a new node u in N and set $\lambda(r) = u$.*
 (iv) *For each edge e in N: let $C' \in C'$ be the cluster represented by e. If $C' \cup \{r\} \in C$ then mark e.*
  (v) *For each lowest marked edge e: Insert a new node v into e, create an edge from v to $\lambda(r)$.*
 (vi) *Let h be the lowest stable ancestor of u and let P be the path of tree edges from the root ρ to h. If there exists an edge in P that represents a cluster that does not contain r, then let e be the first such edge in P. Insert a new node w into e and create a new edge from w to r.*

If no such taxon r exists, then C cannot be represented by a galled tree and the algorithm reports fail. *Otherwise, the auxiliary edge below the root (representing $\mathcal{X} - \{r\}$) is contracted and the constructed network N is a galled tree that represents C.*

The algorithm is straightforward to understand, except, perhaps, for step (vi). This step is required because, when representing a cluster by a tree edge e, for each reticulate node u one in-edge must be on. The construction in step (vi) ensures that this is always possible. The construction in step (vi), together with the fact that the tree constructed in step (i) contains an edge that represents the cluster $\mathcal{X} - \{r\}$,

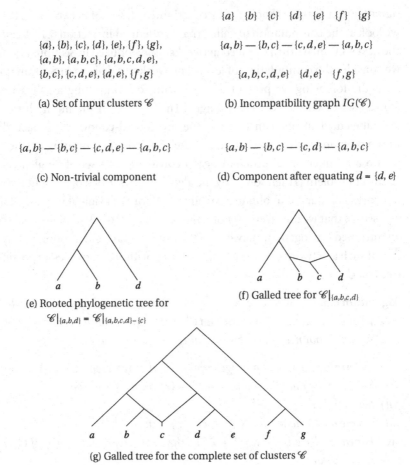

$\{a\}$ $\{b\}$ $\{c\}$ $\{d\}$ $\{e\}$ $\{f\}$ $\{g\}$

$\{a\}, \{b\}, \{c\}, \{d\}, \{e\}, \{f\}, \{g\},$
$\{a,b\}, \{a,b,c\}, \{a,b,c,d,e\},$
$\{b,c\}, \{c,d,e\}, \{d,e\}, \{f,g\}$

$\{a,b\} — \{b,c\} — \{c,d,e\} — \{a,b,c\}$

$\{a,b,c,d,e\}$ $\{d,e\}$ $\{f,g\}$

(a) Set of input clusters $\mathscr{C}$

(b) Incompatibility graph $IG(\mathscr{C})$

$\{a,b\} — \{b,c\} — \{c,d,e\} — \{a,b,c\}$

$\{a,b\} — \{b,c\} — \{c,d\} — \{a,b,c\}$

(c) Non-trivial component

(d) Component after equating $d = \{d, e\}$

(e) Rooted phylogenetic tree for
$\mathscr{C}|_{\{a,b,d\}} = \mathscr{C}|_{\{a,b,c,d\}-\{c\}}$

(f) Galled tree for $\mathscr{C}|_{\{a,b,c,d\}}$

(g) Galled tree for the complete set of clusters $\mathscr{C}$

Figure 8.3   (a) A set of 14 input clusters C. (b) The corresponding incompatibility graph $IG(C)$. (c) The single non-trivial connected component of $IG(C)$. (d) The incompatibility graph for $C|_{\{a,b,c,d\}}$, obtained after equating $d = \{d, e\}$, which is done because these two taxa are not separated by any cluster in the component. Note that removal of c results in a compatible set of clusters $C|_{\{a,b,d\}}$. (f) The rooted phylogenetic tree associated with $C|_{\{a,b,d\}}$. (g) The galled tree that represents $C|_{\{a,b,c,d\}}$, obtained by attaching c to the previous tree. (i) The galled tree that represents the complete set of clusters C.

also ensures that we can represent the cluster $\mathcal{X} - \{r\}$, if required. The algorithm is correct.

Lemma 8.3.2 (Tangled clusters and galled trees)   *A set of tangled clusters C on $\mathcal{X}$ can be represented by a galled tree N if and only if Algorithm 8.3.1 succeeds.*

Proof   Let C be a set of tangled clusters on $\mathcal{X}$ and assume that we have a galled tree N that represents C. The algorithm is based on the fact that removal of the one reticulate node u from N always results in a rooted phylogenetic tree T for $C' = C|_{\mathcal{X}-\{r\}}$, where u is the taxon that labels u. Because the algorithm tries all possible choices for r, this configuration will be considered and found. Note that step (vi) is required because we need to be able to turn all edges leading to the reticulate node u off when representing a cluster C that does not contain r by an edge e that lies above v. To ensure that this can be done, we create a new reticulate edge so as to move the lowest stable ancestor of u above e.   □

Any network N obtained from C by using the divide-and-conquer approach and Algorithm 8.3.1 as a base method is a galled tree. This is because all local networks are galled trees and merging the local networks to obtain N does not affect the tree cycles present in them, by Lemma 8.2.5. Since Algorithm 8.3.1 computes all possible galled trees on a tangled set of clusters, Lemma 6.11.3 ensures us that we can use the divide-and-conquer approach to obtain the minimum galled tree on a not necessarily tangled set of clusters.

Algorithm 8.3.1 is formulated for galled trees that are not necessarily bicombining, that is, may contain reticulate nodes of indegree ≥ 3. However, as the algorithm considers *all possible* multicombining rooted phylogenetic networks, in particular it also considers all that are bicombining, too. Hence, the algorithm can be easily modified so that it only outputs bicombining networks, or reports *fail*, if no such network exists. We illustrate the computation of a galled tree for a set of clusters in Figure 8.3.

8.3.1 Application

Just like a cluster network (see Section 8.1), a galled tree can be computed from the clusters of a set of rooted phylogenetic trees on $\mathcal{X}$ so as to show which parts of a phylogeny the trees agree upon and which parts are resolved in different ways.

For the three rooted phylogenetic trees reported in Figure 8.1, we show a galled tree that represents all clusters contained in the trees in Figure 8.4. Note that this network is simpler than the cluster network computed for the same data, which is shown in Figure 8.1d. This is because, as a softwired representation, a galled tree generally uses fewer reticulations that a cluster network to represent the same set of clusters.

8.4 Galled networks

The concept of a galled network is introduced in Section 6.11.3 as a generalization of galled trees. Recall that a rooted phylogenetic network N on $\mathcal{X}$ is said to be a *galled network*, if every reticulation in N has a tree cycle.

Let C be a set of clusters on $\mathcal{X}$. The goal of this section is to provide an algorithm for computing a galled network N that represents C using a minimum number of

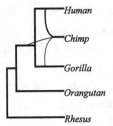

Figure 8.4   A (multicombining) galled tree that represents all clusters contained in the three rooted phylogenetic trees shown in Figure 8.1. The line width of each edge is proportional to the number of trees that contain the cluster represented by it.

(possibly multicombining) reticulations. To address this, we use the divide-and-conquer approach described in Section 8.2. In consequence, only such networks that have the decomposition property are considered.

In the following, we describe an algorithm that computes a galled network N that contains a minimum number of reticulate nodes for a given set of tangled clusters C on X [131]. Note that the problem of finding such a minimum galled network is NP-hard, and so it is not surprising that the algorithm that we present needs to solve two computationally hard problems. However, the practical performance of this algorithm, as implemented in the program Dendroscope, is acceptable in many cases.

The algorithm has two stages. First we must determine a minimum set of "reticulate taxa" R to remove from the taxon set X such that the set of induced clusters C_{X-R} is compatible and can be represented by a rooted phylogenetic tree T. This is the maximal compatible subset problem, which we discuss next in Section 8.4.1. Then we need to determine how to attach the associated set of reticulate nodes to the induced tree T in the best possible way. This is the minimum attachment problem, which we discuss later in Section 8.4.2.

8.4.1 Determining a minimum set of reticulations

The first step in the computation of a galled network is to determine a minimum set of taxa whose removal results in a set of compatible clusters.

Problem 8.4.1 (Maximal compatible subset) *Let C be a set of clusters on X. The maximal compatible subset (MCS) problem is to remove a minimum set of taxa R from X so that the set of clusters $C|_{X-R}$ induced by $X - R$ is compatible.*

For a general set of clusters, this is known to be NP-hard, as already mentioned in Section 6.2.1. However, in the context of the divide-and-conquer strategy, we are only interested in sets of clusters that are tangled. Is the MCS problem difficult for this restricted set of instances? The answer is yes [131].

Lemma 8.4.2 (MCS is hard for tangled clusters) *Let C be a set of clusters on X with properties (C1) and (C2), see Definition (8.3.2). Solving the MCS problem is NP-hard for such input.*

Proof We consider the decision problem of whether there exists a set of k taxa that "removes" the incompatibilities in a set of clusters, that is, whose removal results in a set of induced clusters that are compatible. We then show by reduction of this decision problem for general cluster sets that the decision problem for a set of clusters with properties (C1) and (C2) is NP-complete, which implies the NP-hardness of the MCS problem for such input. Let C be a set of clusters on X, not necessarily fulfilling properties (C1) and (C2). Set $X' = X \cup \{o\}$, where o is some new taxon not contained in X, and let C_1 denote the set of all trivial clusters, that is, clusters that each contains only one taxon.

Define $C' = (C - C_1) \cup \{X\} \cup \{\{o, x\} \mid x \in X\}$. Note that by construction, the set of clusters C' on X' has properties (C1) and (C2). We prove that there is a solution R' of size $k + 1$ of the restricted problem for C' if and only if there is a solution R of size k of the general problem for C.

$\Rightarrow$: Let R' be a solution of the restricted problem for C' of size $k + 1$, there are two cases to consider.

Case (1): $o \in R'$. Then $R = R' - \{o\}$ is of size k and removes all the incompatibilities from C.

Case (2): $o \notin R'$. In this case, the set R' must contain all but one element of X, that is, $|R'| = k + 1 = |X| - 1$, because otherwise $C'|_{X'-R'}$ would contain two clusters of the form $\{o, x\}$ and $\{o, y\}$, which are incompatible. Then any subset $R \subset X$ of size $|X| - 2 = k$ is a valid solution for C, because $|C| - |R| = 2$ and a collection of clusters on only two taxa cannot be incompatible.

$\Leftarrow$: Let R be a solution of the general problem for C of size k. We consider $R' = R \cup \{o\}$. This set has size $k + 1$, removes all incompatibilities between clusters $\{o, x\}$ as it removes taxon o. All other incompatibilities are removed because $R \subset R'$. $\square$

In the following we describe the *seed-growing algorithm* that addresses this problem and often runs fast in practice [131]. For each pair of incompatible clusters $A, B \in C$ we define an *incompatibility statement* consisting of the three terms $(A - B, A \cap B, B - A)$, and we use

$$H = \{(h_{11}, h_{12}, h_{13}), \ldots, (h_{m1}, h_{m2}, h_{m3})\} \tag{8.2}$$

to denote the list of all such statements for C (see Figure 8.5a–b).

A set of taxa $R \subset X$ is said to *resolve* an incompatibility statement (h_{i1}, h_{i2}, h_{i3}) if we have $h_{i1} \subseteq R$, $h_{i2} \subseteq R$ or $h_{i3} \subseteq R$. To solve the MCS problem, we must find a minimum set of taxa $R \subset X$ that resolves all incompatibility statements in H.

{d,e}

|

{e,f}

|

{f,g,h}

|

{d,e,f}

(a) Incompatibility
graph

({d},{e},{f})
({e},{f},{g,h})
({d,e},{f},{g,h})

(b) Incompatibility
statements

seed	size	rank	order
{d}	1	1	1
{e}	1	1	2
{f}	1	1	4
{d,e}	2	2	
{d,f}	2	2	
{d,g,h}	3	2	
{e}	1	2	3
{d,e}	2	3	
{e,f}	2	3	
{e,g,h}	3	3	
{f}	1	2	5
{f}	1	3	6

(c) Seeds

Figure 8.5   The four clusters occurring in the incompatibility graph indicated in (a) give rise to the three incompatibility statements listed in (b). In (c) we list the seeds considered by the seed-growing algorithm. The column entitled "order" shows the order in which certain seeds are chosen as R in the seed-growing algorithm. The algorithm halts after considering the sixth seed and returns $\{f\}$ as an optimal solution.

The algorithm maintains a list S of candidate solutions called *seeds*. Each seed $R \in S$ is assigned a *rank*, denoted by rank(R), which is defined as the number of incompatibility statements that it has been shown to resolve in succession, starting from the beginning of the list H.

Initially, the three parts of the first incompatibility statement (h_{11}, h_{12}, h_{13}) are chosen as seeds and we set rank(h_{11}) = rank(h_{12}) = rank(h_{13}) = 1.

The algorithm then proceeds by repeatedly choosing among all seeds of minimum size a seed R of maximal rank. If rank(R) = $|H|$, then R is an optimal solution of the MCS problem and the algorithm halts. Otherwise, if R resolves the next incompatibility statement (h_{j1}, h_{j2}, h_{j3}), with $j =$ rank(R) + 1, then we set rank(R) = j. Otherwise, we replace R by three new seeds $R_1 = R \cup h_{j1}$, $R_2 = R \cup h_{j2}$ and $R_3 = R \cup h_{j3}$, with rank(R_1) = rank(R_2) = rank(R_3) = rank(R) + 1. (See Figure 8.5c.)

Here is a more formal description of the algorithm:

Algorithm 8.4.3 (Seed-growing algorithm) *Assume that we are given a set of incompatibility statements $H = \{(h_{11}, h_{12}, h_{13}), \ldots, (h_{m1}, h_{m2}, h_{m3})\}$ on $\mathcal{X}$. To compute a minimum set of taxa $R \subset \mathcal{X}$ that resolves all statements in H, do the following:*

Initialization:
   Set $S = \emptyset$
   Set $\text{rank}(h_{11}) = \text{rank}(h_{12}) = \text{rank}(h_{13}) = 1$
   Append h_{11}, h_{12} and h_{13} to S
Main loop:
repeat
   Select $R \in S$ of minimum cardinality and then of maximum rank
   if $\text{rank}(R) = k$ **then return** *R*
   Set $i = \text{rank}(R) + 1$
   if *R resolves (h_{i1}, h_{i2}, h_{i3})* **then** *Set $\text{rank}(R) = i$*
   else
     Remove R from S
     for *j = 1 to 3* **do**
       Set $R' = h_{ij} \cup R$, set $\text{rank}(R') = i$ and append R' to S
end

Both the correctness and the performance of the seed-growing algorithm can be analyzed as follows [131]:

Lemma 8.4.4 (Correctness of the seed-growing algorithm) *The seed-growing algorithm solves the MCS problem.*

Proof Let R^* be a minimum solution of the MCS problem. We will show by induction that at the beginning of each iteration of the main loop the list S contains at least one set R' that is a subset of R^*. This implies that the final result of the algorithm has the same cardinality as R^*.

To start the induction, note that S initially contains all three parts of the first incompatibility statement, and one of these must be contained in R^*.

Now, assume that S contains a subset R' of R^* at the beginning of a repeat of the main loop. If R' is not selected as R in the main loop, then R' will not be changed during execution of the main loop and S will still contain R' at the beginning of the next iteration. The same is true if R' is chosen and resolves the next incompatibility statement. Finally, if R' is chosen and does not resolve the next incompatibility statement (h_{i1}, h_{i2}, h_{i3}), then the algorithm adds $R' \cup h_{i1}$, $R' \cup h_{i2}$ and $R' \cup h_{i3}$ to the set of all seeds. Because the solution R^* must resolve this incompatibility statement, too, one of these three unions will be contained in R^*. □

Lemma 8.4.5 (Performance of the seed-growing algorithm) *If a solution to the maximal compatible subset problem has size k for a given set of incompatibility statements H, then the seed-growing algorithm will find it by considering at most 3^{k+1} seeds, and the algorithm has a worst-case time complexity of $O(k|H|3^k)$.*

Proof Consider the enumeration tree of seeds generated by the algorithm. For the purpose of this proof, let the *level* of a node v be the number of edges in the path from the root of the enumeration tree to v. At the beginning of an iteration, the algorithm chooses a seed R of minimum cardinality d. By construction, the level of the corresponding node will be at most d. If R solves the next incompatibility, then no new nodes are added to the enumeration tree. Otherwise, three new nodes are added, each representing some seed whose cardinality is strictly larger than that of R. Thus, the enumeration tree will have at most $k + 1$ different levels, and each level will have at most three times as many nodes as the previous level. This implies that the number of seeds considered in S is at most 3^{k+1}. Moreover, each seed is considered at most $|H|$ times. Finally, each seed has size at most k and so processing of each seed can be done in $O(k)$ steps. □

8.4.2 Solving the minimum attachment problem

Let $\mathcal{C}$ be a tangled set of clusters on $\mathcal{X}$ and $R \subsetneq \mathcal{X}$ a minimum subset of taxa such that the restriction of $\mathcal{C}$ to $\mathcal{X} - R$, denoted by $\mathcal{C}|_{\mathcal{X}-R}$, is compatible. Let T be the rooted phylogenetic tree on $\mathcal{X} - R$ that represents $\mathcal{C}|_{\mathcal{X}-R} \cup (\mathcal{X} - R)$, and let $L(e)$ denote the cluster in $\mathcal{C}|_{\mathcal{X}-R}$ represented by an edge e of T. For each edge e of T, let $C(e) = \{C \in \mathcal{C} \mid C - R = L(e)\}$ denote the set of all clusters in $\mathcal{C}$ that are mapped to $L(e)$ under the restriction of $\mathcal{C}$ to $\mathcal{X} - R$, and let $R(e) = \{r \in R \cap C \mid C \in C(e)\}$ denote the set of all reticulate taxa that map to the cluster $L(e)$. We refer to T as the *top* part (see Figure 8.6d).

Let $\mathcal{C}|_R$ denote the restriction of $\mathcal{C}$ to the taxon set R and let $\hat{\mathcal{C}}|_R$ denote the set of all maximal clusters in $\mathcal{C}|_R$. We now define a graph B associated with $\hat{\mathcal{C}}|_R$, as follows: each cluster $C \in \hat{\mathcal{C}}|_R$ is represented by a node $v(C)$ and each taxon $r \in R$ is represented by a node $v(r)$ and we place an edge from $v(C)$ to $v(r)$ for all taxa r contained in the cluster C. We refer to B as the *bottom* part (see Figure 8.6d).

Problem 8.4.6 (Attachment problem) *The* Attachment *problem is to define a set of link edges from nodes in the top part T to nodes in the bottom part B such that the resulting graph is a galled network that represents the input set of clusters $\mathcal{C}$.*

More precisely, we aim at representing all clusters in $C(e)$ by the edge e in T and all clusters in $\mathcal{C}|_R$ by the in-edges of the nodes of the form $v(C)$, with $C \in \hat{\mathcal{C}}|_R$. To ensure this, the *link set* must fulfill the following properties:

(A1) For every edge e of T and every taxon $r \in R(e)$ there exists a link from some descendent node of e in T to either $v(r)$ or to some node of the form $v(C)$ in B, where $C \in \hat{\mathcal{C}}|_R$ contains r.

$\{a, b\}, \{a, b, x\},$
$\{a, x\}, \{b, x\},$
$\{b, y\}, \{c, d\},$
$\{c, d, x, y\}, \{c, x\},$
$\{d, y\}, \{x, y\}$

(a) Set of clusters $\mathscr{C}$

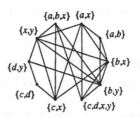

(b) Incompatibility graph $IG(\mathscr{C})$

$(\{b\}, \{a\}, \{x\})$ $(\{x\}, \{b\}, \{y\})$
$(\{a\}, \{b\}, \{x\})$ $(\{b\}, \{x\}, \{c, d, y\})$
$(\{a\}, \{b\}, \{y\})$ $(\{b\}, \{x\}, \{c\})$
$(\{a, x\}, \{b\}, \{y\})$ $(\{b\}, \{x\}, \{y\})$
$(\{a, b\}, \{x\}, \{c, d, y\})$ $(\{b\}, \{y\}, \{c, d, x\})$
$(\{a, b\}, \{x\}, \{c\})$ $(\{b\}, \{y\}, \{d\})$
$(\{a, b\}, \{x\}, \{y\})$ $(\{b\}, \{y\}, \{x\})$
$(\{a\}, \{x\}, \{b\})$ $(\{d\}, \{c\}, \{x\})$
$(\{a\}, \{x\}, \{c, d, y\})$ $(\{c\}, \{d\}, \{y\})$
$(\{a\}, \{x\}, \{c\})$ $(\{c\}, \{x\}, \{y\})$
$(\{a\}, \{x\}, \{y\})$ $(\{d\}, \{y\}, \{x\})$

(c) All incompatibility statements

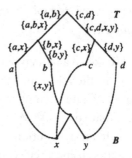

(d) Galled network for $\mathscr{C}$

Figure 8.6   (a) A set of clusters C on $X = \{a, b, c, x, y\}$, (b) the incompatibility graph $IG(C)$, (c) the corresponding list L of incompatibility statements solved by choosing $\{x, y\}$ as the solution of the MCS problem, and (d) a galled network N obtained by solving the minimum attachment problem. Here, the top and bottom parts of the network are labeled T and B, respectively. The edges of the top part are labeled by the non-trivial clusters that they represent and the leaves are labeled by their taxa. Link edges are shown as dashed lines.

(A2) For every node of the form $v(C)$ in B, with $C \in \hat{C}|_R$, there exists exactly one link from some node in T to $v(C)$.

(A3) For every edge e in T and $r \in R$ such that $C(e)$ contains some $C \in C$ that does not contain r, there exists a path from some node in T, that is not a descendant of e, to $v(r)$.

Property (A1) ensures that we can reach $v(r)$ from any edge e in T that has a cluster $C \in C(e)$ that contains r. Property (A2) ensures that all nodes of the form $v(C)$ in B have indegree 1. We do not allow an indegree larger than 1 to ensure that only nodes of the form $v(r)$ for $r \in R$ will be reticulate nodes. For example, in Figure 8.6d, the highest node in B is only attached to the node labeled b, and not to the node labeled c or to the parent of the two nodes labeled c and d, and so forth. Finally, property (A3) ensures that the node $v(r)$ can be avoided from any edge e in T that must represent a cluster that does not contain r.

Our goal is to minimize the number of edges used to solve the Attachment problem:

Problem 8.4.7 (Minimum attachment problem) *Find a collection of links that has all properties (A1)–(A3) and is of minimum size.*

In general, this is hard to solve [131]:

Lemma 8.4.8 (Minimum attachment problem is hard) *The decision problem whether there exists a solution to the Attachment problem that uses at most k edges is NP-complete.*

Proof We sketch a reduction of the *set cover* problem to the minimum attachment problem. In the NP-complete set cover problem we are given a collection of sets C on a set X and a number k. The question is: does there exist a subset $S \subseteq C$ of k sets that covers X? We shall construct an instance of minimum attachment problem that has a solution using $m + k$ edges if and only if the set cover problem has a solution using k sets, with $m = |C|$. Let d be an auxiliary taxon not contained in X and define $X' = X \cup \{d\}$. We construct a phylogenetic tree T with root ρ and $m + 1$ leaves. Set $R(e_0) = X'$ for the first leaf edge $e_0 = (\rho, v_0)$ of T and set $R(e_i) = \{d\}$ for all other leaf edges $e_1 = (\rho, v_1), \ldots, e_m = (\rho, v_m)$ of T. This defines the top part of the graph used in the attachment problem. For the bottom part, B, define one *cluster node* $v(C \cup \{d\})$ for each set $C \in C$ and one reticulate node $v(r)$ for each $r \in X'$. Place an edge from $v(C \cup \{d\})$ to $v(r)$ for each taxon r contained in $C \cup \{d\}$. Note that we can assume that each $r \in X$ is contained in at least two different sets in C, otherwise we can reduce the instance of the set cover problem to a smaller one. This observation ensures that any solution to the minimum attachment problem fulfills property (A3). We now prove the following claim: if there exists a solution of this instance of the minimum attachment problem using $m + k$ edges, then there is one in which the node v_0 is only connected to nodes of the form $v(C \cup \{d\})$, with $C \in C$. It follows from properties (A1) and (A2) that there are k edges that connect v_0 to B and m edges that connect nodes $v_1, \ldots, v_m$ to B. Since all edges departing from v_0 lead to cluster nodes in B, then this set of cluster nodes defines a subset of clusters S that covers X', and thus, also X, providing a solution to the set cover problem of size k. Now, to prove the claim, assume one of the edges departing from v_0 leads directly to a node of the form $v(r)$ in B, for some taxon $r \in X'$. If there is a cluster node $v(C \cup \{d\})$ connected to v_0 with $x \in C \cup \{d\}$, then we can remove the edge from v_0 to $v(r)$, as it is superfluous. Otherwise, there exists such a cluster node $v(C \cup \{d\})$ that is attached below some node v_i in T, with $i > 0$. In this case, we modify the solution as follows: redirect the edge from v_0 to $v(r)$ so that it leads to $v(C \cup \{d\})$ and redirect the edge from v_i

to $v(C \cup \{d\})$ so that it leads to $v(r)$. This is repeated until all children of v_0 are cluster nodes.

Vice versa, it also not difficult to see that any solution with k sets of an instance of the set cover problem leads to a solution with $m + k$ edges of this simplified case of the minimum attachment problem.   □

The instances of this problem that are of interest in practice are often quite small, in which case a branch-and-bound approach can be adequate. The above reduction also proves that the minimum attachment problem is W[2]-hard, remaining intractable even when parameterized by k, and so it is unlikely that an $O(f(k) \times poly(n))$ algorithm can be obtained. Alternatively, the problem can be posed as an ILP (Integer Linear Program) with binary variables that determine whether a possible link edge is used or not and inequalities that ensure that the properties (A1)–(A3) hold. The optimization goal is then to minimize the sum of the binary variables.

As in the computation of galled trees, the following complication must also be taken into account: After solving the minimum attachment problem for a set of clusters C on $\mathcal{X}$ and a given reticulation set R, there may exist a tree edge $e = (v_e, w_e)$ along the path from the root of the top part T to a node $v(r)$, with $r \in R$, that is supposed to represent a cluster C that does not contain r. Assume that e is the highest such edge. If e lies above the lowest single ancestor of $v(r)$, then an additional reticulate edge into $v(r)$ is required that joins some node above e to $v(r)$, so that C can be represented by e. To this end, we create a new edge from v_e to $v(r)$. (This modification can also be used to represent the cluster $\mathcal{X} - R$, because the tree T is defined so as to contain an edge that represents $\mathcal{X} - R$.)

Any network N obtained from C by using the divide-and-conquer approach and the algorithm presented in this section to solve the subproblems is a galled network. This is because all local networks are galled networks and merging the local networks to obtain N does not affect the tree cycles present in them, by Lemma 8.2.5.

8.4.3 Application

As in the case of cluster networks and galled trees, one possible use of a galled network is to summarize a set of different rooted phylogenetic trees. In Figure 8.7, we show the use of a galled network in the context of resolving the phylogeny of *Triticeae*, a tribe of grasses for which single-copy gene studies have failed to reconstruct a reliable phylogeny. In [69], the analysis of phylogenetic trees for 27 different genes identifies several well-supported clades.

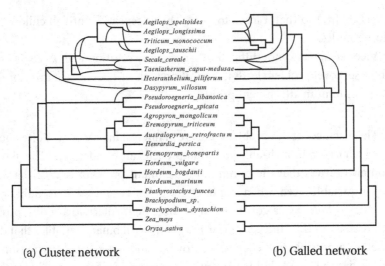

(a) Cluster network   (b) Galled network

Figure 8.7   A cluster network (a) and galled network (b), computed for a set of clusters obtained from 27 gene trees on *Triticeae* [69].

In Figure 8.7 we show both a cluster network and a minimum galled network that represents all clusters contained in the 27 gene trees, extracted after preprocessing the trees so as to reduce incompatibilities among them, as described in [212]. Both networks show that the modified trees are are still in conflict, but also that they agree on several clades. The galled network provides a slightly clearer picture of the conflicts than the cluster network.

8.5 Level-*k* networks

Let C be a set of tangled clusters on $\mathcal{X}$. In this section, we describe an algorithm for computing a level-k network (as defined in Section 6.11.2) that represents C for a given value of k, if such a network exists. Used as the base method for the divide-and-conquer approach introduced in Section 8.2, this allows us to compute a decomposable level-k network for an unrestricted set of clusters C, as described below.

Informally, the algorithm operates as follows. If $k = 0$, then it simply returns the tree that represents C, if such a tree exists. If $k > 0$, then for each taxon x in $\mathcal{X}$, the algorithm first removes x from C. The set of clusters $C' = C_{\mathcal{X}-\{x\}}$ is then collapsed (as described on page 195) so as to obtain a set of collapsed clusters C'' on a set of composite taxa $\mathcal{X}''$. Next, the set of level-$(k-1)$ networks that represent C'' is recursively computed. For each network N'' found, a level-$(k-1)$ network N' that represents C' is obtained by decollapsing all composite taxa in N''. Finally, for each pair of edges in N', the algorithm determines whether placing a reticulation below

$\{a,g\}, \{b,c\}, \{d,e\}, \{a,b,f\}, \{b,c,f\},$
$\{c,d,e\}, \{b,c,d,f\}, \{a,b,f,g\}$

(a) Set of tangled input clusters $\mathscr{C}$

$\{a,g\}, \{d,e\}, \{a,b,f\}, \{b,f\},$
$\{b,d,f\}, \{a,b,f,g\}$

(b) Set $\mathscr{C}' = \mathscr{C}|_{\mathcal{X}-\{c\}}$

$\{a,g\}, \{d,e\}, \{a,\{b,f\}\},$
$\{\{b,f\},d\}, \{a,\{b,f\},g\}$

(c) Set $\mathscr{C}''$ obtained by
collapsing $\mathscr{C}'$

$\{a,g\}, \{d,e\}$

(d) Result of removing
$\{b,f\}$ from $\mathscr{C}''$

(e) Rooted tree T for
set of clusters in (d)

(f) Level-1 network for
set of clusters $\mathscr{C}''$ in (c)

(g) Level-1 network for
set of clusters $\mathscr{C}'$ in (b)

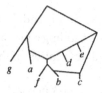

(h) Final level-2 network
for set $\mathscr{C}$ in (a)

Figure 8.8   (a) A tangled set of eight clusters $\mathcal{C}$. (b) The set $\mathcal{C}'$ obtained from $\mathcal{C}$ by removing the taxon c and then all trivial clusters. (c) The set of clusters $\mathcal{C}''$ obtained by collapsing $\mathcal{C}'$. The only non-trivial composite taxon is $\{b, f\}$. (d) The set $\mathcal{C}''$ after the removal of $\{b, f\}$. (e) The rooted phylogenetic tree for the clusters in (d) with an auxiliary edge inserted above the root. (f) A level-1 network for the set of clusters in $\mathcal{C}''$ (together with all trivial clusters). This network is obtained from (e) by adding a reticulate node connecting the two bold edges in (e), with a pendant leaf labeled by the composite taxon $\{b, f\}$. (g) A level-1 network for the set of clusters in $\mathcal{C}'$. This network is obtained from (f) by replacing the composite leaf $\{b, f\}$ by the rooted phylogenetic tree that represents $\{b, f\}$. (h) A level-2 network for the set $\mathcal{C}$. This network is obtained from (g) by adding a reticulate node below the two bold edges in (g), with a pendant leaf labeled by taxon c. The auxiliary edge below the root has been removed.

the two edges, together with a pendent leaf labeled by x, results in a level-k network N that represents $\mathcal{C}$. We illustrate the computation of a level-2 network for a set of tangled clusters in Figure 8.8.

During this process, we also need to be able to add a reticulation to the network that does not have a taxon directly beneath it, as discussed further below with the

$\{a,b\},\{a,e\},\{b,e\},$
$\{c,d\},\{c,e\},\{d,e\},$
$\{a,b,e\},\{c,d,e\}$

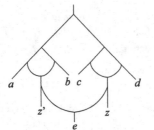

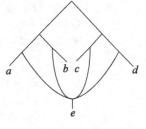

(a) Tangled clusters   (b) Level-3 network with   (c) Final level-3 network
                                  two auxiliary taxa

Figure 8.9   (a) A tangled set of clusters C. (b) A corresponding level-3 network recursively constructed by Algorithm 8.5.1, using $x = e$ (for $k = 3$), $x = z$ (for $k = 2$) and $x = z'$ (for $k = 1$) in line 9 of the algorithm. (c) The final level-3 network N for C, obtained from the previous one by removal of the two nodes labeled by auxiliary taxa, the auxiliary edge below the root and contracting all edges connecting two reticulate nodes.

help of Figure 8.9. To this end, in the main loop of the algorithm (line 9) we also consider the case of adding a new auxiliary taxon z that does not belong to $\mathcal{X}$. Here is a more detailed description of the so-called *Cass algorithm* [136]:

Algorithm 8.5.1 (Level-k networks for tangled clusters)   *Let $k \geq 0$ be a fixed number. Input is a set of clusters C on $\mathcal{X}$, which is assumed to be tangled in the initial call to the algorithm. Recursively construct a set of level-k networks N for C in the following steps (if any such network exists):*

Set $\mathcal{N} = \emptyset$
if $k = 0$ **then**
   if C *is compatible* **then**
      Compute the rooted phylogenetic tree T for C (and all trivial clusters on $\mathcal{X}$)
5   *Insert an auxiliary edge that separates the root from the rest of the tree T*
      Set $\mathcal{N} = \{T\}$
else *(comment: $k > 0$)*
   Create a new auxiliary taxon z
9 **for each** *taxon $x \in \mathcal{X} \cup \{z\}$* **do**
      Set $C' = C|_{\mathcal{X}-\{x\}}$, remove all trivial clusters and then collapse to obtain C''
      on $\mathcal{X}''$
      Recursively compute the set $\mathcal{N}''$ of level-$(k-1)$ networks for C'' on $\mathcal{X}''$
      for each *network N'' in $\mathcal{N}''$* **do**
         Replace each leaf of N'', which is labeled by a composite taxon $\bar{A} \in \mathcal{X}''$,
         by the rooted phylogenetic tree $T(C'|_A)$ that represents $C'|_A$
         Let N' be the resulting network

> **for each** *pair of (not necessarily distinct) edges e_1 and e_2 in a new copy of N'* **do**
>> *Create two nodes r and u, add an edge from r to u and set $\lambda(x) = u$*
>> *Insert a new node v_1 into e_1 and connect v_1 to r*
>> *Insert a new node v_2 into e_2 and connect v_2 to r*
>> **if** *the resulting network N represents C (disregarding all auxiliary taxa)*
>>> **then** *add N to $\mathcal{N}$*
> **return** $\mathcal{N}$

If the returned set $\mathcal{N}$ is empty, then the algorithm reports fail.

If there exists a level-k network that represents C, then this algorithm is guaranteed to find such a network, if $k \leq 2$ [136]. For larger values of k, it is unknown whether failure to find a level-k network by the algorithm always implies that no such network exists. Hence, for $k > 2$, this algorithm is a heuristic for finding a level-k network.

If the initial call to the algorithm is successful and returns a non-empty set of rooted phylogenetic networks $\mathcal{N}$ then, in a post-processing step, in each level-k network $N \in \mathcal{N}$ returned by the algorithm, the auxiliary edge below the root is contracted. In addition, any edge that connects two reticulate nodes is contracted. Moreover, each leaf labeled by an auxiliary taxon is removed. After these modifications, every network N in $\mathcal{N}$ is a level-k network that represents C.

The reason why we contract every edge that connects two reticulate nodes above is that such an edge contributes toward a resolution of an otherwise multicombining node in a way that is not explicitly supported by the input data (see Figure 8.9).

The role of the auxiliary edge inserted below the root node of the tree T in line 5 of the algorithm is to allow the case that a particular taxon x is completely avoided in the network as required in the representation of certain clusters. (A similar construction is also used in the computation of a galled tree and of a galled network, as discussed above.)

The role of the auxiliary taxa created in line 8 and considered in line 9 of the algorithm is to facilitate the computation of level-k networks for C that have nested reticulations. To discuss this, consider the set of tangled clusters C shown in Figure 8.9a. A level-3 network that represents this set of clusters is shown in Figure 8.9c. To generate this network using Algorithm 8.5.1, in the initial call to the algorithm (with $k = 3$), in line 9 we choose $x = e$, then in the first recursive call (with $k = 2$), we choose $x = z$ and then, in the second recursive call (with $k = 1$), we choose $x = z'$. In the final recursive call (with $k = 0$), the tree T on the taxon set $\{a, b, c, d\}$ is built and returned. In Figure 8.9b we indicate the positions of the auxiliary taxa z and z' in the constructed network.

This example also shows that the algorithm really must recurse k times before checking whether C is compatible and then building the tree, because, in this example, we first remove e and from then on, all considered clusters are compatible.

The insertion of auxiliary taxa during the algorithm is actually slightly more complicated than described above. Assume that we are running the algorithm for some value of k. When making a recursive call to the algorithm for $k - 1$, there really are two different cases that must be distinguished, as follows. The first case is that in the recursive call, we intend to only consider values of x that are contained in the taxon set $\mathcal{X}$, in line 9 of the algorithm. In this case, the algorithm proceeds exactly as described above. The other case is that we intend to consider $x = z$, where z is a new auxiliary taxon, in the recursive call. In this case, the algorithm proceeds in a slightly modified manner, namely we refrain from collapsing the set C' in line 10 of the algorithm before making the recursive call. The reason for this change is that when adding an auxiliary taxon to the network we need to have access to the edges associated with the uncollapsed network, as otherwise certain networks will be missed.

Let C be a set of clusters on $\mathcal{X}$ and let $K > 0$ be a fixed number. In practice, the Cass algorithm is used together with the divide-and-conquer algorithm (Algorithm 8.2.4) as follows. For each of the subproblems C' considered in the divide-and-conquer algorithm, we first call the Cass algorithm with $k = 1$. If the algorithm reports fail, we then run it with $k = 2$, etc. We repeat this until either we have successfully computed a level-k network for some value of $k \leq K$, or we assume that no such network exists.

In this way, we seek to compute a rooted phylogenetic network N that minimizes the number of reticulations among all rooted phylogenetic networks that represent C and have the decomposition property. If the number of reticulations found for each subproblem is either 1 or 2, then by the result mentioned above, the network N is indeed guaranteed to be minimal in this sense. For higher values of k, then, this may not be the case.

Exercise 8.5.2 (Level-2 network)   *Using the Cass algorithm with $k = 2$ as the base method M, apply the divide-and-conquer-approach to the following set of clusters (together with all trivial clusters):*

$$\{\{a, b\}, \{b, c\}, \{d, e\}, \{f, g\}, \{d, f\}, \{e, g\}, \{h, i\}, \{a, b, c\}, \{d, e, f\},$$
$$\{e, f, g\}, \{d, e, f, g\}, \{a, b, c, d, e, f, g\}, \{d, e, f, g, h, i\}\}.$$

Finally, as an example of a level-k network, consider the galled network depicted in Figure 8.7b. This is a level-2 network. Note that it contains a multicombining node of indegree 3, which arises from two bicombining reticulations by contracting the edge between them, and thus may be counted as two reticulations, as discussed in Section 6.11.2.

Chapter notes

The algorithms presented in this chapter are all very new. However, easy-to-use implementations of them are available in the Dendroscope program (see Section 14.4) and it remains to be seen whether they will be widely used by biologists in practical studies.

A comparison of the galled network method and the Cass algorithm for level-*k* networks shows that the former usually runs very much faster that the latter, while the latter often obtains a network with fewer reticulations [136].

9

Phylogenetic networks from sequences

In this chapter we describe a number of methods that require sequences, or more precisely, a multiple sequence alignment, as input. To this end, we first introduce the useful concept of a condensed alignment and briefly discuss the relationship between binary sequences and splits. We then discuss a number of methods that can be used to compute unrooted phylogenetic networks from sequences, including the median network, quasi-median network and median-joining methods. Finally, we look at two different ways of computing rooted networks from sequences, both aimed at explaining differences in terms of mutations and recombinations.

9.1 Condensed alignments

For computational reasons, it is sometimes useful to assume that a given multiple sequence alignment M on $\mathcal{X}$ is *condensed* in the sense that no two sequences are identical and no two columns display the same *pattern* of states, that is, induce the same partitioning of the taxa. Additionally, we assume that a condensed alignment does not contain any constant columns (in which all sequences have the same state). We shall sometimes refer to the sequences of a condensed alignment as *haplotypes*.

If the original input multiple sequence alignment M_0 is not condensed, then we will usually *condense* it in a preprocessing step as follows: first, each set of identical sequences is pooled into a single haplotype, then all constant columns are removed and, finally, every set of columns that have the same pattern is replaced by a single column i that is assigned a weight $\omega(i)$ that equals the number of columns in the represented set. Also, in some applications, one may want to consider the haplotypes as weighted by the number of original sequences that they represent. An example is shown in Figure 9.1.

9.2 Binary sequences and splits

Some of the algorithms discussed in this chapter assume that the input consists of a multiple alignment M of *binary* sequences on $\mathcal{X}$. However, in practice, the original

	1	2	3	4	5
a	A	A	G	T	T
b	C	A	C	A	T
c	C	A	C	A	A
d	T	A	T	C	A
e	A	A	G	T	T

	1	2
a	A	A
b	B	A
c	B	B
d	C	B

(a) Alignment M_0 (b) Condensed M

Figure 9.1   In the multiple alignment of DNA sequences M_0 shown in (a), the columns 1, 3 and 4 have the same pattern and column 2 is constant. The sequences for a and e are identical. (b) The resulting condensed alignment M contains four sequences of length two. The sequence for a represents the sequences for both a and e in the original alignment and has weight 2. The first column represents the three original columns 1, 3 and 4 and has weight $\omega(1) = 3$. The second column corresponds to the fifth original column and has weight $\omega(2) = 1$.

input data will usually be a multiple alignment of DNA sequences M_0 on $\mathcal{X}$. In this section, we assume that the given multiple sequence alignment does not contain any gaps. This assumption can be met by ignoring all columns that contain a gap.

There are a number of different ways that such data can be converted into binary sequences. One possibility is to extract all those columns in M_0 that contain precisely two different character states and then to replace the two character states by 0 and 1, perhaps in such a way that the first sequence is always given the state 0, say. A justification for discarding all columns that contain more than two characters states could be that such sites are "hypervariable" and so may fail to be homologous or may have been saturated by multiple substitutions [46].

Another possibility is to use all columns of M_0 and to replace any occurrence of either of the two nucleotides A and G (the two purines) by 0, and to replace any occurrence of the other two nucleotides C and T (the two pyrimidines) by 1. This approach is based on the observation that transitions (changes between A and G or between C and T) are more frequent than transversions (all other possible changes) and the latter are more informative.

If M is a multiple alignment of binary sequences on $\mathcal{X}$, then each non-constant column i of M corresponds to a split $S = \frac{A}{B}$ of $\mathcal{X}$. It is defined by setting A equal to the set of all taxa whose state in column i agrees with that of the first sequence and setting $B = \mathcal{X} - A$. If M is a condensed alignment, then the split S is unique and it is assigned a weight $\omega(S) = \omega(i)$ that equals the number of original columns that were condensed into the column i. We use $\mathcal{S}(M)$ to denote the set of all splits obtainable in this way.

The columns of a multiple sequence alignment M on $\mathcal{X}$ are also called *characters* or *positions*. We say that two characters or positions of a multiple alignment of binary sequences are *compatible* if and only if the corresponding splits are, and *incompatible* otherwise.

Exercise 9.2.1 (Incompatibility of binary characters) *Show that two characters i and j in a multiple alignment M of binary sequences are incompatible if and only if there exist four different sequences in the alignment that equal* 00, 01, 10 *and* 11, *when restricted to the characters i and j.*

9.3 Parsimony splits

The split decomposition algorithm discussed in Section 5.9.4 takes as input a distance matrix D on $\mathcal{X}$ and produces as output a set of weakly compatible splits $\mathcal{S}$ on $\mathcal{X}$. A main feature of the algorithm is that it employs a distance-based "isolation index" that evaluates to a positive number for, at most, two of the three possible non-trivial splits on any quadruple Q of taxa in $\mathcal{X}$.

Let M be a multiple sequence alignment on $\mathcal{X}$. To obtain a similar algorithm that operates on a multiple sequence alignment rather than on a distance matrix, we must replace the computation of the isolation index (which requires a distance matrix) by an analogous calculation that is based directly on the alignment.

In the *parsimony splits* method [8], the basic idea is to use an index that selects, at most, two of the three most parsimonious bifurcating tree topologies for any given quadruple, as the name suggests. In more detail, for each quadruple of taxa $Q = \{w, x, y, z\}$ on $\mathcal{X}$ with corresponding aligned sequences $a = a_1 \ldots a_L$, $b = b_1 \ldots b_L$, $c = c_1 \ldots c_L$ and $d = d_1 \ldots d_L$ in M, we define

$$p_M\left(\frac{\{w, x\}}{\{y, z\}}\right) = |\{i : a_i = b_i \text{ and } c_i = d_i\}| \tag{9.1}$$

as the number of positions $i \in \{1, \ldots, L\}$ in the alignment M at which the character states are the same for the sequences a and b and also for c and d, but not necessarily for all four sequences.

To set things up so for any quadruple of taxa $Q = \{w, x, y, z\}$, we select the quartet topology $\frac{\{w, x\}}{\{y, z\}}$ if and only if

$$p_M\left(\frac{\{w, x\}}{\{y, z\}}\right) > \min\left(p_M\left(\frac{\{w, y\}}{\{x, z\}}\right), p_M\left(\frac{\{w, z\}}{\{x, y\}}\right)\right). \tag{9.2}$$

As this condition is a proper inequality, at most two quartet topologies are chosen for any given quadruple of taxa.

To obtain a set of splits that induce only the selected quartet topologies, we define the *parsimony index* of a split $S = \frac{A}{B}$ as

$$p_M(S) = \min_{\substack{w, x \in A \\ y, z \in B}} \left(p_M\left(\frac{\{w, x\}}{\{y, z\}}\right) - \min\left(p_M\left(\frac{\{w, x\}}{\{y, z\}}\right), \right. \right.$$

$$\left. \left. p_M\left(\frac{\{w, y\}}{\{x, z\}}\right), p_M\left(\frac{\{w, z\}}{\{x, y\}}\right)\right)\right) \geq 0. \tag{9.3}$$

A split S is called a *p-split*, if its parsimony index $p_M(S)$ is positive. The set of all *p*-splits associated with M is weakly compatible by Lemma 5.9.4. It can be computed using a simple modification of the split decomposition algorithm as follows:

Algorithm 9.3.1 (Parsimony splits) *Given a multiple sequence alignment M on $\mathcal{X} = \{x_1, \ldots, x_n\}$, compute the set of all p-splits on $\mathcal{X}$ as follows:*

Initially, set $\mathcal{X}_1 = \{x_1\}$ and $\mathcal{S}_1 = \emptyset$. Now, assume that we have computed the set of all p-splits $\mathcal{S}_i$ on the first i taxa $\mathcal{X}_i = \{x_1, \ldots, x_i\}$. To obtain $\mathcal{S}_{i+1}$ on $\mathcal{X}_{i+1} = \{x_1, \ldots, x_{i+1}\}$, for each split $A \mid B \in \mathcal{S}_i$ do:

(i) *If $p_M\left(\frac{A \cup \{x_{i+1}\}}{B}\right) > 0$, then add $\frac{A \cup \{x_{i+1}\}}{B}$ to $\mathcal{S}_{i+1}$.*

(ii) *If $p_M\left(\frac{A}{B \cup \{x_{i+1}\}}\right) > 0$, then add $\frac{A}{B \cup \{x_{i+1}\}}$ to $\mathcal{S}_{i+1}$.*

(iii) *If $p_M\left(\frac{x_i}{\{x_{i+1}\}}\right) > 0$, then add $\frac{x_i}{\{x_{i+1}\}}$ to $\mathcal{S}_{i+1}$.*

The result is given by $\mathcal{S}_n$.

The splits computed by the algorithm are then usually processed by a method such as the convex hull algorithm (see Section 7.1) to produce a split network.

9.3.1 Application

The *parsimony-splits* method has not been used much in the literature, probably because the resulting set of splits is usually very similar to the one obtained by the more widely known split-decomposition method. An example of the application of the method can be found in a study that compares the phylogeny of different Hox clusters in sharks and mammals [200]. The parsimony-splits algorithm is used together with a number of other phylogenetic methods to support the hypothesis that the shark *HoxN* cluster is orthologous to the mammalian *HoxD* cluster.

9.4 Median networks

Median networks were developed as a means of visualizing the variation in mitochondrial DNA (mtDNA) sequences in the study of human evolution [11]. They are usually constructed for closely related sequences that have evolved without recombinations. In a median network N, every sequence of a given multiple sequence alignment M on $\mathcal{X}$ is represented by a node and additional nodes are said to represent further *unobserved* sequences. Two nodes are connected by an edge if and only if they differ by exactly one mutation (under the assumption that the alignment is condensed).

In more detail, assume that we are given a multiple alignment of DNA sequences M_0 on $\mathcal{X}$. As we shall see, the median network algorithm works only for binary sequences. Thus, the first step is to convert the original input data into a condensed

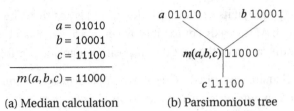

$$a = 01010$$
$$b = 10001$$
$$c = 11100$$
$$\overline{}$$
$$m(a,b,c) = 11000$$

(a) Median calculation   (b) Parsimonious tree

Figure 9.2   (a) The median $m(a, b, c)$ of three binary sequences a, b and c. The median $m(a, b, c)$
minimizes the parsimony score of the unrooted phylogenetic tree on $\{a, b, c\}$ shown in (b),
which in this case is five.

multiple alignment of binary sequences M on $\mathcal{X}$, as described in Section 9.2. Recall
that condensed means that no two columns of M exhibit the same pattern, there
are no constant columns and no two taxa have the same sequence.

Consider three binary sequences $a = a_1 a_2 \ldots a_L$, $b = b_1 b_2 \ldots b_L$ and $c =
c_1 c_2 \ldots c_L$. We define their *median* as the sequence $m(a, b, c)$ whose i-th char-
acter state is the majority state of a_i, b_i and c_i, which is 1, if $a_i + b_i + c_i \geq 2$ and
0, else. An example is shown in Figure 9.2a.

The median sequence $\mu = m(a, b, c)$ has the property that it minimizes the sum
of pairwise Hamming distances (see Section 3.12) $H(a, \mu) + H(b, \mu) + H(c, \mu)$
and provides a most parsimonious sequence for the internal node of the unrooted
phylogenetic tree on $\{a, b, c\}$, as illustrated in Figure 9.2b.

Let M be a multiple alignment of binary sequences on $\mathcal{X}$. The *median closure* of
M is defined as the set $\bar{M}$ of all binary sequences that can be obtained by repeatedly
taking the median of any three sequences in the set and adding it to the set, until
no new sequences can be produced. The median network is defined as follows:

Definition 9.4.1 (Median network)   *Let M be a condensed multiple alignment of
binary sequences on $\mathcal{X}$. The* median network *associated with M is a phylogenetic
network $N = (V, E, \sigma, \lambda)$ whose node set is given by the median closure $V = \bar{M}$ and
in which any two nodes a and b are connected by an edge e of color $\sigma(e) = i$ in E,
if any only if they differ exactly in their i-th position (as haplotypes). An associated
taxon labeling $\lambda : \mathcal{X} \to V$ maps each taxon x onto the node $\lambda(x)$ that represents the
corresponding sequence.*

An example of a median network is shown in Figure 9.3.

Let M be a condensed multiple alignment of binary sequences on $\mathcal{X}$. Definition 9.4.1 suggests the following *naive* algorithm for computing the median network
N for M: First compute the median closure $\bar{M}$. Then construct the graph N that
has node set $\bar{M}$ and in which any two nodes are connected by an edge e of color
$\sigma(e) = i$ if any only if they differ exactly in their i-th position (as haplotypes).
Alternatively, one can compute the canonical split network $N(\mathcal{S}(M))$ for the set

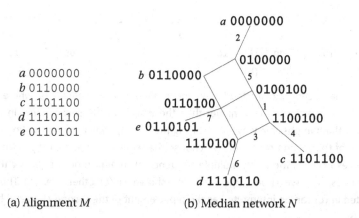

(a) Alignment M (b) Median network N

Figure 9.3   (a) A condensed multiple alignment M of binary sequences on $\mathcal{X} = \{a, \dots, e\}$. (b) The median network N that represents M. The nodes are labeled by haplotypes and the edges are labeled by the corresponding columns in M.

of splits $\mathcal{S}(M)$ associated with M using the convex hull algorithm, based on the following result [6]:

Theorem 9.4.2 (Median network is split network) *Let M be a condensed multiple alignment of binary sequences on $\mathcal{X}$. The naive median network algorithm and the convex hull algorithm both produce the same network.*

An important consequence of this result is the following:

Corollary 9.4.3 (Median network is connected) *Let M be a condensed multiple alignment of binary sequences on $\mathcal{X}$. The median network N associated with M is connected.*

Let M be a condensed multiple alignment of binary sequences on $\mathcal{X}$. If all pairs of columns in M are compatible, it follows from Theorem 9.4.2 that the resulting median network will be a phylogenetic tree T. Trivially, this tree will be the most parsimonious tree for M (see Section 3.7). More generally we have the following result [11]:

Theorem 9.4.4 (All most parsimonious trees) *Let M be a condensed multiple alignment of binary sequences on $\mathcal{X}$. The median network N associated with M contains all phylogenetic trees T on $\mathcal{X}$ whose parsimony score $PS(T, M)$ is minimum.*

9.4.1 Reduced median networks

The main application of a median network is to provide a useful visualization of probable evolutionary pathways. Unfortunately, a median network can have a very

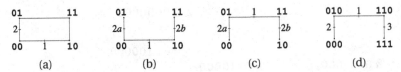

Figure 9.4   (a) A median network restricted to two incompatible binary characters, where the weight of the first character is much higher than the weight of the second one. In (b) and (c) we show the two possible ways of resolving the incompatibility by postulating two parallel changes of the second character. We choose the reduction shown in (b), if the two lower sequences occur in the given multiple sequence alignment more frequently that the two upper ones, and we choose the reduction shown in (c) otherwise. (d) The resolution indicated in (c) can be interpreted as the replacement of the second character by two new characters.

large number of nodes and edges, even for a small number of taxa and characters, if there are a lot of incompatibilities among the characters. In the absence of recombination, many of the incompatibilities will be due to parallel- or back-mutations. Hence, one idea to reduce the complexity of the network is to try and resolve incompatibilities by identifying likely parallel mutations and making them explicit. The resulting network is called a *reduced median network*.

In Figure 9.4a, we display a section of a median network restricted to four binary sequences 00, 01, 10 and 11. Let us assume that the first binary character has a substantially higher weight than the second one, for example $\omega(1) \geq 2\,\omega(2)$. In this case we will assume that the mutation represented by the second character occurred twice, in parallel, as indicated in parts (b) and (c) of the figure. To decide which of the two depicted scenarios is more likely, we count the frequencies of common occurrences of the patterns 00 or 10, and of the patterns 01 or 11, in the given multiple sequence alignment M. If the former two patterns occur with a higher frequency than the latter two, then we will resolve the parallel mutations as shown in (b), otherwise we will resolve them as shown in (c). One way to implement this step computationally is replace the character in which a parallel mutation is assumed to occur by two new characters that each mutate only once. For example, for the resolution indicated in (c), we replace the second character in the sequence by two new characters to obtain the four sequences 000, 010, 111 and 110 in which all characters are compatible with each other, as illustrated in (d).

When one character in the multiple sequence alignment M is incompatible with a whole set of other characters that are all compatible with each other, then again the median network can be simplified by postulating an appropriate parallel mutation, as indicated in Figure 9.5. We do not describe in detail how to systematically detect all opportunities for reduction in a median network.

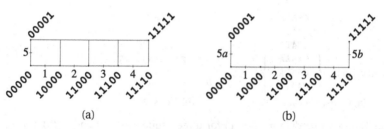

Figure 9.5   (a) This median network shows that the fifth position of the given multiple sequence alignment M is incompatible with the other four positions. (b) The reduced median network obtained after postulating a parallel mutation at position five of the alignment.

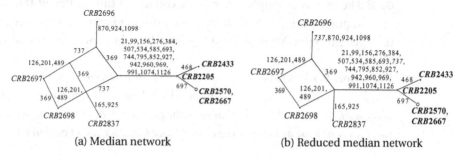

Figure 9.6   (a) The median network for eight specimens of *Callicebus lugens*, based on cytochrome b sequences. Specimens from the right bank of Rio Negro are shown in plain font and those from the left bank are shown in bold font. (b) The reduced median network obtained by postulating a parallel mutation at position 737.

9.4.2 Application

One application domain of median networks is phylogeographic studies. For example, in [44] the distribution of *Callicebus lugens* (Platyrrhini, Primates) at the Rio Negro, in Brazil, is reported. The study focuses on eight specimens, one group of four taken from the left bank of the river and another group of four taken from the right bank. It is based on a multiple alignment M of cytochrome b DNA sequences of length 1140. A median-network analysis shows a clear separation of the two groups, see Figure 9.6. Note that only 35 columns are retained in the condensed version of M (the ones labeling the edges).

9.5 Quasi-median networks

In Section 9.4, we show that median networks are designed for representing binary sequences. To apply them to DNA (or other multi-state sequences), one must first convert the sequences into binary sequences as described in Section 9.2. Because such a conversion usually discards phylogenetic information, the question arises:

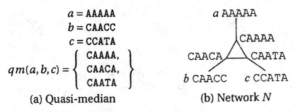

(a) Quasi-median   (b) Network N

Figure 9.7   (a) The quasi-median $qm(a, b, c)$ for three sequences a, b and c. (b) The corresponding quasi-median network N.

How to modify the median calculation so that it works directly with multi-state data? The answer is simple: whenever a column of three different states occurs in the computation of the median, produce three median sequences, one for each of the different states. More formally, we introduce the concept of a quasi-median:

Definition 9.5.1 (Quasi-median)   *Consider three DNA sequences $a = a_1 \ldots a_L$, $b = b_1 \ldots b_L$ and $c = c_1 \ldots c_L$. The quasi-median of a, b and c is the set $qm(a, b, c)$ of all sequences $d = d_1 \ldots d_L$ that have the property that the state d_i occurs in the set $\{a_i, b_i, c_i\}$ at least as many times as any other state, for each position $i = 1, \ldots, L$.*

In other words, $d = d_1 \ldots d_L$ is a member of $qm(a, b, c)$ if and only if either the i-th column (of the alignment of a, b and c) contains at most two different states and d_i is the majority state, or the column contains three different states and d_i is any one of them, for all i. If the number of columns containing three different states is k, then $qm(a, b, c)$ will contain 3^k different sequences. An example with $k = 1$ is shown in Figure 9.7.

Let M be a multiple alignment of DNA sequences on $\mathcal{X}$. We define the *quasi-median closure* $\bar{M}$ as the set of all sequences that can be obtained by repeatedly taking the quasi-median of any three sequences in the set and then adding the result to the set.

A *quasi-median network* is a generalization of the concept of a median network that takes *all* non-constant columns of a multiple alignment of DNA sequences [10] into account:

Definition 9.5.2 (Quasi-median network)   *Let M be a condensed multiple alignment of DNA sequences on $\mathcal{X}$. The quasi-median network associated with M is a phylogenetic network $N = (V, E)$ whose node set is given by the quasi-median closure $N = \bar{M}$ and in which any two nodes are joined by an edge in E if and only if they differ in exactly one position (as haplotypes). The associated taxon labeling $\lambda : \mathcal{X} \to V$ maps each taxon x onto the node $\lambda(x)$ that represents the corresponding sequence.*

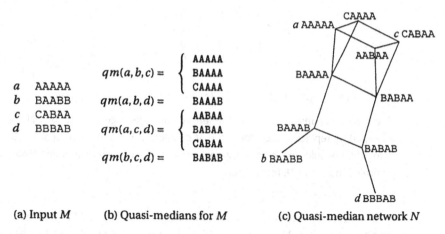

		$qm(a,b,c) =$	$\begin{cases} \text{AAAAA} \\ \text{BAAAA} \\ \text{CAAAA} \end{cases}$
a	AAAAA		
b	BAABB	$qm(a,b,d) =$	BAAAB
c	CABAA		
d	BBBAB	$qm(a,c,d) =$	$\begin{cases} \text{AABAA} \\ \text{BABAA} \\ \text{CABAA} \end{cases}$
		$qm(b,c,d) =$	BABAB

(a) Input M (b) Quasi-medians for M (c) Quasi-median network N

Figure 9.8   For the multiple sequence alignment M of multi-state characters in (a) we list all quasi-medians in (b). Note that taking medians of any of the sequences listed in (a) and (b) does not produce any new sequences, and so in this example the quasi-median closure $\bar{M}$ equals the set of listed sequences. In (c) we show the corresponding quasi-median network N.

The quasi-median network of a condensed multiple sequence alignment M is uniquely defined. Moreover, one can show that it is connected [10]. An example is shown in Figure 9.8. Recall that each column in a condensed alignment corresponds to a *pattern* in the original DNA alignment, which we describe using letters A, B, ... Note that the depicted network contains cycles of length three ("triangles"), thus illustrating that a quasi-median network is *not* a split network in general.

9.5.1 Algorithm

In theory, the quasi-median network can be constructed by first computing the quasi-median closure $\bar{M}$ of the given input dataset M and then producing the phylogenetic network whose node set is given by $\bar{M}$ and whose edges are given by all pairs of haplotypes in $\bar{M}$ that differ in precisely one position, as shown in Figure 9.8.

In practice, it is more efficient to employ a modification of the strategy used for median networks [5]. The basic idea is to encode each column of multi-state data as a group of binary columns, to then compute the median network for the alignment of binary sequences, and then finally to convert back to multi-state characters, being careful to expand each encountered "virtual median" into a set of sequences that display all appropriate multi-state characters.

In more detail, assume that we are given a condensed multiple alignment M of DNA sequences on $\mathcal{X}$. For ease of exposition, assume that the characters in M are labeled A, B, ..., E, representing the four DNA bases and the gap character.

```
A  0      A  0 0 0      A  0 0 0 0      A  0 0 0 0 0
B  1      B  1 1 0      B  1 1 0 0      B  1 1 0 0 0
          C  1 0 1      C  1 0 1 0      C  1 0 1 0 0
                        D  1 0 0 1      D  1 0 0 1 0
                                        E  1 0 0 0 1
```

(a) 2 states   (b) 3 states   (c) 4 states   (d) 5 states

Figure 9.9   The binary expansion M_1 of a condensed multiple alignment M of DNA sequences is obtained by replacing each column that contains 2, 3, 4 or 5 different states by a group of 1 to 5 binary columns, converting each state A–E into a corresponding binary pattern, as indicated in (a) to (d), respectively.

The first step is to compute the *binary expansion M_1* of M in which every column of M is represented by a *group* of binary columns in M_1. For each column C of M, we compute the corresponding group of columns in M_1 as follows: If the column C contains only two different states, then the corresponding group of columns in M_1 contains only one column, which is obtained from C by changing all occurrences of A by 0 and all occurrences of B by 1. If the number of different states contained in the column C is $d > 2$, then the column C gives rise to a group of d binary columns, in which each of the letters A–E is represented by a specific *binary pattern* as prescribed in Figure 9.9.

The second step is to compute the median closure $M_2 = \bar{M}_1$. For computational reasons, one should first collapse all identical columns in M_1 before computing $\bar{M}_1$ and then expand them all again to obtain M_2.

The third step is to convert M_2 back to a representation M_3 of multi-state data. To do this, each group of columns is converted back to a single column using the tables shown in Figure 9.9. Note that a group of three or more columns may contain binary patterns that are not listed in the tables. Such binary patterns are produced by the median operation and we refer to them as *virtual medians*, as they are not the medians of binary characters but rather of our binary encoding of multi-state characters. Virtual medians are represented in M_3 by the state '*'.

By definition, a virtual median is obtained as the median of three different binary patterns that represent three different multi-state characters, according to the tables shown in Figure 9.9. (Please convince yourself that the binary patterns listed in each of the tables have the property that the median of any three different ones is not contained in the table.)

In the fourth step we obtain a new multiple alignment M_4 of multi-state characters from M_3 by expanding each sequence M_3 that contains a virtual median * into a set of new sequences, each one obtained by replacing each virtual median by any one of the states A–E that occur in the column. Remove any duplicate columns from M_4.

Finally, the quasi-median network N associated with the given multiple alignment of DNA sequences M is constructed as the graph $N = (V, E)$ that has node set $V = M_4$ and in which any two nodes are connected by an edge in E if and only if the two nodes differ in exactly one position (as haplotypes). Here is a summary of the algorithm [5].

Algorithm 9.5.3 (Quasi-median network) *Let M be a condensed multiple alignment of DNA sequences on $\mathcal{X}$. The quasi-median network N associated with M is computed in the following steps:*

(i) *Compute the binary expansion M_1 for M.*
(ii) *Compute the median closure M_2 for M_1.*
(iii) *Compute the multi-state representation M_3 for M_2.*
(iv) *Compute M_4 by expanding all virtual medians in M_3 and then removing any duplicate copies.*
(v) *Construct the graph $N = (V, E)$ that has node set $V = M_4$ and in which any two nodes v and w are connected by an edge e in E with label i, if and only if v and w differ only in their i-th position (as haplotypes).*

An example illustrating this algorithm is presented in Figure 9.10.

9.6 Median-joining

The number of nodes of the quasi-median network associated with a condensed multiple alignment of DNA sequences M on $\mathcal{X}$ can be very large, even for a small number of short sequences. Thus, the quasi-median network is rarely useful in practice. In this section, we describe the *median-joining* algorithm [10] that aims at computing a phylogenetic network that is as informative as a quasi-median network, but usually much smaller. As we will see, the algorithm has a parameter Δ that is used to control how complex the resulting phylogenetic network will be.

Let M be a condensed multiple alignment of DNA sequences on $\mathcal{X}$. The median-joining method brings together two different algorithmic ideas. On the one hand, it repeatedly uses the concept of a "(relaxed) minimum spanning network", and on the other hand, it repeatedly employs the quasi-median calculation. While the former construction, on its own, will produce too few nodes to be useful, the latter construction, on its own, will produce too many nodes. Using both together, the median-joining method attempts to provide a useful network of intermediate size.

9.6.1 The relaxed minimum spanning network

Let M be a condensed multiple alignment of DNA sequences $\mathcal{X}$. To be able to employ the concept of a minimum spanning network, we need to define a distance

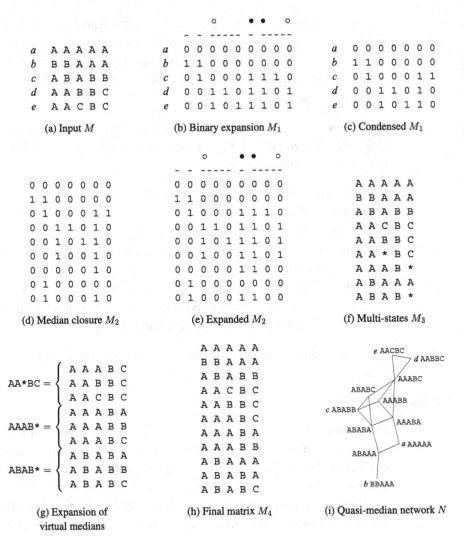

(a) Input M

(b) Binary expansion M_1

(c) Condensed M_1

(d) Median closure M_2

(e) Expanded M_2

(f) Multi-states M_3

(g) Expansion of virtual medians

(h) Final matrix M_4

(i) Quasi-median network N

Figure 9.10 For the condensed multiple sequence alignment M of multi-state characters in (a) we first compute the binary expansion M_1 (b). We then generate the median closure $M_2 = \bar{M}_1$, first collapsing all identical columns in M_1 (c) before computing $\bar{M}_1$ (d). We record which columns are collapsed together (marked ○ and ● in (b)) and then expand them all again to obtain M_2 (e). We then compute the multi-state representation M_3 for M_2 (f). Finally, we obtain the matrix M_4 by expanding all virtual medians in M_3 (g) and then removing any duplicate copies (h). In (i) we show the quasi-median network N for M.

matrix D on $\mathcal{X}$. A natural choice is to use the (unnormalized) Hamming distance $H(a, b)$ between any two sequences a and b in M.

In practice, the multiple sequence alignment M will usually be obtained by condensing some original input multiple sequence alignment M_0. To take this into account, whenever we compare two sequences in M we must weight each position in M by the number of original positions in M_0 that it represents. This gives rise to a distance matrix D in which the distance $d(a, b)$ between any two sequences a and b in M is given by the sum of weights of all positions at which a and b differ.

Consider the graph $G = (V, E, \omega)$ that has node set $V = \mathcal{X}$ and whose edge set E contains all possible edges between any two nodes in V. Moreover, let $\omega : E \to \mathbb{R}^{\geq 0}$ be an edge weighting that reflects D, that is, with $\omega(e) = d(v, w)$ for every edge $e = \{v, w\}$ in E. We will call G the *distance graph* associated with M.

A *minimum spanning tree* for G is any subtree T of G that connects all nodes of G and that minimizes the sum of edge weights given by $\omega(T) = \sum_{e \text{ in } T} \omega(e)$. There are two different algorithms for computing a minimum spanning tree, Prim's algorithm [199] and Kruskal's algorithm [155].

The latter starts with a subgraph T that contains only the nodes of G. It then considers each edge e of G in ascending order of weight and adds e to T, if e joins two different connected components of T. The algorithm terminates as soon as the subgraph T becomes connected.

In the algorithm all edges that have the same weight are processed consecutively in some order. This order serves as an implicit "tie-breaking" rule and different orderings may give rise to different minimum spanning trees.

The *minimum spanning network* for M is defined as the subgraph N of the corresponding distance graph G whose edge set is obtained as the union of the edge sets of all minimum spanning trees of G. It can be computed as follows [10]:

Algorithm 9.6.1 (Minimum spanning network) *For a given edge-weighted, connected graph $G = (V, E, \omega)$, the minimum spanning network N is computed as follows: Initially, let N be the subgraph of G that contains all nodes of G and no edges. Let $w_1 < w_2 \ldots < w_t$ denote the different edge weights that occur in G. For $i = 1, \ldots, t$ repeat the following steps:*

(i) *Add all edges of weight w_i to N that join different connected components.*
(ii) *If N is connected, terminate.*

In this algorithm, all edges of the same weight are treated equally, that is, either they are all put into the minimum spanning network, or they are all left out. For real data, it can be useful to introduce a tolerance Δ up to which we do not distinguish between different distance values. Let G be a distance graph and let N be the corresponding minimum spanning network. We define the *relaxed minimum spanning network* N' with parameter $\Delta > 0$ as the graph that contains all edges of

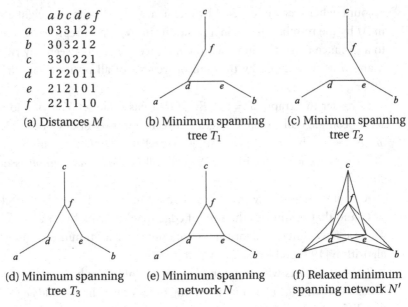

	$a\,b\,c\,d\,e\,f$
a	$0\,3\,3\,1\,2\,2$
b	$3\,0\,3\,2\,1\,2$
c	$3\,3\,0\,2\,2\,1$
d	$1\,2\,2\,0\,1\,1$
e	$2\,1\,2\,1\,0\,1$
f	$2\,2\,1\,1\,1\,0$

(a) Distances M

(b) Minimum spanning tree T_1

(c) Minimum spanning tree T_2

(d) Minimum spanning tree T_3

(e) Minimum spanning network N

(f) Relaxed minimum spanning network N'

Figure 9.11 The distance matrix M on $\mathcal{X} = \{a, \ldots, f\}$ shown in (a) gives rise to three different minimum spanning trees T_1, T_2 and T_3, shown in (b), (c) and (d), respectively. The corresponding minimum spanning network N is shown in (e). The relaxed minimum spanning network N' for $\Delta = 1$ is shown in (f).

N, and, in addition, also those heavier edges of G whose weights do not exceed the heaviest weight in N by more than Δ.

Exercise 9.6.2 *Modify Algorithm 9.6.1 so that it computes the relaxed minimum spanning network for parameter Δ.*

An alternative method of computing the minimum spanning network can be found in [72].

In Figure 9.11, we show a distance matrix D on $\mathcal{X} = \{a, \ldots, f\}$ that gives rise to three different minimum spanning trees T_1, T_2 and T_3. The displayed minimum spanning network N clearly contains all edges of the three trees. As the heaviest weight in N is one, the relaxed minimum spanning network N' with $\Delta = 1$ additionally contains six edges of weight two.

9.6.2 The median-joining algorithm

Let M be a condensed multiple alignment of DNA sequences on $\mathcal{X}$. The median-joining algorithm uses the quasi-median operation to compute new sequences from existing ones. To avoid computing the full quasi-median closure, the algorithm is guided by the relaxed minimum spanning network N_Δ of the distance graph G

associated with M, for a given user-specified tolerance $\Delta \geq 0$. The idea is to apply the quasi-median calculation only to sets of sequences that lie close to each other in the relaxed minimum spanning network N_Δ and are not already connected to each other by chains of short edges [10]:

Algorithm 9.6.3 (Median-joining) *Let M be a condensed multiple alignment of DNA sequences on $\mathcal{X}$ and $\Delta \geq 0$. The associated median-joining network N is computed in the following steps:*

(i) *Set S equal to the set of sequences contained in M.*

(ii) *Compute the relaxed minimum spanning network N_Δ for the current set of sequences S. Determine the set of all couples $\{a, b\}$ of sequences a and b in S that are connected by an edge e in N_Δ and are not connected by any chain of edges $e_1, \ldots, e_t$ in which each edge has a weight of at most $\omega(e) - \Delta$.*

(iii) *Any node contained in N_Δ that does not represent one of the original input sequences in M and is only contained in at most two couples is deemed obsolete and is removed from S. If any removals occur, return to step (ii).*

(iv) *Decide which new sequences to add to S in the following steps.*

    (a) *Let T be the set of all triples $\{a, b, c\}$ of sequences in S for which at least two of the three subsets $\{a, b\}, \{b, c\}, \{a, c\}$ are couples. Compute $\mu = qm(a, b, c)$ for all $\{a, b, c\}$ in T.*

    (b) *For each sequence f in μ that is not already contained in S, compute its connection cost as $d(f, a) + d(f, b) + d(f, c)$, where $d(x, y)$ is computed as the weighted number of differences between sequences x and y.*

    (c) *Let λ be the minimum connection cost seen in (b) over all triples in T.*

    (d) *For each triple $\{a, b, c\}$ in T, compute $\mu = qm(a, b, c)$, and then, for every sequence f in μ, add f to S, if the connection cost for f does not exceed $\lambda + \Delta$.*

    (e) *If any new sequences are added to S in (d), return to step (ii).*

(v) *In this step we generate the final network. Compute the minimum spanning network N for S (with parameter $\Delta = 0$) and determine the set of couples, as in step (i). If any obsolete nodes are present, remove them and return to step (iv). Otherwise, the set of couples defines the edge set of the final network N.*

9.6.3 Application

Median-joining is a widely used method. An interesting application of this method can be found in [151], where a median-joining network on human populations is computed for mitochondrial DNA samples. The network (displayed in Figure 9.12) is consistent with the out-of-Africa model of human origins, which suggests that all non-African populations are derived from one African lineage.

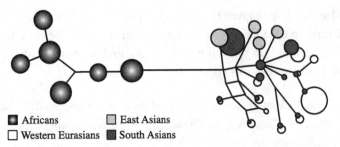

■ Africans   ▢ East Asians
▢ Western Eurasians   ■ South Asians

Figure 9.12 A median-joining network on human populations computed from mtDNA (adapted from [60, 151]). Each disk in the tree represents a cluster of human mitochondrial types and its diameter is proportional to the number of sequences represented. For African sequences, edges between individual types are collapsed and not shown. The cluster containing all non-African sequences is shown here in a non-collapsed view. Note that it attaches to only one of the clusters of African lineages. This network is consistent with the out-of-Africa model of human origins, suggesting that all non-African populations are derived from one African lineage.

9.7 Pruned quasi-median networks

As mentioned above, the number of nodes and edges in a quasi-median network can be unmanageably large. We saw that the median-joining method [10] tries to avoid constructing the full quasi-median closure using a heuristic based on the concept of a relaxed minimum spanning network. A more recent method [5] aims at computing a pruned version of the full quasi-median network by considering only those sequences that lie on a *geodesic* between two of the original input sequences. The result is called a *geodesically pruned quasi-median network*.

The definition and computation of geodesics require some additional concepts.

Definition 9.7.1 (Restricted quasi-median) *Consider three DNA sequences a, b and c of length L. Moreover, assume that we are given two* reference *sequences g and h of length L. The* restricted quasi-median *of a, b and c, relative to g and h is defined as the set $qm_{g,h}(a, b, c)$ of all sequences d of length L that have the property that d_i is the majority state in (a_i, b_i, c_i), if there is one, and otherwise d_i is contained in $\{a_i, b_i, c_i\} \cap \{g_i, h_i\}$, for all positions $i = 1, \ldots, L$.*

Let M be a condensed multiple alignment of DNA sequences of length L. For each character state c and each position i in M we define the score $s(c, i)$ as the fraction of sequences in M that have state c at position i. The score of an arbitrary sequence a of length L is then defined as

$$s(a) = \log\left(\sum_{i=1}^{L} s(a_i, i)\right). \tag{9.4}$$

If a and b are two sequences of length L that are nodes of some graph G whose nodes are all sequences of length L, then a *geodesic* between a and b is any path $P = (v_1, v_2, \ldots, v_t)$ of nodes from a to b in G that has minimum total score $s(P) = \sum_{i=1}^{t} s(v_i)$. The geodesically-pruned quasi-median network can be computed using the following algorithm [5]:

Algorithm 9.7.2 (Geodesically pruned quasi-median network) *Let M be a condensed multiple alignment of DNA sequences on $\mathcal{X}$. Output is the associated geodesically pruned quasi-median network N. Process each pair of sequences a and b in M as follows:*

(i) *Extract the sub-alignment M_{ab} from M that consists of all columns of M for which the states of a and b differ.*

(ii) *Compute the closure $\bar{M}_{ab}$ of the restricted quasi-median operation on M_{ab} relative to a and b.*

(iii) *Construct the graph G_{ab} with node set $\bar{M}_{ab}$ in which any two nodes are connected by an edge if and only if they differ by exactly one position (as haplotypes).*

(iv) *Remove all sequences from $\bar{M}_{ab}$ that do not lie on a geodesic from a to b in G_{ab}.*

(v) *Let S_{ab} be the result of expanding all sequences in $\bar{M}_{ab}$ back to their original length by reinserting all positions at which a and b have the same state.*

The final network $N = (V, E)$ has node set $V = \bigcup_{a \neq b} S_{ab}$ and an edge set E that consists of all pairs of nodes that differ by exactly one position (as haplotypes). An associated taxon labeling $\lambda : \mathcal{X} \to V$ maps each taxon x onto the node $\lambda(x)$ that represents the corresponding sequence.

9.7.1 Application

An application of the geodesically pruned quasi-median network algorithm can be found in [5], where it is applied to 110 sequences from the spacer regions of the 5S ribosomal DNA loci of sea beet (*Beta vulgaris* ssp. *martima*). The resulting phylogenetic network (reproduced in Figure 9.13) contains 648 nodes. It shows a separation between harbor and cliff top populations, in agreement with previous studies.

9.8 Recombination networks

A recombination network is a rooted phylogenetic network that is used to describe the evolution of a set of closely related sequences (usually from different individuals of a population) in terms of mutations (along edges of the network), speciation events (at tree nodes) and recombination events (at reticulate nodes).

In a simple recombination event, two sequences a and b of the same length L give rise to a new sequence c of length L that consists of a prefix of the sequence a

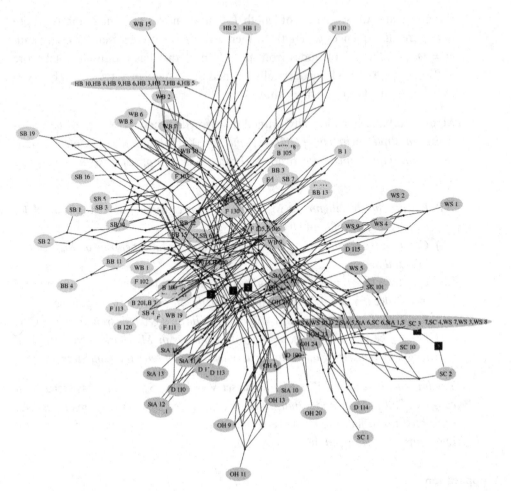

Figure 9.13 The geodesically pruned quasi-median network of 110 sequences from the spacer regions of the 5S ribosomal DNA loci of sea beet (*Beta vulgaris* ssp. *maritima*) [5]. It shows a separation between harbor and cliff top populations, which can be found at the top half and bottom half of the network, respectively. Reproduced from BMC Bioinformatics 9:115 (2008) under the Creative Commons Attribution License.

and a suffix of the sequence b (or vice-versa), see Figure 9.14. This is known as a *single crossover* event. In a *multiple crossover* event, the resulting sequence c consists of a mosaic of segments from a and b. In either type of event, the character state c_i is either a_i or b_i for every position $i = 1, \ldots, L$ of the sequence c.

Let M be a multiple sequence alignment on $\mathcal{X}$. In the following we assume that the sequences in M are binary sequences, as discussed in Section 9.2. However, the ordering of characters along the sequences matters in the context of recombinations

```
a    ACGTTGGCC AGTTGA
b    ccattgaat gtttca
```

```
c    ACGTTGGCC gtttca
```

Figure 9.14 The sequence *c* is obtained from the sequences *a* and *b* by a single crossover between positions 9 and 10. The first nine character states of *c* equal the first nine character states of *a* (shown in upper-case) and the latter six equal those of sequence *b* (shown in lower-case).

and so we usually do *not* assume that the multiple sequence alignment *M* has been condensed. Here is the definition of a recombination network [89, 98, 128]:

Definition 9.8.1 (Recombination network)   *Let M be a multiple alignment of binary sequences of length L, on X. A recombination network N representing M is given by a bicombining rooted phylogenetic network on X, together with two additional labellings:*

(i) *Each node v of N is labeled by a binary sequence $\sigma(v)$ of length L.*
(ii) *Each tree edge e is labeled by a set of positions $\delta(e) \subseteq \{1, \ldots, L\}$.*

These two labellings must fulfill the following compatibility conditions:

(i) *The sequence $\sigma(v)$ assigned to any leaf v must equal the sequence in M that is given for the taxon associated with v.*
(ii) *If r is a reticulate node (often called a recombination node in this context) with parents v and w, then the sequence $\sigma(r)$ must be obtainable from the two sequences $\sigma(v)$ and $\sigma(w)$ by a crossover.*
(iii) *If $e = (v, w)$ is a tree edge, then the set of positions at which the two sequences $\sigma(v)$ and $\sigma(w)$ differ must equal $\delta(e)$.*

For computational reasons, the following condition is usually also required [150]:

(iv) *Any given position may mutate at most once in the network. In other words, for any given position i there exists at most one edge e with $i \in \delta(e)$.*

This condition is usually referred to as the *infinite sites* assumption because for sequences of infinite length it holds that the probability of the same site being hit by a mutation more than once is zero, under a uniform distribution.

A simple example of a recombination network is shown in Figure 9.15.

Exercise 9.8.2 (Verify recombination network)   *Verify that the network N shown in Figure 9.15 is a valid recombination network for the given alignment M. Does it fulfill the infinite sites assumption? Can the sequence at node r be obtained by a single-crossover recombination from the parental sequences?*

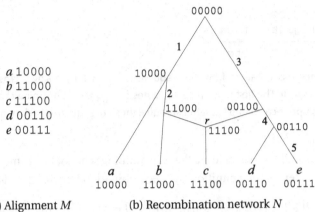

(a) Alignment M (b) Recombination network N

Figure 9.15 (a) A multiple alignment of binary sequences of length five on $\mathcal{X} = \{a, \ldots, e\}$. (b) A recombination network N for M. For each node v we show the sequence $\sigma(v)$ and for each edge e we show the set of positions $\delta(e)$. In this example, that set is either empty (unlabeled edges) or contains only one position (from 1 to 5).

Evolution in the presence of recombination events is usually studied in population genetics, rather than phylogeny, focusing on the statistical analysis of the inheritance and prevalence of genes in populations. Under the "coalescent-with-recombination" model, a description of the history of n sampled sequences going backward in time gives rise to a graph that is called an *ancestral recombination graph* (*ARG*) [90, 108].

9.8.1 The local-tree parsimony approach

Let M be a multiple alignment of binary sequences of length L and let N be a corresponding recombination network, on $\mathcal{X}$. Consider a position i in the alignment. For any particular reticulate node r in N the character state at position i of the sequence $\sigma(r)$ is copied from exactly one of the two parents of r. In consequence, each individual character in M evolves along some rooted phylogenetic tree embedded in N. In the recombination network shown in Figure 9.15, the first two characters evolve along the tree in which the leaf labeled c is a child of the node labeled 11000, whereas the last three characters evolve along the tree in which the leaf labeled c is a child of the node labeled 00100.

So, in consequence, the evolutionary history of any sufficiently short segment of sequence is a rooted phylogenetic tree. This insight leads to the following approach:

(i) Determine a suitable rooted phylogenetic tree T_i for each position i in the alignment M. We will refer to any such tree T_i as a *local* tree.
(ii) Combine all the trees into a suitable rooted phylogenetic network N.

When doing this, there is a trade-off to be made between the number of incompatibilities between a character and its associated local tree on the one hand, and the "recombination cost" of switching from one tree topology to a different when going from position i to $i + 1$, on the other.

The first part of the approach can be formulated as a dynamic programming problem [106]. To this end, we define a *local-tree graph* $G = (V, E)$ for M and X as follows. For each position i in M and each possible rooted phylogenetic tree T on X let $v(i, T)$ be a node in V. Any two nodes $v(i, T)$ and $v(i + 1, T')$ at adjacent positions in the alignment are connected by a directed edge $e = (v(i, T), v(i + 1, T'))$.

Each node in G is assigned a weight $\omega(v(i, T))$ that equals the minimum number of substitutions required for the i-th character on the rooted phylogenetic tree T. Each edge $e = (v(i, T), v(i + 1, T'))$ is assigned a weight $\omega(v(i, T), v(i + 1, T')) = \omega(e)$ that reflects the recombination distance between T and T' (see below for more details).

Now, in the local-tree graph G any path

$$P = (v(1, T_1), v(2, T_2), \ldots, v(L, T_L)) \tag{9.5}$$

that starts at some node $v(1, T_1)$ and ends at some node $v(L, T_L)$ selects a rooted phylogenetic tree T_i for each position i in M. The total weight of such a path P is obtained by adding all node weights and all edge weights along the path:

$$W(P) = \sum_{i=1}^{L} \omega(v(i, T_i)) + \sum_{i=1}^{L-1} \omega(v(i, T_i), v(i + 1, T_{i+1})). \tag{9.6}$$

A *most parsimonious* set of trees for M is given by a path P of minimum weight $W(P)$, which can be computed by standard dynamic programming using the following recursion:

$$W(i, T) = \begin{cases} \min_{T'} \{W(i - 1, T') + \omega((v(i - 1, T'), v(i, T))) + \omega(v(i, T))\}, \\ \qquad \text{if } i > 1, \\ \omega(v(1, T)), \quad \text{else.} \end{cases}$$

$$\tag{9.7}$$

Before this recursion can be used, we need to specify how to determine the weight of a node $\omega(v(i, T))$ and the weight of an edge $\omega(v(i - 1, T'), v(i + 1, T))$. Since the weight of a node is simply the parsimony score for the i-th character on the tree T, it can be efficiently computed as described in Section 3.7.1. The weight of an edge is the minimum number of reticulate nodes in a rooted phylogenetic network that contains both trees T and T'. An algorithm that addresses this NP-hard problem is discussed in Section 11.5.

The described approach is not practical for three reasons. First, the number of possible rooted phylogenetic trees for n taxa is $(2n - 3)!!$ (see Section 3.3), and thus is very large even for relatively small values of n. Second, the computation of each edge weight in the graph is NP-hard. Finally, the second part of the approach, computation of a suitable rooted phylogenetic network that represents all the trees, is an NP-hard problem, see Section 11.5.

Although the approach is not practical, it is conceptually appealing because it explicitly addresses the *mosaic* nature of aligned sequences: a long multiple sequence alignment consists of stretches of sequence that have evolved along a common rooted phylogenetic tree and these stretches are separated by crossover positions at which recombinations have occurred.

We illustrate the approach in Figure 9.16. For the given multiple alignment M of binary sequences of length 11 on $\mathcal{X} = \{a, b, c, d, e\}$ we indicate the corresponding local-tree graph G. The graph has 11 columns of nodes, each column containing one node for each of the possible rooted phylogenetic trees on $\mathcal{X}$. All nodes of a column are connected to all nodes of the next column by edges. The nodes and edges are weighted as described above. In this example, the first five characters of M are all compatible with the tree T_1, whereas the last six characters are all compatible with the tree T_2, and so in this case a path through G that chooses T_1 for the first five positions and then T_2 for the remaining six positions is optimal. The depicted rooted phylogenetic network N contains both T_1 and T_2 and is a recombination network for M. More precisely, N is a recombination network if one adds the appropriate labeling of the nodes by sequences and edges by mutations.

Exercise 9.8.3 (Labeling of recombination network)   *Label the nodes and edges of the rooted network N depicted in Figure 9.16e so that the network becomes a complete recombination network for the alignment M shown in part (a) of the figure.*

9.8.2 A heuristic for the local-tree approach

Let M be a multiple alignment of binary sequences of length L on $\mathcal{X}$. We now look at a heuristic for computing a recombination network N for M that is based on the ideas just discussed [106]. A first simplification in this approach is to use unrooted phylogenetic trees rather than rooted ones. The main simplification is to consider only a small part of the full local-tree graph.

The heuristic starts by computing a phylogenetic tree T from the whole dataset M using the maximum parsimony method. This tree will be used as a start tree and is assigned to position 1 of the alignment. Next, we compute the *SPR-neighborhood S* of T consisting of T and all phylogenetic trees that are exactly one SPR modification away from T. The main recursion of the previous section is repeatedly applied to the local-tree graph restricted to nodes that contain a tree from S until a position

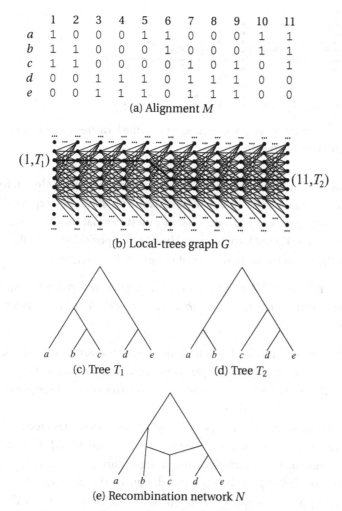

	1	2	3	4	5	6	7	8	9	10	11
a	1	0	0	0	1	1	0	0	0	1	1
b	1	1	0	0	0	1	0	0	0	1	1
c	1	1	0	0	0	0	1	0	1	0	1
d	0	0	1	1	1	0	1	1	1	0	0
e	0	0	1	1	1	0	1	1	1	0	0

(a) Alignment M

$(1,T_1)$ $(11,T_2)$

(b) Local-trees graph G

(c) Tree T_1 (d) Tree T_2

(e) Recombination network N

Figure 9.16 (a) A multiple alignment M of binary sequences of length 11 on $\mathcal{X} = \{a, b, c, d, e\}$. (b) The local-tree graph G associated with M. Each column of nodes represents all possible rooted phylogenetic trees for the corresponding column in M. All nodes in one column are connected to all nodes in the next column (but not all these edges are shown, for clarity). The path highlighted in bold chooses the tree T_1 (c) for the first five positions of M and then the tree T_2 (d) for the remaining six positions. In (e) we show the resulting recombination network N.

is reached at which a jump to a different tree T' in S is indicated by the recursion. We then determine the SPR-neighborhood S' of T' and continue the calculation with T' and S' in place of T and S. See Figure 9.17.

The resulting set of trees might depend quite strongly on the initial tree used. Hence, after completing the first pass of the algorithm, a second pass is performed

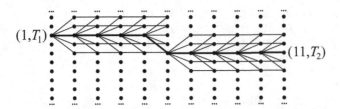

Figure 9.17   The heuristic algorithm considers only those trees in the local-tree graph that differ at most by one SPR from the current tree.

in the opposite direction, using the local tree that was previously chosen for the last position of the alignment as the new start tree. This should be repeated a couple of times to completely erase any dependency on the initial choice of start tree. The final pass produces a set of local trees and breakpoint positions along the sequences that reflect the recombination history of the given set of sequences.

Exercise 9.8.4 (Single SPR neighborhood)   *Design an algorithm that quickly generates all unrooted phylogenetic trees that are one SPR away from a given unrooted phylogenetic tree T on $\mathcal{X}$.*

This heuristic is an example of a general approach to studying recombination that operates by scanning along a multiple sequence alignment, building trees for local segments of the sequence and then comparing the trees in an attempt to determine recombination crossover positions.

   As mentioned above, once a set of phylogenetic trees has been constructed for the different stretches of a multiple sequence alignment M, then in a second step these trees may be combined into a recombination network to represent the evolutionary history of the data. By design of the heuristic, the local trees of neighboring positions will differ by at most one SPR and so in this case the calculation of a corresponding network should be easier than in the general case. However, this step is still challenging in practice and it is not explicitly described in [106].

9.9 Galled trees

In the previous section we reported that the problem of constructing a minimum recombination network is a computationally hard problem. To address this, we presented a heuristic approach. A different way to ease the computational difficulties is to restrict ones attention to phylogenetic networks that have a very simple topology.

   Recall that a rooted phylogenetic network N on $\mathcal{X}$ is called a *galled tree*, if for any two reticulations r and r' and any two undirected cycles that contain r and r', respectively, it holds that the two cycles are edge-disjoint (see Section 6.11.1). This is the topologically simplest form of a non-tree network.

We are interested in the following problem: given a multiple alignment of binary sequences M on $\mathcal{X}$, how to determine whether M can be represented by a recombination network N that is a galled tree using only single crossovers, and, if this is the case, how to construct such a network?

To address this question, we first consider a restricted instance of the problem in which the alignment M is condensed and the corresponding incompatibility graph $IG(\mathcal{S}(M))$ is connected. We then discuss how the solution to the problem in the restricted case can be used to obtain a solution when $IG(\mathcal{S}(M))$ has more than one connected component. Finally, we show how to modify the approach further so as to be able to drop the restriction to condensed alignments, too, thus obtaining a solution for *any* multiple alignment of binary sequences.

9.9.1 Restricted case of condensed alignments and single galls

In the context of recombination it is generally not appropriate to assume that the given alignment M has been condensed because the order of columns and crossover positions is important. However, for ease of exposition we first assume that the given alignment M *is* condensed. Later, in Subsection 9.9.3 we show that only a small modification of the described approach is required to accommodate the more general case of non-condensed alignments.

Let M be a condensed multiple alignment of binary sequences on $\mathcal{X}$. Consider the incompatibility graph $IG(\mathcal{S}(M))$ for the set of splits associated with M. For simplicity, in this section we assume that $IG(\mathcal{S}(M))$ consists of exactly one non-trivial component. This assumption implies that the corresponding recombination network N is biconnected and contains exactly one reticulate node r, if it is a galled tree. In Section 9.9.2 we will discuss how to handle the more general case.

A first necessary condition for the existence of a galled tree that represents M is that removal of a single taxon h (associated with the reticulate node r) must resolve all incompatibilities between columns in M. In other words, we require that a taxon h in $\mathcal{X}$ exists such that we have:

$$\bigcap_{(i,\,j)\in IC(M)} Q(i,\,j) = \{h\}, \tag{9.8}$$

where $IC(M)$ is the set of all pairs of columns i and j of M that are incompatible and $Q(i,\,j) = \{A_i \cap A_j,\, A_i \cap B_j,\, B_i \cap A_j,\, B_i \cap B_j\}$ denotes the set of all four possible intersections of split parts of the two splits $S_i = \frac{A_i}{B_i}$ and $S_j = \frac{A_j}{B_j}$ associated with columns i and j.

Now, let i and j be any two columns of M that are incompatible. As the two associated splits S_i and S_j cannot be represented by the same phylogenetic tree,

	1	2	3
a	0	0	0
b	1	0	0
c	1	0	1
d	0	1	0
e	1	1	1

(a) Alignment M

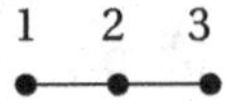

```
  1   2   3
  ●———●———●
```

(b) $IG(\mathcal{S}(M))$

$Q(1,2) = \{\{a\}, \{b,c\}, \{d\}, \{e\}\}$
$Q(2,3) = \{\{a,b\}, \{c,\}, \{d\}, \{e\}\}$

$I(1,2) = \{1\}$
$I(2,3) = \{2\}$

(c) Values for Q and I

Figure 9.18 For the multiple sequence alignment M shown in (a), we display the associated incompatibility graph $IG(\mathcal{S}(M))$ in (b), and the resulting values for Q and I in (c) as used in Lemma 9.9.1. While the values for $Q(1,2)$ and $Q(2,3)$ imply that both d and e satisfy condition (i) of the lemma, the values for $I(1,2)$ and $I(2,3)$ imply that no single-crossover position exists that satisfies condition (ii). Hence, M cannot be represented by a galled tree.

it follows that any recombination network N that represents M must postulate a crossover somewhere between positions i and j, that is, immediately after one of the positions in the set $I(i, j) = \{k \mid i \le k < j\}$. Because we only allow one single-crossover recombination, there must exist one single-crossover position p that takes care of all pairs of incompatibilities. This second necessary condition can be expressed as follows:

$$\bigcap_{(i,j)\in IC(M)} I(i, j) \neq \emptyset. \tag{9.9}$$

These two necessary conditions are also sufficient [221]:

Lemma 9.9.1 (Existence of galled tree recombination network) *Let M be a condensed multiple alignment of binary sequences on $\mathcal{X}$ whose incompatibility graph $IG(\mathcal{S}(M))$ consists of exactly one non-trivial connected component. Then, there exists a recombination network N that represents M and is a galled tree using a single crossover, if and only if the following two conditions hold:*

(i) *There exists a taxon h in $\mathcal{X}$ whose removal resolves all incompatibilities, in other words, $\bigcap_{(i,j)\in IC(M)} Q(i, j) = \{h\}$.*
(ii) *There exists a corresponding single-crossover location, in other words, $\bigcap_{(i,j)\in IC(M)} I(i, j) \neq \emptyset$.*

For an application of the lemma, consider the multiple sequence alignment M shown in Figure 9.18a. The associated incompatibility graph $IG(\mathcal{S}(M))$ shown in part (b) has three nodes and two edges. The intersection of the two sets $Q(1, 2)$ and $Q(2, 3)$ (shown in part (c)) contains two sets, namely $\{d\}$ and $\{e\}$. Hence, condition (i) of the lemma holds. This means that either taxon d or taxon e might be associated with a recombination event. However, the intersection of the sets

$I(1, 2)$ and $I(2, 3)$ (also shown in part (c)) is empty, and thus a recombination involving only a single crossover is not possible in this example.

The lemma is the basis of an algorithm [98] for determining a galled tree (single crossover) recombination network, if one exists, which we now describe. Let M be a condensed multiple alignment of binary sequences on $\mathcal{X}$ whose associated incompatibility graph $IG(\mathcal{S}(M))$ is connected.

First, check the conditions of the lemma. If both are fulfilled, then let h be the taxon identified in the first condition. Determine the set of clusters $\mathcal{C}_h = \{S(h) - \{h\} \mid S \in \mathcal{S}(M)\} \cup \{\mathcal{X} - \{h\}\}$ obtained by considering each split in $\mathcal{S}(M)$, taking the split part $S(h)$ that contains h and removing h from it. Then we construct a graph $H(\mathcal{C}_h) = (V, E)$ with node set $V = \mathcal{C}_h$ and edge set E that is essentially the Hasse diagram for $(\mathcal{C}_h, \subseteq)$, but in which the directions of all edges have been reversed. In more detail, for any two nodes v and w we have that (v, w) is an edge in E if and only if $w \subsetneq v$ (as clusters) and there exists no third node u with $w \subsetneq u \subsetneq v$.

We label each node v by the taxon that is contained in v but not in any child of v (as clusters). At least one such taxon must exist by definition of the graph. The fact that not more than one such taxon can exist follows from the assumption that M is condensed. For the sake of contradiction, let us suppose that two such taxa a and b exist. This implies that no cluster in $\mathcal{C}_h$ separates them. This in turn implies, from the definition of $\mathcal{C}_h$, that a and b have exactly the same sequence, a contradiction.

We extend the graph $H(\mathcal{C}_h)$ by creating an additional node r with label h. Because all pairs of incompatible columns are separated by a single crossover position, there are only two possible layouts of the graph $H(\mathcal{C}_h)$. The first possible layout is that H consists of a single directed path, in which case we attach the node r to the first and last node of the path. The other possible layout is that $H(\mathcal{C}_h)$ consists of two directed paths that both issue from a common node (the root), in which case we attach the node r to the last node of each path.

Strictly speaking, the resulting network does not fulfill the definition of a rooted phylogenetic network because the nodes labeled by taxa are not leaves (in fact the network constructed in this subsection does not have any leaves). This is easily fixed by simply adding all missing leaf edges at the very end of the computation.

If the alignment M can be represented by some galled tree, then there may be more than one solution. For example, in the simplest case of four different binary sequences of length 2 there exist four possible galled trees $N_1, \ldots, N_4$ that represent M, as shown in Figure 9.19.

Here is a summary of the algorithm [98]:

Algorithm 9.9.2 (Galled tree for single component) *Let M be a multiple alignment of binary sequences on $\mathcal{X}$. Assume that M is condensed and that the incompatibility*

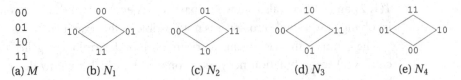

00	00	01	10	11
01				
10				
11				
(a) M	(b) N_1	(c) N_2	(d) N_3	(e) N_4

Figure 9.19 The multiple alignment of binary sequences M shown in (a) can be represented by any of the four different recombination networks $N_1, \ldots, N_4$ indicated in (b)–(e).

graph $IG(\mathcal{S}(M))$ consists of a single non-trivial connected component. If and only if both conditions of Lemma 9.9.1 are fulfilled, then a galled-tree recombination network N representing M can be computed as follows:

(i)   Let h be the taxon identified in condition (i) of Lemma 9.9.1.
(ii) Determine the set of clusters

$$\mathcal{C}_h = \{S(h) - \{h\} \mid S \in \mathcal{S}(M')\} \cup \{\mathcal{X} - \{h\}\}. \tag{9.10}$$

(iii) Construct the network $H(\mathcal{C}_h)$ and a new node r labeled h.
(iv)   If $H(\mathcal{C}_h)$ consists of a single directed path, then attach r to the first and last node of the path.
(v)   Otherwise, $H(\mathcal{C}_h)$ must consist of two directed paths that both issue from a common node (the root). Attach r to the last node of each path.
(vi)   The crossover location associated with r can be chosen as any number that satisfies condition (ii) of Lemma 9.9.1.

For an example of how the algorithm operates, see Figure 9.20. The multiple sequence alignment M depicted there gives rise to values of the functions Q and I that satisfy the conditions of Lemma 9.9.1 if one chooses $h = e$, for example. For this choice, we obtain the following set of clusters:

$$\mathcal{C}_e = \{\{a, b, c, d\}, \{b, c\}, \{c\}, \{d\}\}. \tag{9.11}$$

The Hasse diagram $H(\mathcal{C}_e)$ is shown in Figure 9.20e. By attaching an additional node (labeled e) to the bottom nodes of the Hasse diagram, and then adding all necessary leaf edges, we obtain a galled tree N on $\mathcal{X}$ that represents M, as shown in part (g). The tree edges of N are labeled by the positions at which the sequences change along those edges. The sequence for e obtains its first two positions from c and its last position from d, by a single crossover located between positions 2 and 3.

Exercise 9.9.3 (Alternative galled tree) *In the example just described, choose $h = d$ rather than $h = e$ and then construct the corresponding galled tree.*

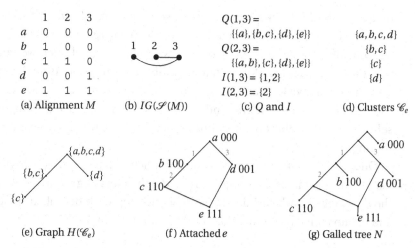

	1	2	3
a	0	0	0
b	1	0	0
c	1	1	0
d	0	0	1
e	1	1	1

(a) Alignment M

(b) $IG(\mathcal{S}(M))$

$Q(1,3) =$
$\quad \{\{a\},\{b,c\},\{d\},\{e\}\}$
$Q(2,3) =$
$\quad \{\{a,b\},\{c\},\{d\},\{e\}\}$
$I(1,3) = \{1,2\}$
$I(2,3) = \{2\}$

(c) Q and I

$\{a,b,c,d\}$
$\{b,c\}$
$\{c\}$
$\{d\}$

(d) Clusters $\mathcal{C}_e$

(e) Graph $H(\mathcal{C}_e)$

(f) Attached e

(g) Galled tree N

Figure 9.20 For the multiple sequence alignment M shown in (a), we display the associated incompatibility graph $IG(\mathcal{S}(M))$ in (b), and the resulting values for Q and I in (c). Choosing e to be the taxon involved in a recombination, we list the associated set of clusters C_e in (d). The associated Hasse diagram $H(C_e)$ is displayed in (e). (f) The network obtained by attaching the node labeled e. The final galled tree N is shown in (g).

9.9.2 Allowing multiple galls

Let M be a condensed multiple alignment of binary sequences on $\mathcal{X}$. In the previous subsection, we discussed how to construct a recombination network that is a galled tree in the case that the incompatibility graph $IG(\mathcal{S}(M))$ consists of a single non-trivial connected component. Now, let us assume that the incompatibility graph $IG(\mathcal{S}(M))$ has more than one connected component.

For each non-trivial connected component C of $IG(\mathcal{S}(M))$, which we refer to as a *subproblem*, we construct the multiple sequence alignment M' induced by the set of splits contained in the component C. Is M' condensed? Clearly, M' will not contain any constant column and also no two columns in M' will display the same pattern of states, because M is condensed. However, it may happen that two sequences in M' are identical. In the following, we disregard all duplicate sequences and can therefore assume that the alignment M' associated with any subproblem C is condensed.

By definition, the incompatibility graph $IG(\mathcal{S}(M'))$ associated with M' consists of a single non-trivial connected component. Thus we can apply Lemma 9.9.1 to determine whether a galled tree exists for M' and then use Algorithm 9.9.2 to construct such a network in the affirmative case.

Once we have established that each subproblem can be solved by a galled tree, then the next task is to merge all these galled trees using a minor extension of the decomposition-based divide-and-conquer strategy described in Section 8.2. As described, the divide-and-conquer strategy assumes that the input data are clusters and then uses a rooted backbone tree to guide the merging process.

If one of the input sequences a is known to be ancestral, then we can use it to root the backbone tree, thus fulfilling the above assumption. If none of the input sequences is known to be ancestral, then one can proceed by considering all possible rootings of the backbone tree and for each such rooting, determining whether a solution to each subproblem can be chosen in a way that is compatible with the rooting.

Once we have chosen one recombination network for each non-trivial component of the incompatibility graph $IG(S(M))$ then we merge all networks following the decomposition-based divide-and-conquer approach described in Section 8.2 and obtain a final recombination network N for M on $\mathcal{X}$.

9.9.3 Handling non-condensed alignments

In the previous two subsections we make the simplifying assumption that the considered multiple alignment of binary sequences M on $\mathcal{X}$ is condensed. To be able to drop this assumption, we must discuss how to handle

(i) duplicate sequences,
(ii) constant columns, and
(iii) multiple columns corresponding to the same split.

The first two cases are easy to deal with. To accommodate duplicate sequences, first run the described algorithm using one representative taxa for each set of taxa that have the same sequence and then add the other taxa as additional labels to the network at the end of the computation. The constant columns are simply ignored in all calculations.

The third case requires that we modify the method of choosing an appropriate crossover location [221]. To this end, consider two columns i and j of a multiple alignment of binary sequences M on $\mathcal{X}$ and assume that $i < j$. We define

$$I'(i, j) = \left\{ k \mid \max S_i^{-1} \leq k < \min S_j^{-1} \right\}, \tag{9.12}$$

where S_i^{-1} denotes the set of all positions in M that give rise to the same split S_i as column i does, for any column i. To ensure that a suitable single crossover position exists, we require that the last occurrence of a column contained in S_i^{-1} comes before the first occurrence of a column contained in S_j^{-1}. This can be taken into account by replacing condition (ii) of Lemma 9.9.1 by the following more general condition:

(ii') There exists a single-crossover location: $\bigcap_{(i, j) \in IC(M)} I'(i, j) \neq \emptyset$.

Strain	Non-constant sites of alignment

```
28436   gaccatcacgatgtgggtgggctcctgaacccccaactactttcagacccacctggttgtggcg
28723   ................................................................
29010   ................................................................
2903    ....g......c....................................................
28585   ....g......c....................................................
28718   ....g......c....................................................
25797   t.t.....t..c...a.....................................t.a........
29148   t.t.....t..c...a.....................................t.a.......a
29020   .g...g.....c.........tt....a...............c.tt...tt..t.....a.ca..
26916   .g...g.....c.........tt....a...............c.tt...tt..t.....a.ca..
29011   .g...g.....c.........tt....a...............c.tt...tt..t.....a.ca..
29105   .g...g.....c.........tt....a...............c.tt...tt..t.....a.ca..
26752   ......g...gc.......................t.....................t.......
26754   ......g..a.c......................t....................tc.......
26755   ......g...gc..a...................t.....................t.......
6101    .......g...c.....c....g...........a...................tt..c.....
13818   .......g...c.....c....g...........a...................tt..c.....
26156   .......g...c.....c....g...........a...................tt..c.....
28720   .......g...c.....c....g...........a...................tt..c.....
28721   .....................................................tt..c.....
5883    ...t.......ca......a..g.t....t.................t....c........
6394    ...t.......ca......a..g......t.................t....c........
13383   ...t.......ca......a..g......t.....g...........t....c........
28063   ...t.......c.......a..g......t.g...............t....c........
28336   ...t.......ca......a..g......t.................t....c........
28439   ...........c.......a..g......t.................t....c........
29169   ...t.......c.......a..g......t.g...............t....c........
O13393  ..........c.c..a.a....t.a.gg.t...g.tcggc.c..cgtt.c.t....c.c..a.
```

Figure 9.21 The 64 non-constant sites of a multiple sequence alignment for the *TRI101* gene for 27 different strains of *F. graminearum* and one outgroup sequence O13393 representing *F. lunulosporum* (adapted from [190]).

9.9.4 Application

The phylogeographic structure of lineages of the fungus *Fusarium graminearum* is investigated in [190]. The paper studies 37 strains and uses the DNA sequence of six different genes to infer phylogenetic relationships between the strains.

One result reported is that the nuclear 3-O-acetyltransferase gene (*TRI101*) has undergone intragenic recombination in one of the strains. The data set for this gene consists of a multiple sequence alignment of 28 DNA sequences of length 1336. The DNA sequences represent different strains of *F. graminearum* and are identified by numbers. The strains are partitioned into 7 lineages (1–7) excluding the strain numbered 28721, which is the recombinant strain. The paper reports that the *TRI101* sequence for strain 28721 appears to have arisen by recombination between African lineage 2 and Asian lineage 6.

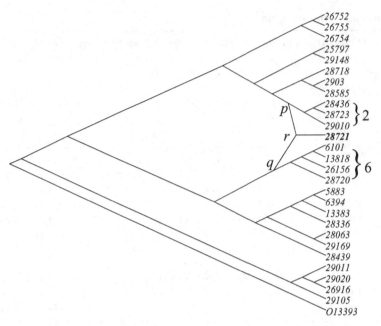

Figure 9.22 A recombination network N computed for the multiple sequence alignment of *TRI101* sequences listed in Table 9.21. This network suggests that the sequence of taxon 28721 arose by recombination of the lineages labeled 2 and 6 (data from [190]).

In Figure 9.21 we list all non-constant positions of the *TRI101* multiple sequence alignment. As each character in this alignment takes on precisely two different states, the data is binary and can be used as input for the galled tree algorithm.

Application of the described galled tree algorithm establishes that this dataset can be represented by a recombination network that is a galled tree, as shown in Figure 9.22. Here we see clearly how strain 28721 arises through recombination from the two lineages 2 and 6. The sequences assigned to the three nodes labeled p, q and r are as follows:

p = gaccatcacga**C**gtgggtgggc**C**cctgaacccccaactactttcagacccatttgcttgtggcg
q = gaccatcacgatgtgggtgggctcctgaacccccaactactttcagaccca**CC**tg**G**ttgtggcg
r = gaccatcacgatgtgggtgggctcctgaacccccaactactttcagacccatttgcttgtggcg

The sequences of the parent nodes p and q differ from the sequence of the reticulate node r at positions 12, 23 and 52, 53, 56, respectively (emphasized in upper-case). Hence, the sequence assigned to the reticulate node r can be obtained from the sequences assigned to p and q by a single-crossover recombination, whose crossover point is located anywhere between positions 23 and 52.

Chapter notes

Bandelt's median network and median-joining method [10, 11, 12] are widely used in population genetics based on mtDNA and these two approaches are among the most widely used phylogenetic network methods. They are implemented in a Windows program called Network and have recently been added to the SplitsTree program. The *netting method* described in [80] computes a network that is similar to a median network. The TCS [52] method is also similar to the median network method in that it attempts to place sequences onto the nodes of a network, infer additional nodes and label edges by the number of differences between different sequences.

Work on recombination in the context of population genetics by Jotun Hein and his co-authors goes far beyond the one approach that we describe in Section 9.8.1, see, for example, [105, 106, 167, 211, 222, 223, 224].

Dan Gusfield's initial papers on galled trees [95, 98] generated a lot of interest in this topic. Other papers in this area include [96, 97, 127, 128, 221]. The description of the algorithm in Section 9.9 is based on [95, 127, 221]. By developing and improving lower and upper bounds for the number of recombinations required by a dataset, Song, Wu and Gusfield have developed a new approach that can be used (in theory) to compute a minimal recombination network [225]. A new approach aiming a computing a putative recombination history in practice is presented in [194].

Phylogenetic networks from distances

Distance methods for phylogenetic trees are widely used because they are usually fast and often perform well in practice. Distance methods for computing phylogenetic networks are also usually fast. In this chapter, we first recall the concept of a minimum spanning network. We then discuss the two most popular distance methods, namely split decomposition and the neighbor-net method, and finally, we describe the T-Rex heuristic.

10.1 Distances and splits

As discussed in Section 3.12, we use the term distance matrix to refer to a distance function that satisfies the symmetry and the triangle inequality conditions, but not necessarily the identity condition.

Let S be a set of splits on $\mathcal{X}$. Any split $S \in \mathcal{S}$ can be used to define a distance matrix D_S on $\mathcal{X}$, by setting:

$$d_S(x, y) = \begin{cases} 1 & \text{if } S \text{ separates } x \text{ and } y, \\ 0 & \text{else.} \end{cases} \tag{10.1}$$

for all taxa x and y in $\mathcal{X}$.

We define the *unweighted split metric* D_S on $\mathcal{X}$ as:

$$d_S(x, y) = \sum_{S \in \mathcal{S}} d_S(x, y) \tag{10.2}$$

for all taxa x and y in $\mathcal{X}$. Thus $d_S(x, y)$ is the number of splits that separate x and y.

Assume we are given a weighting of the splits, $\omega : \mathcal{S} \to \mathbb{R}^{\geq 0}$, then we define the *weighted split metric* $D_{(\mathcal{S}, \omega)}$ as:

$$d_{(\mathcal{S}, \omega)}(x, y) = \sum_{S \in \mathcal{S}} \omega(S) \times d_S(x, y) \tag{10.3}$$

for all taxa x and y in $\mathcal{X}$. To reduce notation, both the unweighted and the weighted split metric are denoted by D_S. In other words, we consider unweighted splits as splits that all have weight one.

Let D be a distance matrix on $\mathcal{X}$. We say that a set of positively weighted splits S on $\mathcal{X}$ *represents* D, if $D_S = D$. Additionally, in this case, if N is a split network representing S, then we also say that N *represents* D. If a split network N represents D, then for any two taxa x and y we have that the distance $d(x, y)$ equals the sum of edge lengths of any shortest path from the node labeled x to the node labeled y in N.

Note that a distance matrix D is additive if and only if it can be represented by a set of of positively weighted splits S that is compatible, which can in turn be represented by an unrooted phylogenetic tree.

10.2 Minimum spanning networks

Let $G = (V, E, \omega)$ be a connected graph with node set V and edge set E, together with an edge weighting $\omega : E \to \mathbb{R}^{\geq 0}$ that assigns a non-negative number to all edges. In Section 9.6.1 we describe an algorithm for computing the minimum spanning network for G (see Algorithm 9.6.1).

Assume that we are given a distance matrix D on $\mathcal{X}$. For example, the distances might be the Hamming distances obtained from a multiple sequence alignment M on $\mathcal{X}$. Let $G = (V, E)$ be the complete graph on $\mathcal{X}$ whose edges are weighted by the given distances between taxa. The minimum spanning network N associated with G can be used to represent the distances between the taxa in $\mathcal{X}$. Although the method is not used in phylogenetics, it is used in population genetics, as described in the following subsection.

10.2.1 Application

In [176], the patterns of inter-population genetic diversity in a migratory bat species, *Miniopterus schreibersii natalensis*, are studied to determine how much population substructure is present. While bats and birds have traditionally shown minimal population substructure, this work shows strong population substructure and philopatry in both sexes, as indicated by concordance of nuclear and mtDNA findings. The authors argue that the genetic structure correlates with local biomes and differentiation in wing morphology. They use a minimum spanning network on 55 unique mtDNA control region sequences to show that the sequences cluster into three regional subpopulations, see Figure 10.1.

10.3 Split decomposition

Given a distance matrix D on $\mathcal{X}$, the split decomposition method produces a set of weighted splits S, as described in detail in Section 5.9. The set of splits is weakly

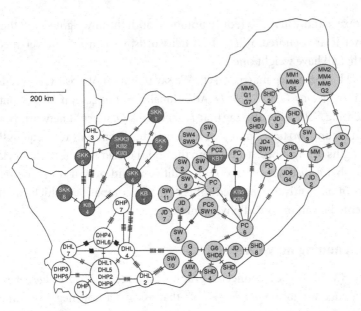

Figure 10.1 Minimum spanning network for 55 unique mtDNA control region sequence haplotypes identified in bat (*M. s. natalensis*) colonies, superimposed over a map of South Africa [176]. Haplotypes are clustered into three regional subpopulations: southern (shown in white), western (black) and northeastern (gray). Differences along an edge are indicated by cross-hatches. Reprinted by permission from Macmillan Publishers Ltd: Nature, 424:187–191, ©2003.

compatible, but not necessarily compatible, and thus gives rise to a split network that is not necessarily a phylogenetic tree. The splits approximate the distances in D from below, that is, we have $d(x, y) \geq \sum_{S \in \mathcal{S}(x,y)} \omega(S)$ for all pairs of taxa $x, y \in \mathcal{X}$, where $\mathcal{S}(x, y)$ denotes the set of all splits in $\mathcal{S}$ that separate x and y. For ease of reference we summarize the algorithm again here [9]:

Algorithm 10.3.1 (Split decomposition) *Given a distance matrix D on $\mathcal{X} = \{x_1, \ldots, x_n\}$, to compute the set of all D-splits on $\mathcal{X}$, proceed as follows:*

Initially, set $\mathcal{X}_1 = \{x_1\}$ and $\mathcal{S}_1 = \emptyset$. Now, assume we have computed the set of splits $\mathcal{S}_i$ on the first i taxa $\mathcal{X}_i = \{x_1, \ldots, x_i\}$. To obtain $\mathcal{S}_{i+1}$ on $\mathcal{X}_{i+1} = \{x_1, \ldots, x_{i+1}\}$, for each split $\frac{A}{B} \in \mathcal{S}_i$ do:

(i) *Consider $S = \frac{A \cup \{x_{i+1}\}}{B}$. If $\alpha_D(S) > 0$, set $\omega(S) = \alpha_D(S)$ and add S to $\mathcal{S}_{i+1}$.*

(ii) *Consider $S = \frac{A}{B \cup \{x_{i+1}\}}$. If $\alpha_D(S) > 0$, set $\omega(S) = \alpha_D(S)$ and add S to $\mathcal{S}_{i+1}$.*

(iii) *Consider $S = \frac{\mathcal{X}_i}{\{x_{i+1}\}}$. If $\alpha_D(S) > 0$, set $\omega(S) = \alpha_D(S)$ and add S to $\mathcal{S}_{i+1}$.*

The result is given by $\mathcal{S}_n$.

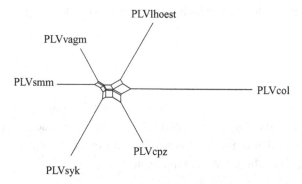

Figure 10.2 A split decomposition network computed for a dataset of primate lentiviruses, as described in the text, adapted from [209]. Names at the tips of the network represent monophyletic groups. Internal edges have a bootstrap support between 85% and 100%.

As discussed in Section 5.9, the isolation index $\alpha_D(S)$ of a split $S = \frac{A}{B}$ is defined as

$$\frac{1}{2} \min_{\substack{w,x \in A \\ y,z \in B}} (\max\{d(w, x) + d(y, z), d(w, y) + d(x, z), d(w, z)$$
$$+ d(x, y)\} - d(w, x) - d(y, z)). \qquad (10.4)$$

A split network representing this set of splits can be computed using the convex hull algorithm (see Algorithm 7.1.4). The resulting split networks are often quite close to being planar, since the set of D-splits is weakly compatible.

10.3.1 Application

Split decomposition has been used in several studies and on a wide range of different biological data.

For example, in [209] the authors use split decomposition (and other methods) to investigate the relationships among six primate lentivirus (PLV) lineages for which full genome sequences are available. For this group, six monophyletic clades have been identified, but uncertainty remains about the phylogenetic relationships among these six major PLV lineages.

Distances were obtained from sequence alignments by the "GTR+Γ+I" model of nucleotide substitution using only first and second codon positions. Taxa belonging to a monophyletic clade were grouped and analyzed as a single taxonomic unit and the average distance from each group to any other was computed using the group option in SplitsTree, see Section 14.1. One thousand bootstrap replicates were generated to assess the reliability of each edge in the network. The resulting network is shown in Figure 10.2. This network confirms that the dataset contains conflicting phylogenetic signals and is consistent with the hypothesis of early recombination events among the major lineages, which is supported in the paper by several other results, as well.

10.4 Neighbor-net

The neighbor-net method takes as input a distance matrix D on $\mathcal{X}$ and produces as output a collection of weighted splits $\mathcal{S}$ on $\mathcal{X}$ [32]. As we shall see, the produced set of splits is circular, that is, there exists a circular ordering $(x_1, x_2, \ldots, x_n)$ of the taxa $\mathcal{X}$ such that every split $S \in \mathcal{S}$ is of the form

$$S = \frac{\{x_p, x_{p+1}, \ldots, x_q\}}{\mathcal{X} - \{x_p, \ldots, x_q\}} \tag{10.5}$$

for some pair of indices p and q with $1 < p \leq q \leq n$. The resulting set of splits $\mathcal{S}$ is then given to the circular network algorithm described in Section 7.2 to compute an actual split network N.

As we saw in Section 5.7, circularity implies that $\mathcal{S}$ can be represented by an outer-labeled planar split network, that is, a split network drawn in the plane in such a way that no two edges cross and all taxon labels occur around the outside of the network. The resulting networks are not overly complicated and that is one reason why neighbor-net is a popular method for constructing phylogenetic networks.

A distance matrix D on $\mathcal{X}$ is called *circular*, if it equals the weighted split metric of some circular set of splits $\mathcal{S}$ on $\mathcal{X}$. The relationship between the input distance matrix D and the splits produced by neighbor-net is given by the following result:

Theorem 10.4.1 (Consistency of neighbor-net) *Let D be a distance matrix on $\mathcal{X}$. The set of weighted splits $\mathcal{S}$ computed by neighbor-net represent D (exactly) if and only if D is circular.*

The proof of this result is quite technical and can be found in [33].

Because compatible splits are always circular, it follows that the neighbor-net method always computes the correct phylogenetic tree, given a distance matrix that is additive.

On real biological data, the obtained distance matrices usually do not fulfill the four-point condition and are also not circular. Nevertheless, just as the neighbor-joining method is used to compute trees from distance matrices even when the four-point condition is not satisfied, it is also common practice to apply the neighbor-net algorithm even when circularity is not given. The hope is that deviations from the required conditions do not distort the result too much.

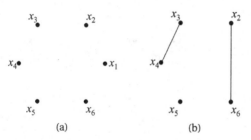

Figure 10.3 (a) Initially, the graph G considered in the neighbor-net algorithm consists of one node for each taxon, illustrated here for $\mathcal{X} = \{x_1, \ldots, x_6\}$. (b) Before and after every iteration of the main algorithm, the graph G consists only of singles (isolated nodes) and couples (pairs of connected nodes).

Given a distance matrix D on $\mathcal{X}$, the *neighbor-net method* proceeds in two steps:

(i) The main computation is to determine a circular ordering Z of $\mathcal{X}$, using an iterative algorithm similar to neighbor-joining.
(ii) Then a set of weighted splits $\mathcal{S}$ is computed that respect the circular ordering Z.

In the following, we discuss each step in detail.

10.4.1 Computation of a circular ordering

Given a distance matrix D on $\mathcal{X}$, neighbor-net uses an iterative algorithm to compute a circular ordering for $\mathcal{X}$. The algorithm maintains a graph G and an auxiliary distance matrix F on the nodes of the graph.

Initialization: The nodes of G are set to $\mathcal{X}$ (See Figure 10.3a) and the distance matrix F is set to D. No new nodes are generated during execution of the algorithm and so nodes in the graph always correspond directly to taxa in $\mathcal{X}$.

Iteration: At the beginning of the i-th iteration, the graph G consists of two types of connected components, namely isolated nodes, which we call *singles*, and pairs of connected nodes, which we call *couples* (see Figure 10.3b). The distance matrix F defines a distance between each pair of nodes x and y.

The algorithm proceeds by choosing two components that have minimum distance from each other and then merging these to produce a new couple. We discuss how the distances between components are defined further below, after we have discussed the process of merging in detail.

Note that there are three possible combinations of pairs of components, namely:

(M1) two singles,
(M2) a single and a couple, or
(M3) two couples.

We shall discuss each case in turn (see Figure 10.4).

(M1) Single–single:

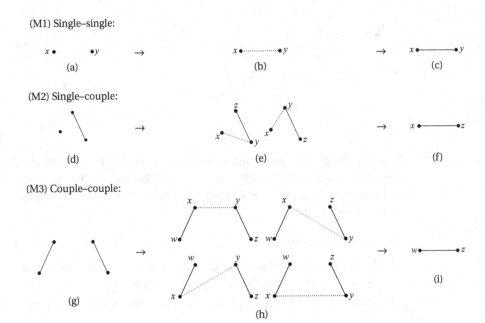

(M2) Single–couple:

(M3) Couple–couple:

Figure 10.4 The three different cases considered in the main iteration of neighbor-net: Case (M1): (a) Two singles are to be merged. (b) There is only one configuration to be considered. (c) The result is a couple. Case (M2): (d) A single and a couple are to be merged. (e) There are two different orientations of the couple to consider. (f) The result is one couple. Case (M3): (g) Two couples are to be merged. (h) There are two different orientations for each component to be considered, giving rise to four possible combinations. (i) Again, the result is one couple. Dotted lines indicate which pairs of taxa are recorded as neighbors.

In case (M1), to merge two singles, x and y, the algorithm simply defines a new edge between x and y. Moreover, the algorithm records that x and y are *neighbors*.

In case (M2), the two components are a single, x, and a couple, containing two nodes y and z. The algorithm deletes the node in the couple that is closest to x, based on a distance calculation that we describe further below. Assume, without loss of generality, that the deleted node is y, then the algorithm constructs a new couple by placing an edge between the two nodes x and z. The algorithm records that the two taxa x and y are *neighbors*.

In case (M3), the two components are both couples, one containing two nodes w and x, and the other containing two other nodes y and z. The algorithm deletes one node from each component, namely the two that are closest to each other, as discussed below, and then constructs a new couple from the remaining two nodes. Assuming, without loss of generality that the two nodes that are deleted are x and y, then the algorithm records x and y as *neighbors*.

The algorithm repeatedly merges pairs of components in this way until all possible components have been merged into a single component, which is a couple. Let z_1 and z_2 denote the two nodes contained in the remaining couple.

Note that each time we merge two components, two taxa are recorded as neighbors. More precisely, each taxon (except z_1 and z_2) is reported as the neighbor of some other node exactly twice. It is first reported as a neighbor when it first enters into a couple, either as the node x or y in case (M1) or as x in case (M2). It is again reported as a neighbor when it is deleted from the graph, either as node y in case (M2) or as node x or y in case (M3). The two remaining taxa z_1 and z_2 each have only one neighbor. Hence, the reported neighbor relations form a chain from z_1 to z_2 containing all taxa in $\mathcal{X}$. This defines a linear ordering Z of $\mathcal{X}$.

Computing the distances

To complete the description of the first part of the algorithm, we need to describe how to define the distances between components of the graph G, how to determine which pair of nodes from two different clusters are closest to each other and how to update the distance matrix F whenever the graph G has changed.

Assume that the graph G has connected components $C_1, C_2, \ldots, C_m$ and consider two such components C_i and C_j. The distance $d(C_i, C_j)$ between C_i and C_j is defined as the average distance between the nodes contained in the two components, given by

$$d(C_i, C_j) = \frac{1}{|C_i||C_j|} \sum_{x \in C_i} \sum_{y \in C_j} F(x, y). \tag{10.6}$$

However, to decide which two components to merge, as in the neighbor-joining method, the algorithm does not pick a pair of components that minimize this distance, but rather it chooses a pair of components that minimize the adjusted distance

$$d'(C_i, C_j) = d(C_i, C_j) - (r_i + r_j), \tag{10.7}$$

with

$$r_i = \frac{1}{m-2} \sum_{k=1, k \neq i}^{m} d(C_i, C_k). \tag{10.8}$$

Suppose that components C_i and C_j minimize the distance d'. In the cases (M2) and (M3) we need to determine which pair of nodes from C_i and C_j are closest to each other. This is done using the same distance calculation as for components, but treating all nodes contained in C_i and C_j as singles. More precisely, we choose a pair of nodes $x_i \in C_i$ and $x_j \in C_j$ that minimize:

$$d'(x_i, x_j) = F(x_i, x_j) - (\hat{r}_i + \hat{r}_j), \tag{10.9}$$

with

$$\hat{r}_i = \frac{1}{\hat{m} - 2} \sum_{k=1, k\neq i}^{\hat{m}} d(x_i, C_k), \tag{10.10}$$

where $\hat{m} = m + |C_i| + |C_j| - 2$ is the number of clusters present when all nodes in C_i and C_j are treated as singles.

After applying one of the three merging steps, the distance matrix F must be updated to reflect the fact that the graph G has changed. In case (M1), an edge is added to the graph, but no nodes are deleted and so in this case the matrix F remains unaltered.

In case (M2), a configuration of three nodes x, y and z is replaced by a couple consisting of x and z. In this case, we compute new distances F' between x or z and every node w still in the graph G as follows:

$$F'(x, w) = \tfrac{2}{3} F(x, w) + \tfrac{1}{3} F(y, w), \tag{10.11}$$

$$F'(z, w) = \tfrac{2}{3} F(z, w) + \tfrac{1}{3} F(y, w) \quad \text{and} \tag{10.12}$$

$$F'(x, z) = \tfrac{1}{3}(F(x, y) + F(x, z) + F(y, z)). \tag{10.13}$$

In case (M3), a configuration of four nodes w, x, y and z is replaced by a couple consisting of w and z. To update the distances in this case, we first apply the update just described to nodes x, y and z and remove y. We then apply the same calculation to nodes w, x and z, before removing x.

10.4.2 Computing the weighted splits

Let D be a distance matrix on $\mathcal{X}$. Above we discussed how the neighbor-net algorithm computes a linear ordering Z of $\mathcal{X}$ from D. In this section, we describe how Z is used to produce a set of weighted splits $\mathcal{S}$.

To begin, construct the set of all splits that respect the linear ordering $Z = (x_1, x_2, \ldots, x_n)$ as

$$\mathcal{S} = \left\{ \frac{\{x_p, x_{p+1}, \ldots, x_q\}}{\mathcal{X} - \{x_p, \ldots, x_q\}} \mid 1 < p \leq q \leq n \right\}. \tag{10.14}$$

The set of splits $\mathcal{S}$ defined in this way is circular and Z is a circular ordering for $\mathcal{S}$ (see Definition 5.7.1).

The weights of these splits are chosen using the *least squares method*. To this end, we represent the distance matrix D as an $\frac{n(n-1)}{2}$-dimensional vector

$$\mathbf{d} = \begin{pmatrix} d_{12} \\ d_{13} \\ \dots \\ d_{(n-1)n} \end{pmatrix} \tag{10.15}$$

of pairwise distances between distinct taxa, where $d_{ij} = d(x_i, x_j)$ is the distance between taxa x_i and x_j.

The set of splits $\mathcal{S} = \{S_1, \dots, S_k\}$ is represented by an $(m \times k)$ matrix A (with $m = \frac{n(n-1)}{2}$) in which each row represents a pair of taxa x_i, x_j and each column represents a split S_t. We set

$$A_{(ij)t} = \begin{cases} 1 & \text{if } x_i \text{ and } x_j \text{ are separated by } S_t, \\ 0 & \text{else.} \end{cases} \tag{10.16}$$

Let $\mathbf{b}$ be the k-dimensional vector representing the weights of the splits. Then the ordinary least square estimate of $\mathbf{b}$ is given by

$$\mathbf{b} = (A^T A)^{-1} A^T \mathbf{d}. \tag{10.17}$$

The ordinary least squares calculation is not ideally suited to the task of estimating the weights of splits because some values of the solution may be negative.

Simply removing all splits that are assigned negative weights does not solve this problem because the weights of the remaining splits may be severely overestimated, leading to a phylogenetic network N that is both overly complicated and also a bad fit to the distance matrix D.

The answer is to impose a non-negativity constraint [32]. There is no closed formula for constrained least square estimates and enforcing the constraint increases the computation time considerably. However, the resulting split network is usually much cleaner and provides a more accurate representation of D.

The output of this part of the neighbor-net method is the set of all weighted splits in $\mathcal{S}$ that have a positive weight.

10.4.3 Constructing an outer-labeled planar split network

As we have seen, given a distance matrix D on $\mathcal{X}$, the output of the neighbor-net method is a set of weighted splits $\mathcal{S}$ and a circular ordering Z of the taxa for $\mathcal{S}$. The set of splits $\mathcal{S}$ and the circular ordering Z are passed to the circular network algorithm (see Section 7.2) to compute an actual network N that is guaranteed to be outer-labeled planar. The neighbor-net method is implemented in SplitsTree (see Section 14.1).

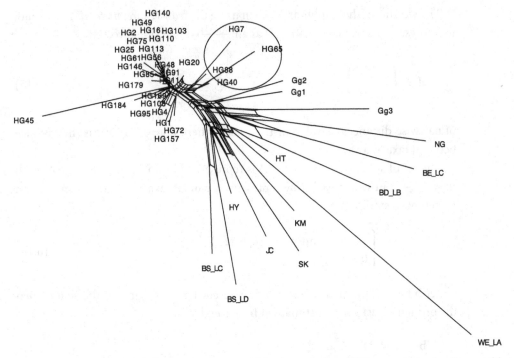

Figure 10.5   A split network for 44 Vietnamese chicken populations [19] computed using the neighbor-net method. Four of the domestic chicken populations (encircled) appear isolated from the other ones and more closely related to wild populations. Reprinted from BMC Genetics 10:1 (2009) under the Creative Commons Attribution License.

10.4.4 Application

In [19] the relationship between wild and domestic chicken is examined. The authors assume that hybridization between wild and domestic chickens may have occurred in Ha Giang, Vietnam, leading to a high genetic diversity. In a split network on 44 Vietnamese chicken populations, computed using Latter's genetic distances and the neighbor-net method, four of the domestic chicken populations (encircled in Figure 10.5) appear isolated from the other domestic populations and more closely related to the wild populations, supporting an hypothesis of admixture between the two.

10.4.5 Discussion

Neighbor-net is an attractive method for computing split networks for the following reasons: First, the resulting networks are outer-labeled planar and thus easy to draw and to read. Secondly, the algorithm is quite fast. Thirdly, it produces resolved networks even for quite large numbers of taxa, unlike the split decomposition method, which rapidly loses resolution as the number of taxa increases.

10.5 T-Rex

Most algorithms presented in this book produce either an unrooted phylogenetic network that is a split network or a rooted phylogenetic network that is a rooted DAG. One of the few exceptions is the *T-Rex algorithm* [162, 172]. This method takes as input a distance matrix D on X and produces as output an unrooted phylogenetic network N that consists of an unrooted phylogenetic tree on X, augmented by a set of *short-cut edges*, which are auxiliary edges that are added to produce a better fit between distances in N and the input distances D. More precisely, the output is a network that fulfills the following definition:

Definition 10.5.1 (Reticulogram) *A reticulogram $R = (V, E, F, \omega)$ is an unrooted phylogenetic network on X that consists of an unrooted phylogenetic tree $T = (V, E)$, a set of short-cut edges F and a map $\omega : E \cup F \rightarrow \mathbb{R}^{\geq 0}$ that assigns a length to every edge in R.*

Let R be a reticulogram on X. The *reticulogram distance* matrix D_R is defined for any two taxa x and y as $d_R(x, y)$, the minimum sum of edge lengths on any path from the node $\lambda(x)$ to the node $\lambda(y)$ in R.

The basic idea of the T-Rex algorithm is to first compute a phylogenetic tree from the given input distance matrix D, using a method such as neighbor-joining, and then to greedily add short-cut edges to the tree so as to improve the least squares fit between the reticulogram distances and the input distances.

Algorithm 10.5.2 (T-Rex algorithm)   *Let D be a distance matrix on X. To construct a reticulogram R on X, proceed as follows. First, compute a phylogenetic tree T for D using the neighbor-joining algorithm and set $R = T$. Second, while the goodness-of-fit between the reticulogram distances D_R and the input distances D can be improved, do the following:*

(i) *For each pair of nodes v and w that are not connected by an edge: Determine the optimal length $\omega(e)$ for the candidate short-cut edge $e = \{v, w\}$ and then compute the improvement $\Delta(e)$ of the least-square fit that would be gained by adding e to R.*

(ii) *Add the candidate edge e that maximizes $\Delta(e)$ to R.*

To fully specify this algorithm, we must answer two questions: (1) How to calculate the optimal length $\omega(e)$ for a prospective short-cut edge e? (2) How to determine the goodness-of-fit between D_R and D?

To answer the first question, let v and w be a pair nodes that are not yet connected by an edge $e = \{v, w\}$ in R. The first step is to determine all pairs of taxa s and t for which the distance $d_R(s, t)$ might possibly change if we were to add the edge e to R. Clearly, the distance between a pair of taxa s and t can only change if

$d_R(s, t) > \omega(e)$. To find all such pairs, for the moment, let us assume that the length of e is $\omega(e) = 0$, as this is the most inclusive choice.

Upon adding e to R, for any pair of taxa s and t the new distance from $\lambda(s)$ to $\lambda(t)$ would be given by:

$$\min\left\{d_R(s, t), d_R(s, v) + d_R(w, t), d_R(s, w) + d_R(v, t)\right\}. \tag{10.18}$$

Hence, the distance can only change for those pairs of taxa s and t for which the following inequality holds:

$$\min\left\{d_R(s, v) + d_R(w, t), d_R(s, w) + d_R(v, t)\right\} < d_R(s, t). \tag{10.19}$$

We use $A(v, w)$ to denote the set of all pairs of taxa (s, t) that satisfy this condition. Our next goal is to partition $A(v, w)$ into a collection of sets $A_1(v, w)$, $A_2(v, w), \ldots$ containing all pairs of taxa whose distances may be affected by adding e to R, for different possible lengths of e.

To this end, we first consider all pairs of taxa $(s, t) \in A(v, w)$ that minimize the following difference:

$$d_R(s, t) - \min\left\{d_R(s, v) + d_R(w, t), d_R(s, w) + d_R(v, t)\right\}. \tag{10.20}$$

We use $\ell_1(v, w)$ to denote the minimum and we define $A_1(v, w)$ as the set of all pairs of taxa in $A(v, w)$ for which the minimum is attained. Note that the reticulogram distance between pairs in this group can only change if the length of e is less than $\ell_1(v, w)$.

Now, assume that we have already defined the sets $A_1(v, w), \ldots, A_j(v, w)$ and that there still is at least one pair of taxa in $A(v, w)$ that is not contained in one of these sets. Then we define $A_{j+1}(v, w)$ as the set of all pairs of taxa in $A(v, w) - \bigcup_{i=1}^{j} A_i(v, w)$ that minimize the expression given in Equation (10.20), with $\ell_{j+1}(v, w)$ defined as the attained minimum.

At the end of this process, $A(v, w)$ is partitioned into sets $A_1(v, w)$, $A_2(v, w), \ldots, A_m(v, w)$, for some number $m \geq 1$, and each such set has an associated value $\ell_1(v, w) < \ell_2(v, w) \cdots < \ell_m(v, w)$. To determine the optimal length $\omega(e)$, we consider each interval of values $I_j = [\ell_j(v, w), \ell_{j+1}(v, w)]$ (with $0 \leq j < m$) in turn and determine the best choice for each interval.

What is the best choice of $\omega(e)$ restricted to such an interval I_j? If the length of the new edge e is between $\ell_j(v, w)$ and $\ell_{j+1}(v, w)$, then only distances between pairs of taxa (s, t) contained in $\bigcup_{i=j+1}^{m} A_i(v, w)$ can change. As the aim of this algorithm is to minimize the least-squares fit between the input distances D and the reticulate distances D_R, we choose $\omega_j(e)$ such that it minimizes

$$\sum_{i=j+1}^{m} \sum_{(s, t) \in A_i(v, w)} (\ell_i(v, w) + \omega_j(e) - D(s, t))^2. \tag{10.21}$$

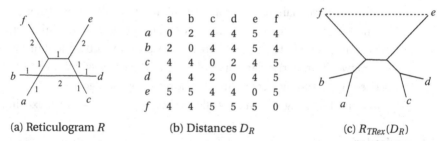

(a) Reticulogram R   (b) Distances D_R (c) $R_{TRex}(D_R)$

Figure 10.6 (a) A reticulogram R on $\mathcal{X} = \{a, \ldots, e\}$ with edges labeled by their lengths. (b) The distance matrix D_R on $\mathcal{X}$ that is defined by R. (c) The reticulogram $R_{TRex}(D_R)$ obtained by applying the T-Rex algorithm to D_R. Solid lines represent the initial unrooted phylogenetic tree and the dashed line indicates the added short-cut edge.

A nontrivial solution is given by

$$\omega_j(e) = \frac{\sum_{i=j+1}^{m} \sum_{(s,t) \in A_i(v,w)} (D(s,t) - \ell_i(v,w))}{\sum_{i=j+1}^{m} |A_i(v,w)|}, \tag{10.22}$$

which we must then also compare with the two boundary values $\ell_j(v, w)$ and $\ell_{j+1}(v, w)$.

Once we have obtained an optimal edge length for every interval I_j, for each such choice $\omega_j(e)$ we compute the total least-square fit between the input distances D and the reticulogram distances D_R, the latter computed in the presence of the edge e with length $\omega_j(e)$. Then $\omega(e)$ is set to the value $\omega_j(e)$ that optimizes the least-squares fit over all $j = 1, \ldots, m - 1$.

How to measure the goodness-of-fit between D and D_R to help decide whether more short-cut edges should be added to the network? This is done using the quadratic difference $Q(D, D_R) = \sum_{a,b \in \mathcal{X}} (d(a, b) - d_R(a, b))^2$ between the two distance matrices, set into relation with the number n of taxa and the number k of edges in the network R, in some way. One possibility is to define the *fit* between D and D_R as

$$fit(D, D_R) = \frac{Q(D, D_R)}{\frac{(2n-2)(2n-3)}{2} - 2k}. \tag{10.23}$$

The aim of the T-Rex approach is to compute a reticulogram R on $\mathcal{X}$ whose distances D_R provide a good fit to a given input distance matrix D on $\mathcal{X}$. Unfortunately, it is easy to construct a reticulogram R on $\mathcal{X}$ such that the T-Rex algorithm will fail to compute the "correct" reticulogram from the distance matrix D_R. In Figure 10.6, we display a reticulogram R on $\mathcal{X} = \{a, \ldots, e\}$ and its distance matrix D_R. We also show the reticulogram $R_{TRex}(D_R)$ computed by the T-Rex algorithm (using the web-server http://www.trex.uqam.ca), which provides only an imperfect fit to the distance matrix D_R.

There are two main reasons for such failures. First, the initial distances computed by the tree reconstruction method (for example neighbor-joining) are aimed at providing a good fit to a tree and are not designed for subsequent fitting of the data to a network. In particular, some of the tree distances will under-fit the distances between taxa, and such errors are not addressed by the algorithm. A second problem is that the T-Rex algorithm only allows short-cut edges between existing nodes.

10.5.1 Application

An application of the T-Rex approach can be found in [85], where the phylogenetic relationships inside the *Fagus sylvatica* species complex in Europe and western Asia are studied. For each of 279 local subpopulations (represented by a sample of 50–280 individuals), data from 12 allozyme loci were collected. Employing a distance matrix inferred from the sequences using Nei's distance transform [187], the paper presents a reticulogram calculated by T-Rex and also a neighbor-net network. The two networks are cited as evidence that a substantial taxonomic revision is necessary for *F. sylvatica*. Moreover, the two networks are said to provide a better insight into the phylogenetic relationships among the components of this group. For example, both networks suggest a strong genetic similarity of the Calabria and Balkan populations, confirming prior studies and readily explained by the existence of a land-bridge over the Otranto Strait, which disappeared at the beginning of the Pliocene. However, in some cases the reticulogram displays dubious edges connecting geographically distant populations for which a contact in the past was hardly possible and which are separated by other regional populations.

Chapter notes

The split decomposition method [9] was the first method for computing unrooted phylogenetic networks from distances and has been used in many studies, particularly for data that are not expected to be tree-like, such as in the evolution of viruses [114, 147], or, to cite a non-biological example, in the "evolution" of manuscripts that were copied by hand [14, 226]. The neighbor-net method [31] is more widely used than the split decomposition method, due to the fact that it shows much better resolution on larger and more conflicting datasets. A recent paper discusses the use of neighbor-net as a tool for metagenome comparison [177]. For a discussion of strengths and weaknesses of the methods presented in this chapter, see [180].

11

Phylogenetic networks from trees

In this chapter, we look at a number of methods that take as input one or more phylogenetic trees and compute as output a phylogenetic network. We first discuss methods for computing consensus trees and *consensus split networks* from unrooted phylogenetic trees, and then briefly touch on doing the same for rooted phylogenetic trees. We then present an algorithm for computing a *minimum hybridization network* for two trees. The two last topics in this section both address the issue of gene duplications. First we discuss how to compute a phylogenetic network for a given *multi-labeled* phylogenetic tree. Then we discuss an approach used for computing an optimal *duplication-loss-transfer scenario* that reconciles a gene tree with a species tree.

11.1 Consensus split networks

Assume that we are given a collection $T = (T_1, \ldots, T_k)$ of unrooted phylogenetic trees on $\mathcal{X}$. They might be different phylogenetic trees inferred for different genes, a collection of trees obtained for a single gene using a number of different tree inference methods or a list of trees produced via bootstrapping or a sampling approach, for example.

Such a collection of unrooted phylogenetic trees T is often represented by a single consensus tree, as discussed in Section 3.17. Recall that two standard approaches to computing a consensus tree both start by determining the splits associated with each tree in T, and then either consider the set $\mathcal{S}_{strict}(T)$ of all splits that occur in every tree, or consider the set $\mathcal{S}_{majority}(T)$ of all splits that occur in more than half of the trees. The unrooted phylogenetic tree T_{strict} associated with the former set of splits $\mathcal{S}_{strict}(T)$ is known as the strict consensus tree, whereas the latter set of splits $\mathcal{S}_{majority}(T)$ gives rise to the majority consensus tree $T_{majority}$.

Now, more generally, let us consider the set $\mathcal{S}_p(T)$ of all splits that are present in more than a proportion p of all unrooted phylogenetic trees $T = (T_1, \ldots, T_k)$

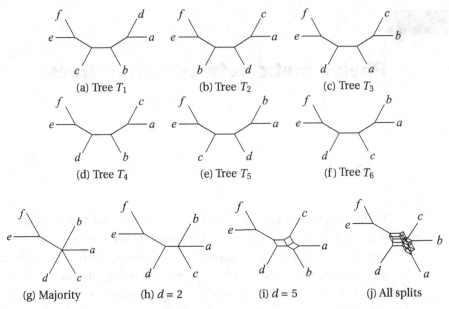

Figure 11.1 (a)–(f) Six different phylogenetic trees $T_1, \ldots, T_6$ on $\mathcal{X} = \{a, b, c, d, e, f\}$. (g) Their majority consensus tree and (h) their consensus split network for $d = 2$, representing all splits that are present in more than $\frac{1}{3}$ of the trees. Note that in this case the network is still a tree, but more resolved than the majority consensus tree. (i) The consensus split network for $d = 5$ and (j) the split network representing all splits present in the six trees.

on $\mathcal{X}$, for some fixed value of p, with $0 \le p < 1$, defined as

$$\mathcal{S}_p(T) = \left\{ S \in \mathcal{S}(T) : \frac{|\{T \in \mathcal{T} : S \in \mathcal{S}(T)\}|}{|\mathcal{T}|} > p \right\}. \tag{11.1}$$

Then, clearly $\mathcal{S}_{\frac{1}{2}}(T) = \mathcal{S}_{majority}(T)$ and thus $\mathcal{S}_{\frac{1}{2}}(T)$ is compatible. Moreover, for $p > \frac{1}{2}$ we have $\mathcal{S}_p(T) \subseteq \mathcal{S}(T)_{\frac{1}{2}}$, and so $\mathcal{S}_p(T)$ is compatible for any choice of p above $\frac{1}{2}$.

Consider $\mathcal{S}_p(T)$ for a value of p that is less than $\frac{1}{2}$. In this case, the splits in $\mathcal{S}_p(T)$ need not be compatible and so the graphical representation of $\mathcal{S}_p(T)$ might not be a phylogenetic tree T but rather a more general split network N, which we call a *consensus split network* in this context [113]:

Definition 11.1.1 (Consensus split network) *Assume that we are given a set $\mathcal{T}$ of phylogenetic trees on $\mathcal{X}$ and an integer $d > 1$. We define the (d-dimensional) consensus split network N for $\mathcal{T}$ as the split network that represents $\mathcal{S}_{(\frac{1}{d+1})}(\mathcal{T})$.*

First, note that the proportion used in the definition is $p = \frac{1}{d+1}$. Moreover, since we assume $d > 1$, we have $0 < \frac{1}{d+1} \le \frac{1}{2}$ and so the proportion p lies between 0 and at most $\frac{1}{2}$. What is the interpretation of the parameter d? It prescribes the *maximal*

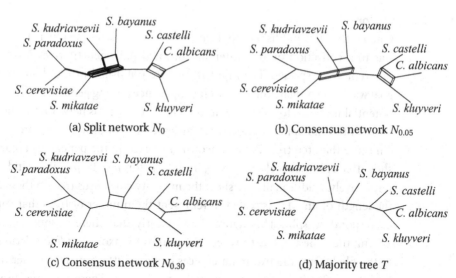

(a) Split network N_0

(b) Consensus network $N_{0.05}$

(c) Consensus network $N_{0.30}$

(d) Majority tree T

Figure 11.2 For a set T of 106 phylogenetic trees on eight yeast species reported in [206], we show three split networks (a) N_0, (b) $N_{0.05}$ and (c) $N_{0.30}$ that represent all splits that occur in at least one input tree, in more than 5% of all trees, and in more than 30% of all trees, respectively. In (d) we show the majority consensus tree for T.

dimensionality of the consensus split network N: For $d = 1$, we have $p = \frac{1}{2}$ and the consensus split network is one-dimensional, namely a phylogenetic tree (the majority consensus tree). For $d = 2$, the consensus split network may be "two-dimensional" and contain "parallelograms", that is, undirected cycles of length 4. For $d = 3$, the consensus split network may contain (the edge skeletons of) cubes of dimension ≤ 3, and so on. In Figure 11.1 we see that the maximal dimensionality is not always attained: for example, for $d = 2$ the displayed consensus split network is a tree, whereas for $d = 5$, the network is only "two-dimensional".

A consensus tree has a nice interpretation as representing those parts of a evolutionary history on which a set of phylogenetic trees agree on. Parts for which the trees provide incompatible accounts are suppressed and are represented by unresolved nodes in the consensus. A consensus split network can be used to represent those incompatible alternative scenarios that are supported by a good proportion of the input trees.

However, in many cases the well-supported alternatives are often caused by simple local rearrangements around individual nodes in the trees. Other, more interesting incompatibilities, for instance, caused by a major horizontal gene transfer event, are often missing from the consensus split network, because such events usually involve only one or a few genes.

11.1.1 Application

One of the first published applications of the consensus split network method was to a collection of 106 different unrooted phylogenetic trees involving eight different yeast species. The input trees were published in [206] where the main goal was to investigate the issue of incongruence among gene trees and to consider potential methods for resolving it. In [112], a consensus network for the 106 trees representing all splits that occur in at least 21% of all trees was presented (not "in more than ten trees", as erroneously reported in the paper). In Figure 11.2 we show tree split networks N_0, $N_{0.05}$ and $N_{0.30}$ for $p = 0$, $p = 0.05$ and $p = 0.30$, respectively. Additionally, we show the majority consensus tree. In these drawings, the length of an edge is scaled to represent the number of trees that support the corresponding split. This figure shows clearly that there is some disagreement among the gene trees as to where the outgroup taxon *Candida albicans* attaches to the phylogeny, the two main choices being an attachment to either the lineage leading to *Saccharomyces castelli* or to *S. kluyveri*. There is some disagreement concerning how to arrange the other five taxa, the largest contention being whether *S. kudriavzevii* and *S. bayanus* are sister taxa.

In the depicted consensus tree, the majority of gene trees place *C. albicans* as a sister of *S. kluyveri* and the information that a significant number of genes prefer an alternative placement is suppressed.

11.2 Consensus super split networks for unrooted trees

The above approach to calculating consensus trees and consensus split networks requires that all input trees are on exactly the same set of taxa $\mathcal{X}$. However, in practice, it often occurs that some (or even all) of the input trees lack some of the members of the total taxon set $\mathcal{X}$. For example, when studying multiple gene trees, particular gene sequences may be absent for certain taxa. In this section, we discuss how to address this problem.

Let $\mathcal{X}$ be a set of taxa. A phylogenetic tree T on $\mathcal{X}'$ is called a *partial tree* on $\mathcal{X}$, if $\mathcal{X}'$ is a subset of $\mathcal{X}$. Similarly, any split on $\mathcal{X}' \subset \mathcal{X}$ is called a *partial split* on $\mathcal{X}$. A split that contains all members of $\mathcal{X}$ is sometimes called a *complete* split on $\mathcal{X}$, for emphasis.

A key question is how to extend partial splits so that they become complete splits. As we shall see, one answer is given by the *Z-closure* method, which takes a set of partial splits on $\mathcal{X}$ as input and infers a set of complete splits on $\mathcal{X}$ as output. Such a set of complete splits can then be used to compute consensus trees and consensus split networks [126].

Consider a set $\mathcal{S}$ of partial splits on $\mathcal{X}$. Two partial splits $S_1 = \frac{A_1}{B_1} \in \mathcal{S}$ and $S_2 = \frac{A_2}{B_2} \in \mathcal{S}$ are said to be in *Z-relation* to each other, if exactly one of the four intersections $A_1 \cap A_2$, $A_1 \cap B_2$, $B_1 \cap A_2$ or $B_1 \cap B_2$ is empty. In this case, if the empty intersection is $A_1 \cap B_2$, say, then we write $\frac{A_1}{B_1} \mathbf{Z} \frac{A_2}{B_2}$, where the three lines in the "Z" connect those pairs of split parts that have nonempty intersections. The definition of the Z-relation is very similar to that of the compatibility of two splits, however it involves partial splits rather than complete splits.

Assume that we have a collection of partial splits $\mathcal{S}$ on $\mathcal{X}$. Consider two partial splits $S_1 = \frac{A_1}{B_1} \in \mathcal{S}$ and $S_2 = \frac{A_2}{B_2} \in \mathcal{S}$ for which the Z-relation $\frac{A_1}{B_1} \mathbf{Z} \frac{A_2}{B_2}$ holds. The Z-operation applied to S_1 and S_2 is defined as the creation of two new splits

$$S_1' = \frac{A_1}{B_1 \cup B_2} \quad \text{and} \quad S_2' = \frac{A_1 \cup A_2}{B_2}. \tag{11.2}$$

It can happen that the two new splits equal the original ones, in which case we say that the pair of splits $\{S_1, S_2\}$ is *unproductive*. (Note that a pair of full splits on $\mathcal{X}$ is always unproductive.) Otherwise, at least one of the two new splits contains more taxa than its predecessor, and in this case the pair of splits is called *productive*. Based on the Z-operation, the following algorithm computes a set of complete splits $\bar{\mathcal{S}}$ called a *Z-closure*:

Algorithm 11.2.1 (Z-closure) *Let $\mathcal{S} = \{S_1, \ldots, S_m\}$ be a set partial splits on $\mathcal{X}$. We obtain a set $\bar{\mathcal{S}}$ of complete splits from $\mathcal{S}$ as follows: While $\mathcal{S}$ contains a productive pair of splits $\{S_i, S_j\}$, apply the Z-operation to obtain two new splits $\{S_i', S_j'\}$ and then replace the former pair by the latter pair in $\mathcal{S}$. Finally, define $\bar{\mathcal{S}}$ as the union of the set of all complete splits in $\mathcal{S}$ and the set of all trivial splits on $\mathcal{X}$.*

This algorithm operates "in place", maintaining the splits in an array of constant size, in which pairs of splits are replaced by pairs of derived ones. The algorithm runs in polynomial time, because at least one of the splits derived from a productive pair will contain more taxa than the split that it replaces in $\mathcal{S}$. However, the result of this algorithm depends on the order in which splits are processed. In our experience, the order-dependency is usually not very severe, with different runs resulting in a difference of very few splits. In practice, the algorithm is run ten times and then the union of all results is taken to be an approximation of the "full" Z-closure. (The latter is defined as the full closure of the Z-closure operation applied to $\mathcal{S}$, obtained by repeatedly applying the Z-closure to all pairs of sequences generated so far, until no new sequences arise.) The complexity of computing the full Z-closure is currently unknown. We illustrate an application of the Z-closure method in Figure 11.3.

Suppose we are given a collection of partial phylogenetic trees $\mathcal{T} = (T_1, \ldots, T_k)$ on $\mathcal{X}$. Let $\mathcal{S}$ be the set of all partial splits represented by the trees in $\mathcal{T}$ and let $\bar{\mathcal{S}}$ be the Z-closure of $\mathcal{S}$ obtained using the above algorithm. How to define the set of all

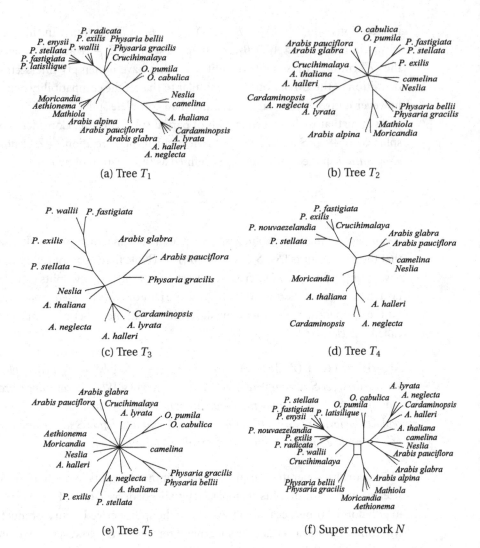

Figure 11.3 (a–e) Five partial gene trees $T_1, \ldots, T_5$ on 13–25 plant taxa. (f) The corresponding super split network N on all 26 taxa, computed using the Z-closure method. The edges in N are scaled to represent the number of input trees that contain the edge. The network N shows that the placement of the pair of taxa *Physaria bellii* and *Physaria gracilis* differs in the five trees.

splits in $\bar{S}$ that occur in a proportion of more than p of the input trees, for a fixed number p?

To answer this, we must define what we mean by the statement that a complete split "occurs" in a given partial tree. For any complete split $S = \frac{A}{B}$ on $\mathcal{X}$ and any partial tree T on $\mathcal{X}$, we say that T is *relevant* for S, if T contains taxa from both parts of S, that is, if we have $\mathcal{X}(T) \cap A \neq \emptyset$ and $\mathcal{X}(T) \cap B \neq \emptyset$, where $\mathcal{X}(T) \subseteq \mathcal{X}$ denotes the set of all taxa that appear as labels in T. If T is relevant for S, then we say that S *occurs* in T, if the *restriction* of S to the taxon set $\mathcal{X}(T)$, in symbols $S|_{\mathcal{X}(T)}$ or $S|_T$, is a split of T, that is, is contained in the split encoding $\mathcal{S}(T)$ of T.

With these concepts in place, for any proportion p, with $0 < p < 1$, we can compute

$$\bar{\mathcal{S}}_p(T) = \left\{ S \in \bar{\mathcal{S}} : \frac{|\{T \in \mathcal{T}(S) : S|_{\mathcal{X}(T)} \in \mathcal{S}(T)\}|}{|\mathcal{T}(S)|} > p \right\}, \tag{11.3}$$

where $\mathcal{T}(S)$ denotes the set of all trees in $\mathcal{T}$ that are relevant for the split S. Based on this, we define:

Definition 11.2.2 (Consensus supertrees and networks) *Let $\mathcal{T}$ be a set of partial trees on $\mathcal{X}$. The set of* strict consensus splits *for $\mathcal{T}$ is given by*

$$\bar{\mathcal{S}}_{strict}(\mathcal{T}) = \{S \in \bar{\mathcal{S}} \mid T \in \mathcal{T}(S) \Rightarrow S \in \mathcal{S}(T)\}, \tag{11.4}$$

that is, the set of all splits that occur in all trees that are relevant for them. The set of majority consensus splits *for $\mathcal{T}$ is given by*

$$\bar{\mathcal{S}}_{majority}(\mathcal{T}) = \bar{\mathcal{S}}_{\frac{1}{2}}(\mathcal{T}). \tag{11.5}$$

The split network used to represent $\bar{\mathcal{S}}_p(\mathcal{T})$ for any value of p from 0 to 1 is called a super split network.

We conjecture that the set of strict consensus splits $\bar{\mathcal{S}}_{strict}(\mathcal{T})$ is always compatible and thus representable by a phylogenetic tree. If this is true, then this construction provides a *supertree* method. However, the set of majority consensus splits $\bar{\mathcal{S}}_{majority}(\mathcal{T})$ is not necessarily compatible, in general. In fact, the latter statement holds for $\bar{\mathcal{S}}_p(\mathcal{T})$ with any choice of p between 0 and 1.

Exercise 11.2.3 (Majority consensus incompatible) *Construct an example that shows that the majority consensus splits for a set of partial trees is not necessarily compatible.*

11.2.1 Inferring split weights

If the partial trees $T_1, \ldots, T_k$ on $\mathcal{X}$ under consideration have weighted edges, then the question arises how to use the given edge weights to assign weights to the splits in the Z-closure of the splits from the trees.

It can be shown that for each split S in the Z-closure $\bar{\mathcal{S}}$ there exists some relevant input tree $T \in \mathcal{T}$ in which S occurs [126]. Based on this result, there are several possible ways to assign a weight $\omega(S)$ to a split $S \in \bar{\mathcal{S}}$. One possibility would be to use the minimum weight over all partial splits of the form $S|_T$, where T is any input tree in which S occurs. A second possibility would be to use the average of

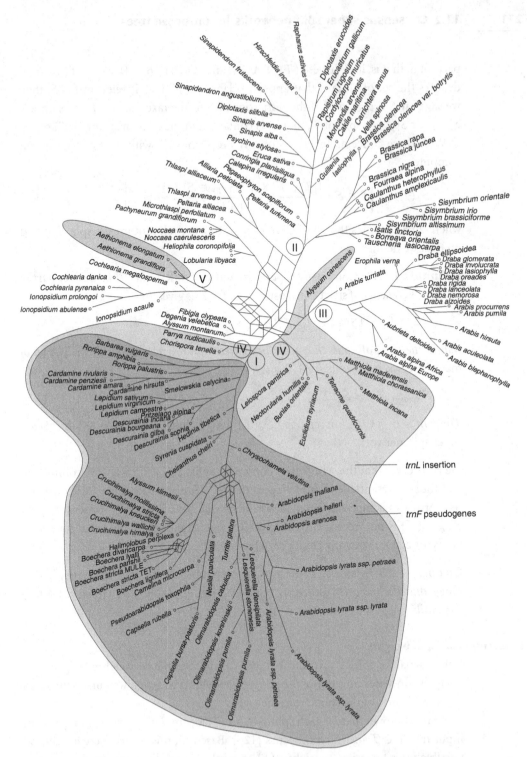

Figure 11.4 A super split network on 71 plant taxa. It was constructed from five different gene trees that were computed using the maximum parsimony method and then processed to remove uncertainties [154]. Reprinted by permission from Oxford University Press: Molecular Biology and Evolution, 24:63–73 ©2007.

these weights. However, both of these choices assume that the edge weights in the different input trees are all on the same scale, which is not always the case.

A more widely applicable approach is to use the *average relative length* of edges in the input set, by setting

$$\omega(S) = \frac{1}{|\mathcal{T}(S)|} \sum_{T \in \mathcal{T}(S)} \frac{\omega_T(S|_{\mathcal{X}(T)})}{\bar{\omega}(T)}, \tag{11.6}$$

where $\mathcal{T}(S)$ denotes the set of all input trees in which S occurs, $\omega_T(S|_{\mathcal{X}(T)})$ is the weight of the split $S|_{\mathcal{X}(T)}$ assigned by the tree T, and $\bar{\omega}(T)$ is the average weight of all edges in tree T.

11.2.2 Applications

In [154], the evolutionary history of nonfunctional *trn*F pseudogenes in flowering plants is studied and the authors identify multiple independent origins for *trn*F pseudogenes in crucifers. The authors present a super split network on five gene trees on the loci *adh, chs, matK, trnL-F,* and ITS containing a total of 71 taxa representing 68 different species (three species were represented by duplicate copies of the *chs* locus). The network was computed using the Z-closure method and is shown in Figure 11.4.

In [243], super split networks are computed for different values of the threshold *p*, to help address some of the common phylogenetic problems that arise from gene tree incongruence. Six individual gene trees on parasitoid wasps from the families Braconidae and Ichneumonidae are inferred using independent Bayesian analysis of six gene partitions. None of these trees is compatible with the expected phylogeny, which was inferred by a mixed-model Bayesian analysis of the combined partitions [159]. The super split network of the Z-closure of all splits found in the six gene trees is shown in Figure 11.5a. By considering only those splits that are contained in more than two of the gene trees, the expected phylogeny is obtained (see Figure 11.5b).

11.3 Distortion-filtered super split networks for unrooted trees

In the previous section, we discussed how to extend the idea of a consensus split network to partial trees, using the number of relevant trees in which a split occurs as a filter to control which splits make it into the set of consensus splits.

In this section, we briefly describe an alternative approach to filtering the splits produced by the Z-closure algorithm. The idea is to assign a *distortion* value to each split that measures how much every relevant input tree would have to be modified in order to accommodate the split [132]. A threshold on the distortion values is then used to filter the splits. This provides a parameter for controlling the complexity

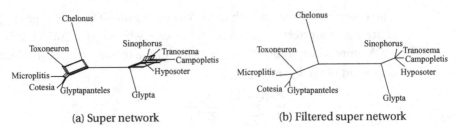

(a) Super network   (b) Filtered super network

Figure 11.5 (a) The Z-closure super network for the six gene trees for parasitoid wasps, and (b) a filtered Z-closure super network for the same dataset, based only on splits found in more than two of the gene trees, both as reported in [243].

of the resulting super split network that is similar to the maximum dimensionality parameter d, which was introduced above for consensus split networks.

Let $\mathcal{T} = (T_1, \ldots, T_k)$ be a set of partial trees on $\mathcal{X}$. Formally, for any complete split S on $\mathcal{X}$ we define the *distortion* of S (with respect to $\mathcal{T}$) as

$$\partial(S) = \sum_{i=1}^{k} \partial(T_i, S), \tag{11.7}$$

where $\partial(T_i, S)$ denotes the *minimal homoplasy score* for S on the input tree T_i. In more detail, this score is defined as the smallest number of (SPR or TBR) branch-swapping operations required to transform some refinement of T_i into a tree that contains the split S. The distortion of a split can be efficiently computed using dynamic programing.

11.4 Consensus cluster networks for rooted trees

Recall that a cluster network N on $\mathcal{X}$ is a rooted phylogenetic network that is used to represent a given set of clusters $\mathcal{C}$ on $\mathcal{X}$ in the hardwired sense, see Section 6.4.

Now, suppose we are given a collection of rooted phylogenetic trees $\mathcal{T} = (T_1, \ldots, T_k)$ on $\mathcal{X}$. Consider the set $\mathcal{C}(\mathcal{T})$ of all clusters contained in $\mathcal{T}$. We can use a cluster network N computed for $\mathcal{C}(\mathcal{T})$ by the cluster-popping algorithm to represent the set of trees $\mathcal{T}$. As is shown in Section 6.7, this network contains a refinement of every tree in $\mathcal{T}$.

In practice, this works well when the number of incompatibilities in $\mathcal{C}(\mathcal{T})$ is small or, more precisely, when the maximum size of the connected components in the associated incompatibility graph $IG(\mathcal{C}(\mathcal{T}))$ is not too large.

In more complicated cases, one possible approach is to use only those clusters that occur in a certain percentage of the input trees, thus obtaining a *consensus cluster network* that is akin to the concept of a consensus split network, described in Section 11.1. However, this suffers from the same draw-back as a consensus split network, namely that major differences that are only supported by one tree (or too few trees) are not represented in the network.

Note that the cluster-popping algorithm, introduced in Section 6.4, takes as input a set of clusters and so it can also be used to compute a rooted *consensus super cluster network* for a given set of rooted phylogenetic trees on overlapping, not necessarily identical, taxon sets. An example is shown in Figure 8.7.

11.5 Minimum hybridization networks

Suppose we are given two rooted, bifurcating phylogenetic trees T_1 and T_2 on $\mathcal{X}$. In this section, we consider the problem of computing a phylogenetic network N on $\mathcal{X}$ that represents ("displays" or "contains") both T_1 and T_2 and has a minimum number of reticulations. One possible application is in the context of hybridization, for example, of plants, where we might be given two gene trees, one based on a nuclear gene and the other based on a chloroplast gene, say. The goal is then to attempt to reconcile the two different trees by postulating hybridization events at the reticulate nodes of a suitable rooted phylogenetic network. Because this is a main application of this type of approach, we refer to such a network N as a *hybridization network* for T_1 and T_2.

In the following, we first introduce and discuss the concept of an *agreement forest*, which plays an important role in the computation of hybridization networks. As a first application of agreement forests, we then look at the problem of computing the *rooted SPR* distance between two rooted phylogenetic trees using *maximum* agreement forests. Finally, we discuss how to compute an optimal hybridization network using a maximum *acyclic* agreement forest.

11.5.1 Agreement forests

Let T be a rooted phylogenetic tree on $\mathcal{X}$ and let $C \subset \mathcal{X}$ be a subset of taxa. Recall that, by Definition 3.16.3, we use $T(C)$ to denote the smallest connected subgraph of T that contains all leaves that are labeled by elements of C, while we use $T|_C$ to denote the *restriction* of T to C, which is the rooted phylogenetic tree on C that is obtained from $T(C)$ by suppressing all suppressible nodes in $T(C)$.

Consider a set of trees $\mathcal{F} = \{F_1, \ldots, F_h\}$, where each tree F_i is a rooted phylogenetic tree on some taxon set $\mathcal{X}_i$. We call $\mathcal{F}$ a *forest on* $\mathcal{X}$, if no two trees share any taxa and $\mathcal{X}$ is the union of all the taxa involved. As such forests are usually constructed from a pair of two input trees, we refer to the members of such a forest as *components* rather than trees or subtrees, to avoid confusion.

Assume that we are given two rooted, bifurcating phylogenetic trees T_1 and T_2 on $\mathcal{X}$. For technical purposes (namely to identify the component of a forest that contains the root of a tree), we assume that the root ρ of either tree is a pendant

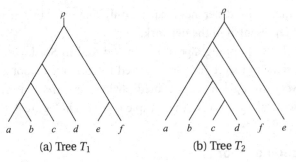

(a) Tree T_1 (b) Tree T_2

Figure 11.6 Two rooted phylogenetic trees T_1 and T_2 on $\mathcal{X} = \{a, \ldots, f\}$. For technical reasons, each tree has a pendant root node ρ.

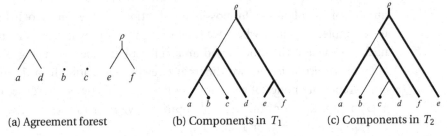

(a) Agreement forest   (b) Components in T_1 (c) Components in T_2

Figure 11.7 (a) An agreement forest $\mathcal{F}(T_1, T_2)$ for the two rooted phylogenetic trees T_1 and T_2 in Figure 11.6. In (b) and (c) we indicate (in bold) how the components of $\mathcal{F}(T_1, T_2)$ appear as subtrees of T_1 and T_2, respectively.

edge that has been adjoined to the original root. Moreover, we sometimes implicitly treat ρ as an additional taxon label and then implicitly consider the two trees as phylogenetic trees on $\mathcal{X} \cup \{\rho\}$. This is illustrated in Figure 11.6. An agreement forest for two trees is defined as follows [107]:

Definition 11.5.1 (Agreement forest)   *An agreement forest for two rooted phylogenetic trees T_1 and T_2 on $\mathcal{X} \cup \{\rho\}$ is a forest $F = \{F_\rho, F_1, \ldots, F_{h-1}\}$ on $\mathcal{X} \cup \{\rho\}$ such that*

   (i) *Each component F_i is a restricted subtree of T_1 and of T_2.*
   (ii) *The root ρ is contained in F_ρ.*
   (iii) *The trees in $\{T_1(\mathcal{X}_i) \mid i = \rho, 1 \ldots, h - 1\}$ and $\{T_2(\mathcal{X}_i) \mid i = \rho, 1, \ldots, h - 1\}$ are node disjoint subtrees of T_1 and T_2, respectively.*

In Figure 11.7, we show an agreement forest for the two rooted phylogenetic trees T_1 and T_2 in Figure 11.6. Any such agreement forest describes one way of simultaneously partitioning the two input trees into a collection of components that occur in both trees.

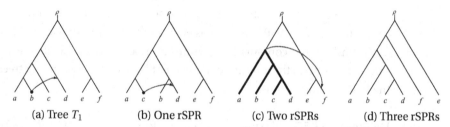

| (a) Tree T_1 | (b) One rSPR | (c) Two rSPRs | (d) Three rSPRs |

Figure 11.8 How to transform the rooted phylogenetic tree T_1 into T_2 using a set of rSPR moves based on an agreement forest $\mathcal{F}(T_1, T_2)$. In each illustration, the component of $\mathcal{F}(T_1, T_2)$ to be moved is show in bold and the intended rSPR move is indicated by an arrow. (a) The initial tree T_1. (b) Result of the first rSPR move, involving taxon b. (c) Result of the second rSPR move, involving taxon c. (d) Result of the the third rSPR move, involving the taxa $\{a, b, c, d\}$. This is the target tree T_2.

11.5.2 Maximum agreement forests and the rSPR distance

In Section 3.8 we introduced branch-swapping methods on unrooted phylogenetic trees. Recall that in a subtree prune and regraft (SPR) move, a subtree is detached ("pruned") from a given phylogenetic tree T and then reattached ("re-grafted") at a different location of the tree T. In the following, we discuss the rooted version of this operation, called a *rooted SPR (rSPR)*, that is performed by detaching a rooted subtree of T and then reattaching it somewhere else in the tree.

Assume that we are given two rooted, bifurcating phylogenetic trees T_1 and T_2 on $\mathcal{X}$ and let $\mathcal{F}(T_1, T_2) = \{F_\rho, F_1, \ldots, F_{h-1}\}$ be an agreement forest for T_1 and T_2. In both trees, the root of any of the components $F_1, \ldots, F_{h-1}$ is attached to the rest of the tree by a single edge. To transform one tree into the other, what is required is that we detach each such edge and then reattach it at the appropriate location. As the latter operation is a rSPR move, this provides an (oversimplified) argument for the following result [25, 107]:

Lemma 11.5.2 (Agreement forest and rSPR)   *Let T_1 and T_2 be two rooted, bifurcating phylogenetic trees on $\mathcal{X}$. One tree can be transformed into the other by a sequence of $h-1$ different rSPR moves if and only if there exists an agreement forest $\mathcal{F}(T_1, T_2)$ for the two trees that contains at most h components.*

To illustrate this result, consider the two rooted phylogenetic trees T_1 and T_2 displayed in Figure 11.6. One possible agreement forest $\mathcal{F}(T_1, T_2)$ for these two trees is given in Figure 11.7. This forest contains four components and so the lemma implies that one can transform T_1 into T_2 using three rSPR moves, as demonstrated in Figure 11.8.

Let T_1 and T_2 be two rooted, bifurcating phylogenetic trees on $\mathcal{X}$. The *rSPR distance*, denoted by $d_{rSPR}(T_1, T_2)$, is defined as the minimum number of rSPR moves required to transform one tree into the other. By the above lemma, the computation of the rSPR distance is equivalent to computing an agreement forest for T_1 and T_2 that contains a minimum number of components. Because such a forest "maximizes the amount of agreement" detected between two trees, we have the following definition:

Definition 11.5.3 (Maximum agreement forest) *Let T_1 and T_2 be two rooted, bifurcating phylogenetic trees on $\mathcal{X}$. A maximum agreement forest $\mathcal{F}(T_1, T_2)$ for T_1 and T_2 is an agreement forest for the two trees that contains a minimum number of components.*

Note that Lemma 11.5.2 implies that $d_{rSPR}(T_1, T_2) = |\mathcal{F}(T_1, T_2)| - 1$ holds.

The problem of computing the rSPR distance between two rooted phylogenetic trees on $\mathcal{X}$ is known to be NP-hard, but fixed-parameter tractable (FPT) [25]. In this context, the FPT property means that the problem can be solved by an algorithm whose runtime depends polynomially on the size of the problem and exponentially on the value of the answer. In the following, we discuss the two main approaches to simplifying the problem of finding a maximum agreement forest, which are also used to obtain the FPT result. These also play a central role in the computation of a minimum hybridization network.

The first approach is based on the following observation: If two rooted phylogenetic trees T_1 and T_2 share a maximal common subtree T' then that subtree is contained in a component of every maximum agreement forest $\mathcal{F}(T_1, T_2)$. So, in a preprocessing step, one can replace any such subtree T' by a new leaf that is labeled by a new formal taxon $x_{T'}$, which represents the subtree T' (and all taxa that label it). Clearly, this *subtree reduction* step decreases the size of input without changing the rSPR distance.

The second reduction method is based on the concept of a "caterpillar chain". Let T be a rooted, bifurcating phylogenetic tree on $\mathcal{X}$. Consider a set of leaves $q_1, \ldots, q_t$ together with their parents $p_1, \ldots, p_t$, such that p_{i+1} is the parent of p_i for $i = 2, \ldots, t$, and either p_2 is the parent of p_1 or $p_2 = p_1$. The subtree C with path $(p_t, \ldots, p_1)$ and leaves $(q_t, \ldots, q_1)$ is called a *caterpillar chain* in T.

For example, consider the two rooted, bifurcating phylogenetic trees T_1 and T_2 displayed in Figure 11.9a and b. The nodes labeled $a, b, \ldots, f$ form the leaf-set of a caterpillar chain in both trees.

Assume that two rooted phylogenetic trees T_1 and T_2 on $\mathcal{X}$ contain the same caterpillar chain C. The *chain reduction* applied to such a shared caterpillar chain C is defined as the operation that replaces the caterpillar chain C by a reduced caterpillar chain C' that has only three parent nodes (p_1, p_2, p_3) and three new leaves labeled by three formal taxa (x_1, x_2, x_3), in both trees.

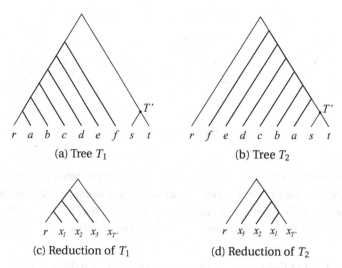

(a) Tree T_1 (b) Tree T_2

(c) Reduction of T_1 (d) Reduction of T_2

Figure 11.9 In (a) and (b) we show two rooted, bifurcating phylogenetic trees T_3 and T_4. The nodes labeled $a, \ldots, f$ form the leaf-set of a maximal caterpillar chain that is contained in both trees, highlighted in bold. In (c) and (d) we show the result of applying a chain reduction and a subtree reduction (involving the leaves labeled s and t) to both trees.

A maximal caterpillar chain C that occurs in both trees must necessarily be a component of every maximum agreement forest $\mathcal{F}(T_1, T_2)$ [25]. Hence, in a pre-processing step we can reduce all such chains using the chain reduction operation, decreasing the size of input and leaving the rSPR distance unchanged.

In Figure 11.9c and d we see the result of reducing the two rooted phylogenetic trees T_1 and T_2. Now, to determine the minimum rSPR distance between the two networks, we use an exhaustive search to find a maximum agreement forest for the two reduced trees.

The two described reduction rules can be used to show that the computation of the rSPR distance is fixed-parameter tractable. First, note that it is easy to tell where to apply which operation, there is no order-dependency and application takes only polynomial time. Once the reductions have all been applied then the two resulting trees cannot be much larger than the number of rSPR moves required to transform one tree into the other. In more detail, one can show that the number of taxa after application of all possible reductions is at most 28 times the number of required rSPR operations [25].

11.5.3 Acyclic agreement forests and hybridization networks

Now we come to the main aim of this section, which is to solve the following problem:

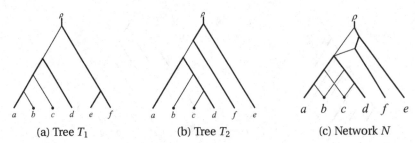

(a) Tree T_1 (b) Tree T_2 (c) Network N

Figure 11.10 (a and b) Two rooted phylogenetic trees T_1 and T_2 on $\mathcal{X} = \{a, \ldots, f\}$ where all components in the agreement forest $\mathcal{F}(T_1, T_2)$ are indicated in bold. In (c) a rooted phylogenetic network N that represents the two trees T_1 and T_2 is shown. This network was obtained by attaching all components in the agreement forest $\mathcal{F}(T_1, T_2)$ exactly as they are attached to each other in both trees T_1 and T_2. The components of $\mathcal{F}(T_1, T_2)$ are highlighted in bold.

Problem 11.5.4 (Minimum hybridization network) *Let T_1 and T_2 be two rooted, bifurcating phylogenetic trees on $\mathcal{X}$. Determine a rooted, bicombining and bifurcating phylogenetic network N that represents both T_1 and T_2 and has a minimum number of reticulations $h(T_1, T_2)$.*

The minimum hybridization network problem is NP-hard, but fixed-parameter tractable [26].

Let T_1 and T_2 be two rooted, bifurcating phylogenetic trees on $\mathcal{X}$. As we have seen, any agreement forest $\mathcal{F}(T_1, T_2)$ represents a partitioning of both trees into components that occur in both trees simultaneously. So far, we have discussed how to use $\mathcal{F}(T_1, T_2)$ to determine a sequence of rSPR moves that can transform one of the two trees into the other. Now we want to use $\mathcal{F}(T_1, T_2)$ to compute a rooted phylogenetic network N that represents both trees.

Figure 11.10a and b shows two rooted phylogenetic trees T_1 and T_2 and how an agreement forest $\mathcal{F}(T_1, T_2)$ appears in the two rooted phylogenetic trees. The root node of each component (except for ρ) is attached to two different locations, one in tree T_1 and the other in tree T_2.

Now, the basic idea is to construct a rooted phylogenetic network N by simply joining all components of $\mathcal{F}(T_1, T_2)$ in the way that they are attached in T_1 and also as they are attached in T_2 (we discuss how to join the components later in this section). The result is illustrated in Figure 11.10c.

However, this simple approach needs to be modified, because otherwise a problem can occur, as demonstrated in Figure 11.11. In this example, there is a circular dependency between the two components of the agreement forest $\mathcal{F}(T_1, T_2)$ associated with the label sets $\{b, e\}$ and $\{c, d\}$, displayed in the two phylogenetic trees T_1 and T_2. While one component attaches to an edge of the other component in tree T_1, in tree T_2 the two components attach to each other in the opposite order.

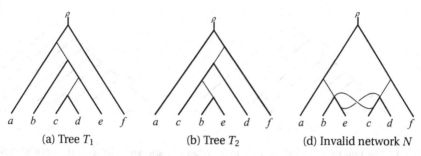

(a) Tree T_1 (b) Tree T_2 (d) Invalid network N

Figure 11.11 In (a) and (b) we show two rooted phylogenetic trees T_1 and T_2 on $\mathcal{X}$. In both trees, the components of a maximum agreement forest $\mathcal{F}(T_1, T_2)$ are highlighted in bold. Connecting all components as prescribed by both trees produces the network N shown in (c). Note that this network contains a directed cycle and is therefore not a valid rooted phylogenetic network.

In consequence, the resulting network N contains a directed cycle and is therefore not a valid rooted phylogenetic network.

So, for the purpose of computing a minimum hybridization network N for two rooted, bifurcating phylogenetic trees T_1 and T_2 on $\mathcal{X}$, we must restrict our attention to such agreement forests that do not exhibit any circular dependencies between components:

Definition 11.5.5 (Acyclic agreement forest) *Let T_1 and T_2 be two rooted, bifurcating phylogenetic trees on $\mathcal{X}$. An agreement forest $\mathcal{F}_a(T_1, T_2)$ for T_1 and T_2 is called acyclic, if the components of $\mathcal{F}_a(T_1, T_2)$ can be numbered in such a way that if the root of one component F is an ancestor of the root of some other component F', then the number assigned to F is lower than the number assigned to F', for all pairs of components F and F' in $\mathcal{F}_a(T_1, T_2)$.*

Exercise 11.5.6 (Maximum acyclic agreement forest) *Determine an acyclic agreement forest for the two trees T_1 and T_2 displayed in Figure 11.11 that is of minimum size.*

Such an acyclic agreement forest for two rooted, bifurcating phylogenetic trees T_1 and T_2 is called a *maximum acyclic agreement forest*, if it contains a minimum number of components. Based on this, we have the following result [15]:

Theorem 11.5.7 (Agreement forests and hybridization networks) *Let T_1 and T_2 be two rooted, bifurcating phylogenetic trees on $\mathcal{X}$. Any acyclic agreement forest $\mathcal{F}_a(T_1, T_2)$ with h components corresponds to a rooted phylogenetic network N on $\mathcal{X}$ that represents both T_1 and T_2, is bicombining and bifurcating, and contains at most $h - 1$ reticulations.*

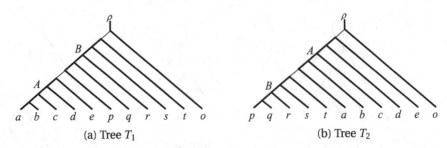

(a) Tree T_1 (b) Tree T_2

Figure 11.12 The two rooted phylogenetic trees T_1 and T_2 shown in (a) and (b) each contain the same two maximal caterpillar chains labeled A and B. Because A lies below B in tree T_1 and above B in tree T_2, these two chains cannot occur together in an acyclic agreement forest for T_1 and T_2.

Based on this result, the minimum hybridization network problem can be solved by determining a maximum acyclic agreement forest and then extending it to a network. We address both problems in the following.

The problem of finding a maximum acyclic agreement forest $\mathcal{F}_a(T_1, T_2)$ is NP-hard, but fixed-parameter tractable [26]. As in the case of computing the rSPR distance, a set of reductions is first applied to the two input trees T_1 and T_2 so as to reduce the size of the problem and then an exhaustive search is performed to find a maximum agreement forest for the two reduced trees.

One possibility is the subtree reduction presented in the previous section, which can be used without modification in the context of acyclic agreement forests. In this case we reduce a subtree T' and we obtain two trees T_1' and T_2' that both contain a leaf labeled by a formal taxa $x_{T'}$. In this case we have that $h(T_1, T_2) = h(T_1', T_2')$. One can compute $\mathcal{F}_a(T_1, T_2)$ from $\mathcal{F}_a(T_1', T_2')$ by expanding $x_{T'}$ back. Because we are now concentrating on *acyclic* agreement forests, we can introduce a new reduction method called the *cluster reduction*. Let T_1 and T_2 be two rooted, bifurcating phylogenetic trees on $\mathcal{X}$. If there exists a cluster C on $\mathcal{X}$ that is represented by both trees, then we can use a divide-and-conquer approach, and decompose the problem into the two problems of finding a maximum acyclic forest for the pairs $(T_1|_C, T_2|_C)$ and $(T_1|_{(\mathcal{X}-C)}, T_2|_{(\mathcal{X}-C)})$. In this case we have $h(T_1, T_2) = h(T_1|_C, T_2|_C) + h(T_1|_{(\mathcal{X}-C)}, T_2|_{(\mathcal{X}-C)})$ and $\mathcal{F}_a(T_1, T_2) = \mathcal{F}_a(T_1|_C, T_2|_C) \cup \mathcal{F}_a(T_1|_{(\mathcal{X}-C)}, T_2|_{(\mathcal{X}-C)})$ [166].

Note that both of these reductions involve whole subtrees of the two input trees and so they do not affect the acylicity of the agreement forests associated with the trees. This is not the case for the chain reduction as defined in Section 11.5.2, because a caterpillar chain can lie above nodes that do not belong to the chain. This is illustrated in Figure 11.12, where one caterpillar chain A lies below a second caterpillar chain B in the rooted phylogenetic tree T_1, and above it in the rooted phylogenetic tree T_2. In consequence, the two components A and B can never occur together in an acyclic agreement forest for the two trees T_1 and T_2. This shows that a maximal caterpillar chain C that occurs in both trees is not necessarily a component of every maximum *acyclic* agreement forest.

To accommodate this issue, we must first modify the definition of the chain reduction slightly. Let C be a maximal caterpillar chain defined by a set of leaves $q_1, \ldots, q_t$ of T with parents $p_1, \ldots, p_t$ that is present in both trees. A *chain reduction* replaces such a chain C by a reduced caterpillar chain C' that has path nodes p_t, p_1 and leaf nodes labeled by two formal taxa (x_1, x_2) together with a new edge (p_t, p_1). We keep track of the new caterpillar chains of length 2, called *2-chains* for short.

When replacing a caterpillar chain C by a single edge $e = (v, w)$ in the reduced trees T_1' and T_2', we assign weight $m(e) = \ell(C) - 2$ to the edge e, where $\ell(C)$ denotes the number of leaves in C. In the presence of overlapping chains or reduced clusters, additional adjustments to this count have to be made by adding all the weights already assigned to edges in the caterpillar.

To find an optimal forest $\mathcal{F}_a(T_1', T_2')$, we restrict our attention to "legitimate" acyclic agreement forests. An acyclic agreement forest $\mathcal{F}_a(T_1', T_2')$ is called *legitimate* if and only if, for each new 2-chain $\{x_1, x_2\}$ created by a chain reduction, one of the following conditions hold:

(i) There exists a component F_i in $\mathcal{F}_a(T_1', T_2')$ such that $\{x_1, x_2\} \subseteq L(F_i)$ hold, or
(ii) there exist two components F_i and F_j in $\mathcal{F}_a(T_1', T_2')$ such that $\{x_1\} = L(F_i)$ and $\{x_2\} = L(F_j)$ hold.

Then, when searching for a maximum acyclic agreement forest for the two reduced trees T_1' and T_2', we aim at minimizing the following score among all possible legitimate acyclic agreement forests:

$$H(\mathcal{F}_a(T_1', T_2')) = (|\mathcal{F}_a(T_1', T_2')| - 1) + \sum m(e), \tag{11.8}$$

where the sum is taken over all edges e that have a positive weight $m(e)$ and are not present in the agreement forest. These edges correspond to those caterpillar chains contained in the original input trees T_1 and T_2 that do not occur as chains in $\mathcal{F}_a(T_1', T_2')$.

See [26] for justification for this formula, where it is shown that for any caterpillar chain C present in the two input trees there are only two possibilities: Either the chain C is contained in the final network N, or every leaf v in C occurs as a reticulate node in N. In other words, every caterpillar chain C that occurs in the input, but does not make it into the output (because of the acyclicity condition), gives rise to $m(e)$ additional reticulations (see Figure 11.13c–d).

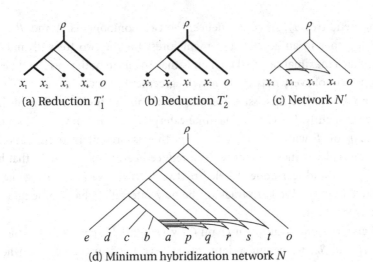

(a) Reduction T_1' (b) Reduction T_2' (c) Network N'

(d) Minimum hybridization network N

Figure 11.13 In (a) and (b) we show the two rooted phylogenetic trees T_1' and T_2' obtained from the trees T_1 and T_2 in Figure 11.12 by two chain reductions. A maximum acyclic agreement forest $\mathcal{F}_a(T_1', T_2')$ containing four components is highlighted in bold. (c) A rooted phylogenetic network N' obtained from T_1' and T_2' using $\mathcal{F}_a(T_1', T_2')$. (d) A minimum hybridization network N for T_1 and T_2 obtained by "undoing" the chain reductions in N'.

Let $\mathcal{F}_{al}(T_1', T_2')$ denote a minimum-weight legitimate acyclic agreement forest. Then we have that $H(\mathcal{F}_{al}(T_1', T_2')) = h(T_1, T_2)$ and we can obtain $\mathcal{F}_a(T_1, T_2)$ from $\mathcal{F}_{al}(T_1', T_2')$ in the following way. For each 2-chain $\{x_1, x_2\}$ created by chain reduction, do the following:

(i) if there exists a component F_i in $\mathcal{F}_{al}(T_1', T_2')$ such that $\{x_1, x_2\} \subseteq L(F_i)$, then undo the reduction of $\{x_1, x_2\}$ in F_i, or,

(ii) if there exist two components F_i and F_j in $\mathcal{F}_{al}(T_1', T_2')$ such that $\{x_1\} = L(F_i)$ and $\{x_2\} = L(F_j)$, then $\mathcal{F}_a(T_1, T_2) = \mathcal{F}_{al}(T_1', T_2') - F_i - F_j \cup \{a_1, \ldots, a_n\}$, where $\{a_1, \ldots, a_n\}$ are the taxa of the caterpillar chain that was reduced into $\{x_1, x_2\}$.

Given an acyclic agreement forest for T_1 and T_2, we can compute a minimum hybridization network for T_1 and T_2 using the following algorithm:

Algorithm 11.5.8 (Minimum hybridization network)   *Let $\mathcal{F}_a(T_1, T_2)$ be an acyclic agreement forest for T_1 and T_2. An associated hybridization network N is computed in the following steps:*

(i) *Find an acyclic ordering, $F_\rho, F_1, F_2, \ldots, F_k$ of $\mathcal{F}_a(T_1, T_2)$.*

(ii) *Set $N_0 = F_\rho$.*

(iii) *For $i = 1$ to k compute a hybridization network N_i as follows:*

> Connect F_i to N_{i-1} using two new edges that each join the root of F_i to
> some (not necessarily distinct) edge of N_{i-1} so that the resulting hybridization
> network N_i displays $T_1|_{L(F_1,...,F_i)}$ and $T_2|_{L(F_1,...,F_i)}$.
>
> (iv) *The result is given by $N = N_k$.*

Let T_1 and T_2 be two rooted, bifurcating phylogenetic trees on $\mathcal{X}$. Assume that
we are given an acyclic agreement forest $\mathcal{F}_a(T_1', T_2')$ for two reductions T_1' and T_2'
of the two input trees, possibly with weighted edges that indicate the presence of
reduced caterpillar chains. To construct a hybridization network for T_1 and T_2, as
indicated in Figure 11.13d we undo all reductions in the opposite order that they
where originally done in. We obtain an acyclic agreement forest $\mathcal{F}_a(T_1, T_2)$ and
then we construct the corresponding network by Algorithm 11.5.8.

11.5.4 Application

In Figure 11.14, we display two trees T_1 and T_2 computed for 14 difference grass
species (*Poaceae*), based on the *phyB* and *waxy* gene, respectively, see [87]. Each
of the two accompanying rooted phylogenetic networks N_1 and N_2 contain both
trees and have a minimum number of reticulate nodes with this property, namely
three. If we assume that differences in topology of the two trees T_1 and T_2 are due
to hybridization events, then, for example, the network N_1 suggests that *glyceria*
is a hybrid of the lineages leading to *melica* and *triticium*. In the case of the two
other putative hybrid species, *lygeum* and *chusquea*, their evolution requires the
postulation of additional lineages to resolve the fact that they appear to be hybrids
of recent and less recent lineages. We emphasize that neither network *proves* that
hybridization is the cause of the incongruence between the trees T_1 and T_2 and
additional biological evidence is required to support suspected cases of speciation
by hybridization.

11.6 Minimum hybridization networks and galled trees

Let T_1 and T_2 be two rooted, bifurcating phylogenetic trees on $\mathcal{X}$. If their rooted SPR
distance $d_{rSPR}(T_1, T_2)$ is one, then all their maximum agreement forests are acyclic
and their hybridization number $h(T_1, T_2)$ is also one. In this case any minimum
hybridization network for T_1 and T_2 has the galled tree property.

In [185], this approach is extended to the case of two rooted, *not necessarily
bifurcating* phylogenetic trees T_1 and T_2 on $\mathcal{X}$ as follows: Determine whether there
exist two rooted, *bifurcating* phylogenetic trees T_1' and T_2' that refine T_1 and T_2,
respectively, and for which $d_{rSPR}(T_1', T_2') = 1$ holds. If such trees exist, then a galled
tree representing T_1' and T_2' is computed. Even this extended approach is very

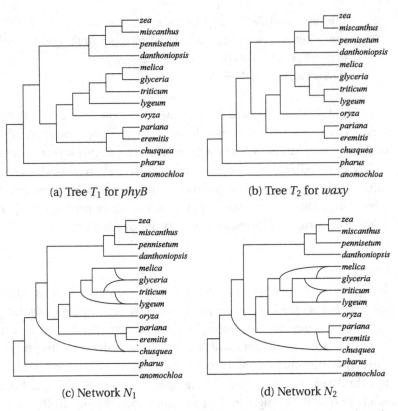

(a) Tree T_1 for *phyB*   (b) Tree T_2 for *waxy*

(c) Network N_1 (d) Network N_2

Figure 11.14 Two rooted phylogenetic trees (a) T_1 and (b) T_2, on 14 grasses, based on the *phyB* gene and *waxy* gene [87]. The two rooted phylogenetic networks N_1 and N_2 shown in (c) and (d) both contain each of the two trees T_1 and T_2, each using only three reticulate nodes. Each network displays a set of of putative hybridization events that may explain the differences between two trees.

limited, as it can only solve the case in which two trees can be represented by a galled tree that has only a single gall.

An alternative and more general approach operates as follows [127]. Let T_1 and T_2 be two rooted, not necessarily bifurcating phylogenetic trees on $\mathcal{X}$, and let $\mathcal{C} = \mathcal{C}(T_1) \cup \mathcal{C}(T_2)$ be the set of clusters contained in the two trees. To obtain a galled tree N that represents both T_1 and T_2, if one exists, compute a galled tree N that represents $\mathcal{C}$, as described in Section 8.3, if one exists, by virtue of Lemma 6.11.4.

When using a method that accepts multifurcating trees as input, one can first contract such edges in the input trees that have low bootstrap values (Section 3.10) or posterior probabilities (Section 3.11) so as to avoid the computation of superfluous reticulations that would only serve to represent uncertainties in the input trees.

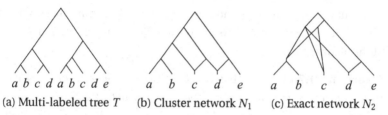

(a) Multi-labeled tree T (b) Cluster network N_1 (c) Exact network N_2

Figure 11.15 (a) A multi-labeled tree T on $\mathcal{X}$. (b) A rooted phylogenetic network N_1 representing all clusters contained in T. (c) A rooted phylogenetic network N_2 that provides an exact representation of T.

In contrast, when using a method that requires the input trees to be bifurcating, such as the one described in the previous section, then the resulting networks can be overly complicated because they must accommodate all topological differences among the trees, whether biologically relevant or not.

11.7 Networks from multi-labeled trees

Gene duplication is a common event in evolution and so many genes are present in multiple copies in a genome. In most phylogenetic studies based on gene sequences, a single instance of the gene is used per taxon. When some taxa are each represented by more than one copy of a gene, then the resulting phylogenetic tree is called *multi-labeled*. If the duplication history of a gene is itself of interest, then it may be helpful to map such a multi-labeled phylogenetic tree onto a rooted phylogenetic network so as to see which parts of the tree are similar and where are there differences [117]. We start with a formal definition of a multi-labeled tree:

Definition 11.7.1 (Multi-labeled phylogenetic tree)   *A multi-labeled phylogenetic tree T on $\mathcal{X}$ consists of a tree $T = (V, E)$, in which all nodes have degree $\neq 2$ (except a possible root node), together with a taxon multi-labeling λ that maps each taxon x to a set $\lambda(x)$ of one or more leaves such that every leaf obtains precisely one label.*

In the following, all multi-labeled phylogenetic trees considered are rooted. Let T be a multi-labeled tree on $\mathcal{X}$. Our goal is to compute a rooted phylogenetic network N on $\mathcal{X}$ that "represents" T and is single-labeled. There are two different efficient approaches to this problem, one is based on clusters and the other uses nested labels, see Figure 11.15.

In the cluster-based approach [22], the aim is to represent all clusters present in the multi-labeled tree T. Each edge e in a multi-labeled tree T defines a cluster $C(e)$ that is given by the set of taxa that label all leaves below e. The set of clusters $C(T)$ obtained in this way contains at most one cluster for every edge, as different edges may give rise to the same cluster (using different instances of the same taxa). Then a cluster network N is computed for $C(T)$ to represent T. Here is a summary of the algorithm:

Algorithm 11.7.2 (Cluster-based network)   *For a given multi-labeled phylogenetic tree T on $\mathcal{X}$, we compute a rooted phylogenetic network N on $\mathcal{X}$ that represents all clusters in T as follows: First, determine the set of clusters $C(T)$ represented by T. Then compute N as a cluster network for $C(T)$ using the cluster-popping algorithm.*

In the nested-labels approach [117], the main idea is to determine and reuse large congruent subtrees of the multi-labeled tree T. First, in a bottom-up traversal of T, we assign a *nested label* $v(v)$ to every node v, as described in Section 6.14.4:

  (i) The nested label of a leaf is its taxon label.
 (ii) The nested label of an internal node is defined as the multi-set of all nested labels of its children.

We then order all nodes by their nested labels, first by decreasing size and second lexicographically, in what we call an *elimination* order O. In such an order O, all nodes that have the same nested-label are adjacent to each other. Moreover, any node v comes before all of its descendants. In the second step the aim is to transform T into a rooted phylogenetic network N. To do so, we inspect all nodes of T in the order O. Whenever we find two adjacent nodes v and w that have the same nested label $v(v) = v(w)$, then we add a new reticulate edge from the parent node of w to v, remove the whole subtree rooted at w from T and remove all its nodes from O. Here is a summary of the algorithm:

Algorithm 11.7.3 (Nested-labels-based network)   *For a given multi-labeled phylogenetic tree T on $\mathcal{X}$, we compute a rooted phylogenetic network N on $\mathcal{X}$ that represents T as follows. First, determine the nested labels of all nodes in T, and then use them to obtain an elimination ordering O of the nodes of T. Second, inspect each node v of T in order O. If the successor w of v in O has the same nested label as v, then join the parent of w to v by a new reticulate edge and then remove the subtree rooted at w from T and all its nodes from O.*

Optionally, at the end of the algorithm, one can insert additional edges to ensure that the resulting network satisfies the reticulation-separation property (see Definition 6.4.1).

By construction, the nested label of the root of the resulting phylogenetic network N equals the nested label of the root of the original multi-labeled tree T. Moreover, as the nested label of T is essentially the Newick representation of T, as discussed in Section 4.5, the multi-labeled tree T can be reconstructed from N and so the network provides a loss-less representation of the multi-labeled phylogenetic tree T. For this reason, we call the rooted phylogenetic network N computed by Algorithm 11.7.3 an *exact* representation of T.

In practice, the cluster-based approach usually produces simpler networks than the nested-labeled approach. Hence, when the represented clusters are of more interest than the actual tree itself, the former approach is more suitable. The latter approach is to be preferred when the whole topology of the multi-labeled tree is of interest and an exact representation is desired.

11.7.1 Application

An analysis of the history of complex genome merging events within a set of high-polyploid species in the plant genus *Cerastium* (Caryophyllaceae) is presented in [34]. The authors of the paper apply the nested-label approach to a multi-labeled phylogenetic tree. This is the majority tree over a set of maximum parsimony trees that were computed for a noncoding region from the second largest subunit of the RNA polymerase gene family (RPB2). The authors present the resulting rooted phylogenetic network and argue that it shows how hybridization and poly-ploidization have acted as key evolutionary processes in creating a plant group where allopolyploids of high ploidy levels outnumber extant parental genomes (see Figure 11.16).

11.8 DLT reconciliation of gene and species trees

A fundamental observation in phylogeny is that different genes give rise to different trees. In more detail, suppose we are interested in the evolutionary history of a set of taxa $\mathcal{X}$. If we compute two phylogenetic trees T_1 and T_2 on $\mathcal{X}$ based on the sequences of two distinct genes g_1 and g_2, then in most cases the two trees will exhibit differences in their topologies. Although some of the differences may be explained by inaccuracies in the tree-inference process, others may be the consequence of evolutionary events such as *gene duplications, gene losses* or *horizontal gene transfers*. We refer to all these three types of events as *DLT* events.

Now let us assume that we know the *species phylogeny* T_{sp} on $\mathcal{X}$, that is, the true evolutionary history of the set of taxa $\mathcal{X}$. Then, any *gene phylogeny* T on $\mathcal{X}$ will usually differ from the species tree T_{sp}, even if it has been correctly inferred, see Figure 11.17. To explicitly accommodate gene duplication events, throughout this section all gene trees considered are multi-labeled trees, as defined in the previous section. Moreover, for computational reasons, we assume that all gene trees are bifurcating.

Under the assumption that the discrepancies between a species tree T_{sp} and a gene tree T on $\mathcal{X}$ are due to gene duplication, loss and transfer events, we can formulate the following computational problem:

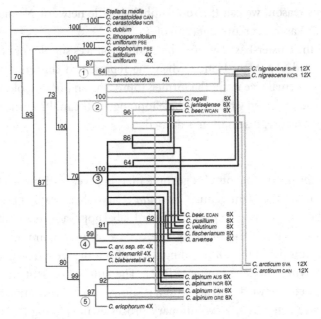

Figure 11.16 Rooted phylogenetic network representing a possible duplication history for a set of high-polyploid species in the plant genus *Cerastium* (Caryophyllaceae), computed from a multi-labeled tree for the RPB2 gene, as reported in [34]. Geographical origin is indicated by prefixes: AUS = Austria, ECAN = East Canada, GRE = Greenland, NOR = Norway, SHE = Shetland, SVA = Svalbard and WCAN = West Canada. Two pseudo genes are labeled PSE. Numbers above the edges of the network are bootstrap values for the multi-labeled tree. Reprinted by permission from Oxford University Press: Systematic Biology, 56(3):467–476 ©2007.

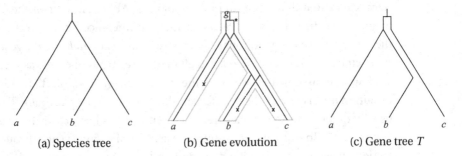

| (a) Species tree | (b) Gene evolution | (c) Gene tree T |

Figure 11.17 (a) A species tree T_{sp} on $\mathcal{X} = \{a, b, c\}$. (b) An evolutionary scenario in which a gene g experiences a duplication event, labeled ∗, and then three subsequent losses, each marked x. (c) The resulting gene tree T.

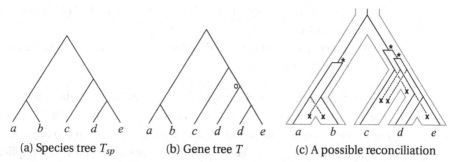

(a) Species tree T_{sp} (b) Gene tree T (c) A possible reconciliation

Figure 11.18 (a) A species tree T_{sp} on $\mathcal{X} = \{a, \ldots, e\}$. (b) A gene tree T on $\mathcal{X}$. (c) A possible reconciliation of the two trees that uses three duplications (marked $*$) and six losses (dashed lines leading to an x).

Problem 11.8.1 (Gene tree–species tree reconciliation)   *Given a species tree T_{sp} and a (multi-labeled) gene tree T on $\mathcal{X}$ that differ topologically, explain the differences by postulating a suitable "optimal" series of gene duplication, loss and transfer events.*

We first discuss the easier problem of reconciling a gene tree and a species tree using only duplication and loss events. Then we tackle the more difficult problem of reconciliation using all three types of DLT events.

11.8.1 Duplication and loss

Let T be a (bifurcating, multi-labeled) gene tree and T_{sp} a species tree, on $\mathcal{X}$. Our aim is to explain or to *reconcile* the differences between T and T_{sp} by postulating a *duplication-loss scenario* consisting of a number of appropriate gene duplication events and gene loss events in T.

In a duplication–loss scenario, each internal node in the gene tree T represents either a speciation event or a duplication event that the gene has experienced during the evolution of the associated species. Each gene-tree node v that represents a speciation event corresponds to a species-tree node w that represents the same event, whereas every gene tree node v that represents a duplication event is associated with a species-tree edge e along which the duplication event took place. (Technically, we map v to a species-tree node in both cases and, in the latter case, the edge e is identified via its target node.) Moreover, the gene tree T can be depicted as evolving completely inside the species tree T_{sp}, as illustrated in Figure 11.18.

Definition 11.8.2 (Duplication–loss scenario)   *Let T be a gene tree and T_{sp} a species tree on $\mathcal{X}$. A duplication–loss scenario for T and T_{sp} is given by an assignment γ that maps the set of nodes of the gene tree T to the set of nodes of the species tree T_{sp}, and has the following properties:*

(i) *If v is a leaf of T with taxon label x, then $\gamma(v)$ is a leaf of T_{sp} and has the same taxon label x.*

(ii) *For every edge $e = (v, w)$ of the gene tree T we have that $\gamma(v)$ is an ancestor of $\gamma(w)$ in the species tree.*

This definition does not completely determine a reconciliation between a gene tree T and a species tree T_{sp} because it does not explicitly say which gene-tree nodes correspond to speciation events and which ones correspond to duplications. So, in principle, many different reconciliations are possible for a single duplication–loss scenario.

To disambiguate this, any internal gene-tree node u is interpreted as a *speciation node*, if its two children v and w are mapped to incomparable species-tree nodes, and as a *duplication node*, otherwise. In the latter case the duplication is assumed to take place along the species-tree edge that leads to $\gamma(u)$. We use V_s and V_d to denote the set of speciation nodes and duplication nodes in T, respectively.

Note that associating a gene-tree node u with a speciation event only makes sense if its two children v and w are mapped to incomparable species-tree nodes. On the other hand, interpreting such a node u as a duplication node is unnecessarily costly, as a duplication event in the gene tree that is located immediately before a speciation event in the species tree implies two losses.

While the speciation and duplication events are explicitly defined in terms of nodes in the gene tree, the losses associated with a duplication–loss scenario are only implicitly given. A loss arises whenever a gene-tree edge e continues through a species-tree node w, thus indicating that the gene did not participate in the speciation event at w, thereby losing a lineage, see Figure 11.18c.

Clearly, the duplication–loss scenario depicted in Figure 11.18c is not optimal in the sense that it contains many more duplications and losses than are necessary to reconcile the depicted species tree and gene tree. There is a straight-forward algorithm for computing a duplication–loss scenario that minimizes the number of duplication and losses required by a duplication scenario, based on the idea of mapping each gene-tree node v onto the lowest common ancestor of the images of the leaves below v [238]:

Algorithm 11.8.3 (Optimal duplication–loss scenario)   *For a given gene tree T and species tree T_{sp} on $\mathcal{X}$, an optimal duplication–loss scenario γ can be computed as follows:*

(i) *For each leaf v of T set $\gamma(v) = w$, where w is the leaf of T_{sp} that is labeled by the same taxon as v.*

(ii) *For each internal node v of T set $\gamma(v)$ equal to the lowest common ancestor of all species-tree nodes $\gamma(u)$, where u is a leaf in T that is a descendant of v.*

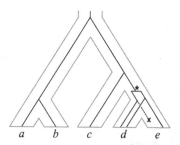

Figure 11.19 An optimal duplication–loss reconciliation for the gene tree T and species tree T_{sp} from Figure 11.18, involving one duplication (marked ∗) and one loss (marked x).

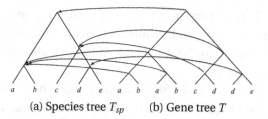

     (a) Species tree T_{sp} (b) Gene tree T

Figure 11.20 (a) A species tree T_{sp} on $\mathcal{X} = \{a, \ldots, e\}$. (b) A gene tree T on $\mathcal{X}$. For each internal gene-tree node v, a dashed arrow indicates to which species tree node $\gamma(v)$ it is mapped by the duplication–loss scenario computed by Algorithm 11.8.3.

Application of this algorithm to the species tree and gene tree shown in Figure 11.18a–b gives rise to a reconciliation that uses only one duplication and one loss, which is caused by the node labeled **o** in Figure 11.18b. The resulting optimal duplication–loss reconciliation is shown in Figure 11.19.

Exercise 11.8.4 (Duplication–loss scenario) *Determine the number of speciations, duplications and losses associated with the duplication–loss scenario depicted in Figure 11.20.*

11.8.2 Duplication, loss and transfer

Let T be a gene tree and T_{sp} a species tree on $\mathcal{X}$. The problem of reconciling the two trees becomes significantly more difficult once we allow horizontal gene transfer (HGT) events. The history of a gene can be very complicated, for example, the gene may transfer back and forth between different lineages many times, as indicated in Figure 11.21. Moreover, there are time constraints to consider, because a gene can only be transferred between two species that exist at the same time.

There are very many possible reconciliations of T and T_{sp} using DLT events. Our goal is to determine a reconciliation that uses only a minimum number of duplications and transfers. This allows us to ignore all those reconciliations that

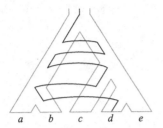

Figure 11.21 A complicated gene history involving many transfers.

cannot be minimum, or are not distinguishable from a simpler one based on the given input data. To this end, we make the following simplifying assumptions:

(i) A transfer edge e in T always starts in one species-tree edge f, connects directly to a second species-tree edge g and terminates there. In other words, we do not allow a transfer edge to continue through multiple species-tree edges, as is the case in Figure 11.22a.

(ii) The two species-tree edges f and g that a transfer edge e starts and ends in must be incomparable in T_{sp}, thus ruling out transfers between ancestors and descendants, such as in Figure 11.22b.

(iii) A transfer event always starts directly at the source node of a transfer edge e, thus excluding the situation illustrated in Figure 11.22c.

(iv) A gene-tree node v may only be the source of at most one transfer edge, ruling out the situation shown in Figure 11.22d.

These assumptions allow us to specify a DLT scenario by determining a mapping of the gene-tree nodes to the species-tree nodes; a partitioning of the gene-tree nodes into three types of nodes, namely speciation, duplication and transfer nodes; and a specification of the transfer edges. Here is a formal definition [237]:

Definition 11.8.5 (DLT scenario) *A DLT scenario Z for a gene tree T and a species tree T_{sp} on $\mathcal{X}$ is given by:*

- *a mapping γ of the gene-tree nodes to the species-tree nodes,*
- *a partitioning of the set of gene-tree internal nodes into three sets V_s, V_d and V_t, called the sets of speciation, duplication and transfer nodes, respectively, and*
- *a set E_t of transfer edges in T.*

These sets must fulfill the following conditions:

(i) *If v is a leaf of T with taxon label x, then $\gamma(v)$ is a leaf of T_{sp} with the same taxon label x.*

(ii) *If u is an internal gene-tree node with children v and w, then:*
   (a) *The node $\gamma(u)$ is not a proper descendant of $\gamma(v)$ or $\gamma(w)$.*
   (b) *At least one of $\gamma(v)$ and $\gamma(w)$ is a descendant of $\gamma(u)$.*

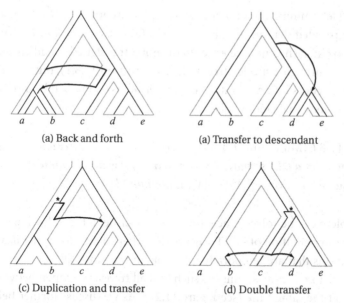

(a) Back and forth   (a) Transfer to descendant

(c) Duplication and transfer   (d) Double transfer

Figure 11.22 Here we show four different kinds of transfers (highlighted in bold) that cannot be identified by the comparison of a gene tree and a species tree and are therefore not represented by a DLT scenario.

(iii) *A gene-tree edge (v, w) is contained in E_t, if and only if $\gamma(v)$ and $\gamma(w)$ are incomparable.*

(iv) *If u is an internal gene-tree node with children v and w, then:*
   (a) *The node u is contained in V_t, if and only if either (u, v) is contained in E_t or (u, w) is contained in E_t.*
   (b) *The node u is contained in V_s, only if $\gamma(u) = \mathrm{lca}(\gamma(v), \gamma(w))$ and the two nodes $\gamma(v)$ and $\gamma(w)$ are incomparable.*
   (c) *The node u is contained in V_d, only if $\gamma(u)$ is an ancestor of $\mathrm{lca}(\gamma(v), \gamma(w))$.*

Let T be a gene tree and T_{sp} a species tree on $\mathcal{X}$. In a DLT scenario Z we represent a gene transfer by a gene-tree edge $e = (u, v)$. How do we determine which two species-tree edges this transfer edge connects?

The transfer edge e "takes off" from the species-tree edge f that leads into $\gamma(u)$. The "landing place" of a transfer edge $e = (u, v)$ in the species-tree is not uniquely defined and we allow it to lie on any species-tree edge on the species-tree path from $\mathrm{lca}(\gamma(u), \gamma(v))$ down to the node $\gamma(v)$. Additional constraints on the possible landing places come into play when we restrict our attention to so-called "acyclic" DLT scenarios further below.

Given a gene tree and a species tree on $\mathcal{X}$, our computational goal is to find a DLT scenario that reconciles the two given trees and uses a minimum number of duplications and transfers. Note that we do not seek to minimize the number of losses. Determining the number of losses that are implied by a DLT scenario is more complicated than in the case of duplication–loss scenarios. Not only do we have to look at each gene-tree edge that is not a transfer edge and check whether it continues through a species-tree node (as in Section 11.8.1), but we also have to take transfer edges into account. The number of losses implied by a transfer edge depends on its actual landing place.

Problem 11.8.6 (Minimum DLT scenario)   *Let T be a gene tree and T_{sp} a species tree on $\mathcal{X}$. Determine a DLT scenario for the two trees for which the number of duplication and transfer nodes given by $|V_d| + |V_t|$ is minimized.*

This problem can be solved efficiently, as we describe in the following subsection. However, our definition of a DLT scenario is perhaps a little too general, as it allows configurations that are temporally infeasible in the sense that one cannot assign time stamps to nodes and transfers such that all transfers happen between species that exist at the same time (see Figure 11.23). As we discuss further below, if we seek to rule out such DLT scenarios by restricting our attention to *acyclic* ones, then the problem of finding a DLT scenario that minimizes duplications and transfers becomes NP-hard.

Computation of a minimum DLT scenario

Let T be a gene tree and T_{sp} a species tree on $\mathcal{X}$. We use T_v to denote the subtree of the gene tree T that is rooted at the node v. A minimum DLT scenario for the two trees can be computed efficiently using dynamic programming, as we now describe.

We define $c(u, p)$ as the minimum cost of any DLT scenario for T_u and T_{sp} such that the gene-tree node u is mapped onto the species-tree node p. Any such DLT scenario classifies the node u either as a speciation, duplication or transfer node, and so the value $c(u, p)$ arises as the minimum of three minimum costs, namely $c_s(u, p)$, $c_d(u, p)$ and $c_t(u, p)$, respectively. These cost functions are defined and linked by the following recursion:

If u is a leaf of T, then we set

$$c(u, p) = \begin{cases} 0 & \text{if } u \text{ and } p \text{ have the same taxon label,} \\ \infty & \text{else.} \end{cases} \tag{11.9}$$

Otherwise, u is an internal node of T with children v and w, and we set

$$c(u, p) = \min\{c_s(u, p), c_d(u, p), c_t(u, p)\}, \tag{11.10}$$

where

$$c_s(u, p) = \begin{cases} \min\{c(v, q) + c(w, r) \mid r \perp q \text{ and } \text{lca}(q, r) = p\} & \text{if } p \text{ internal,} \\ \infty & \text{else,} \end{cases}$$

$$c_d(u, p) = \min\{1 + c(v, q) + c(w, r) \mid p \leq q \text{ and } p \leq r\}, \quad \text{and}$$

$$c_t(u, p) = \min\{1 + c(v, q) + c(w, r) \mid p \leq q \text{ and } p \perp r\}. \tag{11.11}$$

Here, for two species-tree nodes q and r we use the notation $q \perp r$ to indicate that q and r are incomparable in T_{sp} and we use $r \leq q$ to indicate that r is an ancestor of q.

The minimum cost of a DLT scenario for T and T_{sp} is then obtained by determining the best score for the root ρ_T of T, given by

$$\min\{c(\rho_T, p) \mid p \text{ is a node of } T_{sp}\}. \tag{11.12}$$

This recursion can be solved using the following algorithm.

Algorithm 11.8.7 (Minimum DLT scenario) *For a given gene tree T and species tree T_{sp} on X, the minimum cost of any DLT scenario for the two trees is computed as follows.*

Initialize $c(u, p) = \infty$ for all gene-tree nodes u and species-tree nodes p. Set $c(u, p) = 0$ for all pairs of leaf nodes that are labeled by the same taxon.

For each internal gene-tree node u in a post-order traversal, consider all species-tree nodes p in a post-order traversal and perform the following steps:

- *Let v and w be the two children of u.*
- *If p is an internal node, then let q and r be the two children of p and set*

$$c_s = \min\left(\min_{q \leq r'} c(v, q') + \min_{r \leq r'} c(w, r'), \min_{q \leq q'} c(w, q') + \min_{r \leq r'} c(v, r')\right). \tag{11.13}$$

- *Otherwise, p is a leaf and set $c_s = \infty$.*
- *Set $c_d = \min_{p \leq p'} c(v, p') + \min_{p \leq p'} c(w, p') + 1$.*
- *Set $c_t = \min\left(\min_{p \leq p'} c(v, p') + \min_{p \perp p'} c(w, p'), \min_{p \leq p'} c(w, p') + \min_{p \perp p'} c(v, p')\right) + 1$.*
- *Set $c(u, p) = \min\{c_s, c_d, c_t\}$.*

Return $\min_{p \text{ in } T_{sp}} c(\rho_T, p)$.

Once we have determined the minimum possible score, a concrete DLT scenario that attains the optimal score can be determined by performing a recursive *traceback* in which for each internal gene-tree node u we determine which species-tree node p and which of the three functions c_s, c_d or c_t was used to obtain the final minimum.

We then set $\gamma(u) = p$ and classify u as a species, duplication or transfer node, accordingly. Note that this algorithm can be extended to weight duplications and transfers differently.

Acyclic DLT scenarios

Let T be a gene tree and T_{sp} a species tree on $\mathcal{X}$. In the definition of a DLT scenario for T and T_{sp} we explicitly excluded the possibility of a transfer between an ancestor and one of its descendants. This restriction can be justified by arguing that species must exist at the same time for a horizontal gene transfer to be able to happen between them. (While the latter argument looks reasonable at first glance, it can easily be refuted by allowing for the existence of additional species that are not part of the taxon set under consideration. In this case, a transfer between between a node and a descendant should be counted as at least two events.) We refer to this requirement as *temporal feasibility*, which essentially means that one can assign time stamps to the nodes of both trees in such a way that transfers occur only between coexisting nodes.

Unfortunately, temporal feasibility can be affected by a chain of multiple transfers, even if each of the transfers involved is temporally feasible on its own. To address this problem we now introduce the concept of an *acyclic* DLT scenario that ensures that a reconciliation of a gene tree and a species tree is always *temporally feasible*.

Definition 11.8.8 (Acyclic DLT scenario)   *Let T be a gene tree and T_{sp} a species tree on $\mathcal{X}$. A DLT scenario Z for T and T_{sp} is called acyclic, if we can assign a unique number $\pi(v)$ to every node v of the species tree T_{sp} in such a way that we have:*

(i) *The number $\pi(v)$ of any node v is less than the number of any of its children.*
(ii) *If $e = (u, v)$ and $e' = (u', v')$ are two transfer edges in T and if v' is a descendant of v or equal to v (thus allowing $e = e'$), then the number $\pi(p)$ assigned to the parent p of $\gamma(u)$ is less than the number $\pi(\gamma(v'))$.*

In particular, the second condition in the definition states that if one transfer edge e in the gene tree T is followed by a second transfer edge e', then the landing place of e' in the species tree T_{sp} is not allowed to lie above the parent of the take off place of e, as this would give rise to a cycle. In Figure 11.23, we display an example of the kind of configuration that this definition excludes.

Let T be a gene tree and T_{sp} a species tree on $\mathcal{X}$. Unfortunately, the problem of computing an acyclic DLT scenario for T and T_{sp} of minimum cost is known to be NP-hard, as shown in [237]. However, in that paper it is also stated that optimal DLT scenarios are usually acyclic anyway and so in practice it may be sufficient to use Algorithm 11.8.7.

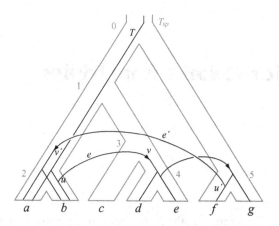

Figure 11.23 A DLT scenario with species tree T_{sp} (in gray) and gene tree T (in black). The internal nodes of the species tree T_{sp} are numbered in accordance with condition (i) of the definition of an acylic DLT scenario, whereas the nodes of two transfer edges e and e' are labeled u, v, u' and v', respectively, fulfilling the prerequisite of condition (ii). Note that the parent p of the node $\gamma(u)$ is equal to $\gamma(v')$ and thus the two nodes p and $\gamma(v')$ have the same number, two. In consequence, condition (ii) is violated and so the depicted DLT scenario is cyclic.

11.8.3 Application

Due to the lack of a suitable implementation, the described algorithm for computing a DLT scenario is currently not used in practice. However, DLT scenarios are of practical interest and in Figure 4.6 we show a published example that is based on the tad locus of the human pathogen *Actinobacillus actinomycetemcomitans* [197]. The network was constructed with the help of the TreeMapper software [48], see Section 14.5.

Chapter notes

Consensus split networks were introduced in [113]. The Z-closure method and filtered super networks were introduced in [126], [132] and [243]. The section on minimum hybridization networks is based on [15, 24, 25, 26]. Multi-labeled trees are discussed in [117] and [22]. The section on DLT reconciliation is based on [100, 237, 238]. Other work in this area includes [48, 86].

The split-based networks described in this chapter were all implemented in SplitsTree [125]. The nested-label algorithm for multi-labeled trees is available in PADRE [167] and also in Dendroscope [120], and the latter program also contains the cluster-based algorithm, too.

Phylogenetic networks from triples or quartets

In this chapter, we look at some methods for computing rooted phylogenetic networks from rooted triples and unrooted phylogenetic networks from unrooted quartet topologies.

12.1 Trees from rooted triples

A *rooted triple* is given by a bifurcating, rooted phylogenetic tree on a set of three taxa $\{x, y, z\}$. We write $\frac{yz}{x}$ to indicate that the most recent common ancestor of y and z is more recent than that of x and y, or x and z. There are three possible rooted bifurcating trees, and consequently rooted triples, on a set of three taxa (see Figure 12.1).

Let T be a rooted phylogenetic tree on $\mathcal{X}$. We use $T|_{\{x,y,z\}}$ to denote the *restriction* of T to the three leaves labeled by $\{x, y, z\}$ (see Definition 3.16.3). We say that a rooted triple t on three taxa x, y, z is *contained* in T, if the restriction $T|_{\{x,y,z\}}$ equals t.

Any rooted tree T can be unambiguously described by its rooted triple encoding $R(T)$ (see, for example, [94]).

Definition 12.1.1 (Rooted triple encoding) *Let T be a rooted phylogenetic tree on $\mathcal{X}$. The set $R(T)$ of all rooted triples contained in T is called the* rooted triple encoding *of T.*

Given a rooted phylogenetic tree T on $\mathcal{X}$, one can easily produce the set $R(T)$ of rooted triples contained in T in a single traversal of the tree: At each node v record all rooted triples $\frac{yz}{x}$ such that x does not label a descendant of v, whereas y and z label descendants of two different children of v.

Now, assume that we are given a set of rooted triples R on $\mathcal{X}$. How to determine whether R is the rooted triple encoding of some unknown rooted phylogenetic tree T on $\mathcal{X}$? Moreover, if such a tree exists, how do we compute it? To address these two questions, we introduce the following:

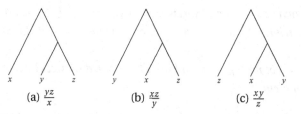

Figure 12.1 The three possible rooted triples on a set of three taxa $\{x, y, z\}$.

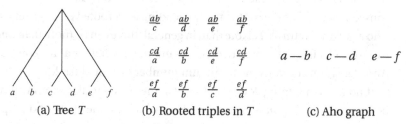

| (a) Tree T | (b) Rooted triples in T | (c) Aho graph |

Figure 12.2 (a) A rooted phylogenetic tree T on $X = \{a, b, c, d, e, f\}$. (b) The set R of all 12 rooted triples contained in T. (c) The Aho graph $AG(R)$ for R.

Definition 12.1.2 (Aho graph)   *Let R be a set of rooted triples on X. The Aho graph $AG(R) = (V, E)$ associated with R has node set $V = X$ and any two nodes y and z are connected by an edge in E if and only if there exists a third taxon x such that the rooted triple $\frac{yz}{x}$ is present in R.*

See Figure 12.2c for an example of the Aho graph.

The *Aho algorithm* [3] below builds a tree top-down from the root to the leaves, in terms of clusters first identifying the largest clusters, then those contained in them, and so on. The algorithm is guided by the Aho graph, operating on smaller and smaller induced sets of rooted triples.

Let R be a set of rooted triples on X and let X' be a subset of X. The set $R|_{X'}$ of rooted triples *induced* by X' consists of all rooted triples $\frac{yz}{x} \in R$ for which all three taxa x, y, z are contained in X'.

Algorithm 12.1.3 (Aho algorithm)   *Given a set of rooted triples R on X, the aim is to construct a rooted phylogenetic tree T on X that contains R, if one exists.*

If X contains only one element, say x, then return the tree consisting of a single node labeled x.

If X contains two elements, say x and y, then return the tree that has two leaves labeled x and y, respectively.

Otherwise, compute the Aho graph $AG(R)$.

If $AG(R)$ has only one connected component, then the algorithm reports fail.

Else, for each node set U of a connected component of AG(R), determine the set of induced rooted triples R|$_U$ and recursively compute the rooted phylogenetic subtree T(R|$_U$).

Finally, create a root node ρ and combine all computed subtrees by connecting ρ to the root of each of them.

This classic algorithm computes a minimal rooted phylogenetic tree T that contains a given set of rooted triples R, if one exists, where minimal means that no internal edge of T can be contracted so that the resulting tree also contains R [213]. In consequence, if R is the rooted triple encoding for a rooted phylogenetic tree T, the Aho algorithm returns T. Note that, in general, there can be more than one minimal tree for a given set of rooted triples. It is unknown whether a tree computed by the Aho algorithm contains the minimum number of rooted triples.

If no tree containing R exists, then at some stage the algorithm will report 'fail' because an *obstruction* is encountered, by which we mean an Aho graph that has only one connected component.

In practice, this algorithm usually reports fail. How to modify it such that it always produces a useful result? Unfortunately, the problem of constructing a rooted phylogenetic tree that contains a maximum subset of R is known to be NP-hard [30, 138, 244]. One heuristic to deal with an encountered obstruction O is to determine a minimum *cut set* of edges in O whose removal will produce a graph with exactly two connected components. After deletion of these edges the algorithm can proceed and there is no need to report fail [192, 215].

12.2 Level-*k* networks from rooted triples

Let R be a set of rooted triples on $\mathcal{X}$. If R is not contained in any rooted phylogenetic tree, then one possibility is to determine a rooted phylogenetic tree that is consistent with as many rooted triples as possible, as mentioned above. A second alternative is that we search for a rooted phylogenetic network N that contains all rooted triples in R, in the following sense:

Definition 12.2.1 (Rooted triples contained in a network) *A rooted triple $t = \frac{yz}{x}$ is said to be* contained *in a rooted phylogenetic network N, if there exist two nodes u and v in N such that the following conditions hold (see Figure 12.3):*

  (i)   *There exists a directed path from u to the node labeled x.*
 (ii)   *There exists a directed path from u to v.*
(iii)   *There exists a directed path from v to the node labeled y.*
 (iv)   *There exists a directed path from v to the node labeled z.*
  (v)   *All four paths are node-disjoint (except at their end-points).*

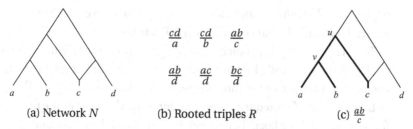

(a) Network N (b) Rooted triples R (c) $\frac{ab}{c}$

Figure 12.3 (a) A rooted phylogenetic network N. (b) The set R of rooted triples contained in N. (c) To show that the rooted triple $\frac{ab}{c}$ is indeed contained in N, here we indicate two nodes u and v that are connected to each other and to the leaves labeled a, b, c by four disjoint paths (highlighted in bold), as required by Definition 12.2.1.

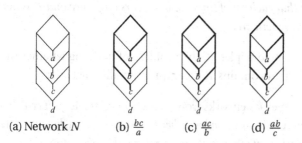

(a) Network N (b) $\frac{bc}{a}$ (c) $\frac{ac}{b}$ (d) $\frac{ab}{c}$

Figure 12.4 (a) A phylogenetic network N containing all possible rooted triples on $\mathcal{X} = \{a, b, c, d\}$. For example, all three possible rooted triples on a, b, c are highlighted in (b), (c) and (d). This construction is easily generalized to a taxon set of arbitrary size by adding more taxa below d.

An equivalent formulation of this definition is the following: we say that a rooted triple t is contained in a rooted phylogenetic network N, whenever t is contained in some rooted phylogenetic tree T that is contained in N.

It is not difficult to find a rooted phylogenetic network N that contains a given set of rooted triples R on $\mathcal{X}$. Indeed, Figure 12.4 shows how to construct a rooted phylogenetic network that contains all possible rooted triples on $\mathcal{X}$. However, a network obtained in this way is not informative, as it contains *all three* possible rooted triples for any given set of three taxa.

To obtain a more useful network, we restrict our attention to phylogenetic networks of limited complexity. More precisely, we consider level-k networks for some small value of k. Recall from Definition 6.11.5 that a bicombining, rooted phylogenetic network has level k, if the number of reticulations in any biconnected component of the network is at most k.

In the previous section we saw that we can use the Aho algorithm to efficiently decide whether a rooted phylogenetic tree exists that contains a given set of rooted triples R on $\mathcal{X}$. Unfortunately, for $k \geq 1$ the problem to decide whether there exists a level-k network N that contains R is NP-hard, in general [139].

To obtain a tractable problem, consider the following definition. A set of rooted triples R on $\mathcal{X}$ is called *dense*, if for every set of three taxa $\{x, y, z\}$ the set R contains at least one of the three possible rooted triples, $\frac{yz}{x}$, $\frac{xz}{y}$ or $\frac{xy}{z}$.

If a dense set of rooted triples R contains more than one of the three possibilities for some set of three taxa, then we can be sure that R is not contained in some tree. However, the converse does not hold. Even if R contains at most one rooted triple for every choice of three taxa in $\mathcal{X}$, it may still be the case that R is not contained in any rooted phylogenetic tree.

Exercise 12.2.2 (Non-treelike dense triples)   *Design a dense set of rooted triples R on four taxa such that each set of three taxa appears only once and R is not contained in any rooted phylogenetic tree.*

For a dense set of rooted triples R, the problem of deciding whether there exists a level-k network N that contains R is computationally tractable:

Theorem 12.2.3 (Level-k networks from dense rooted triples)   *Let R be a dense set of rooted triples on $\mathcal{X}$. For any fixed number $k \geq 0$, to decide whether a level-k rooted phylogenetic network exists that contains R takes only polynomial time.*

We do not prove this result here. A proof for $k = 1$ can be found in [140], for $k = 2$ in [133], and for all k in [236]. In order to give a flavor of the approach, in the next section we discuss the problem of computing a level-1 network for a given set of rooted triples. The key idea is to modify the Aho algorithm (12.1.3) so as to attempt to resolve each encountered obstruction by postulating a single reticulation. To formulate this, we need to introduce the concept of a "collapsible" set of taxa.

Definition 12.2.4 (Collapsible with respect to R)   *Let R be a set of rooted triples on $\mathcal{X}$. A set of taxa $Y \subset \mathcal{X}$ is called* collapsible *(with respect to R), if for every rooted triple $\frac{yz}{x}$ in R we have: if x and y are contained in Y, then so is z.*

Let R be an arbitrary set of rooted triples on $\mathcal{X}$. By definition, every single element subset $\{x\}$ of $\mathcal{X}$ is trivially collapsible, as is $\mathcal{X}$ itself. Now, consider the set R of all rooted triples contained in the rooted phylogenetic network N on $X = \{a, b, c, d\}$ shown in Figure 12.3. Which collapsible sets does R contain in addition to the trivial ones just mentioned? Only $\{a, b\}$. All other proper subsets of R, namely $\{b, c\}, \{b, c, d\}, \ldots, \{b, c, d\}$ are not collapsible. For example, $Y = \{b, c\}$ is not collapsible with respect to R because $\frac{cd}{b} \in R$ implies that Y must also contain d.

Lemma 12.2.5 (Collapsible sets for dense triples are compatible) *Let R be a dense set of rooted triples on $\mathcal{X}$. Any two collapsible sets of R are compatible (in the sense of clusters).*

Proof Let R be a dense set of rooted triples on $\mathcal{X}$. Assume there exist two collapsible sets Y_1 and Y_2 that are not compatible, that is, for which $Y_1 - Y_2 \neq \emptyset$, $Y_2 - Y_1 \neq \emptyset$, and $Y_1 \cap Y_2 \neq \emptyset$ hold. Choose three taxa $x \in Y_1 - Y_2$, $y \in Y_2 - Y_1$, and $z \in Y_1 \cap Y_2$. As R is dense, the set R must contain at least one rooted triple t on $\{x, y, z\}$. There are three possibilities for t, namely $t = \frac{yz}{x}$, $t = \frac{xy}{z}$ or $t = \frac{xz}{y}$. In the first two cases, because $x, z \in Y_1$ and Y_1 is collapsible, we must have $y \in Y_1$, a contradiction. In the third case, because $y, z \in Y_2$ and Y_2 is collapsible, we must have $x \in Y_2$, again obtaining a contradiction. □

If R is not dense, then two different sets that are collapsible (with respect to R) are not necessarily compatible, even if R is contained in a rooted phylogenetic tree. In this case, the tree must contain a node u of outdegree greater than two and any three taxa x, y, z that label leaves contained in different subtrees below u will not be present in a rooted triple in R.

For example, the root node ρ of the tree T displayed in Figure 12.2 has outdegree three. So, the set of rooted triples displayed by T does not contain any rooted triple on x, y, z with $x \in \{a, b\}$, $y \in \{c, d\}$ and $z \in \{e, f\}$. Please convince yourself that $Y_1 = \{a, b, c, d\}$ and $Y_2 = \{c, d, e, f\}$ are two collapsible sets that are incompatible with each other.

12.2.1 Computing a level-1 network

Let R be a dense set of rooted triples on $\mathcal{X}$. We now want to discuss how to compute a level-1 network N that contains R, if one exists (recall that a level-1 network is a galled tree). When running the Aho algorithm, there are only two cases to consider:

Lemma 12.2.6 (Number of components on dense triples) *Let R be a dense set of rooted triples on $\mathcal{X}$. The number of connected components of the Aho graph $AG(R)$ is at most two.*

Proof The Aho graph $AG(R)$ cannot have more than two components, because, by definition of the graph, any three taxa chosen from three different components do not correspond to a rooted triple in R. This contradicts the assumption that R is dense. □

Let R be a dense set of rooted triples on $\mathcal{X}$ and assume that the Aho graph $AG(R)$ has only one component. We shall attempt to resolve this obstruction by postulating a reticulation event, as described in the following.

Let $\mathcal{H} = \{H_1, \cdots, H_k\}$ denote the set of all maximal proper subsets of $\mathcal{X}$ that are collapsible with respect to R. Note that any two sets in $\mathcal{H}$ are disjoint, because $\mathcal{H}$ is compatible by Lemma 12.2.5, and that $\mathcal{H}$ covers $\mathcal{X}$, because $\{x\}$ is collapsible, for any taxon x.

The first step is to *collapse* each collapsible set $H_i \in \mathcal{H}$ by setting $\mathcal{X}' = \mathcal{H}$ and defining the set of rooted triples on $\mathcal{X}'$ as:

$$R' = \left\{ \frac{H_y H_z}{H_x} \mid H_x, H_y, H_z \in \mathcal{X}', \exists x \in H_x, y \in H_y, z \in H_z \text{ with } \frac{yz}{x} \in R \right\}.$$

$$(12.1)$$

Next, for each choice of a taxon $H_r \in \mathcal{X}'$, we do the following. Remove H_r from $\mathcal{X}'$ and remove all rooted triples involving H_r from R'. Run the Aho algorithm on the resulting set of rooted triples R'' to determine whether there exists a rooted phylogenetic tree T that contains R''. If such a tree exists, then attempt to attach H_r to the tree T using two new edges in such a way that all rooted triples in R' that involve H_r are contained in the resulting network N.

If we are unable to identify such a reticulate taxon H_r, then we cannot resolve the obstruction and the algorithm returns *fail*. In this case, no level-1 network exists that contains the set R of rooted triples.

On the other hand, if we are successful, then the original obstruction has been overcome by attaching taxon H_r as a reticulate node. In this case, for each collapsed taxon $H_i \in \mathcal{X}'$, we run the algorithm on the set of taxa $H_i \subsetneq \mathcal{X}$ and set of induced rooted triples $R|_{H_i} \subsetneq R$ to obtain a level-1 network N_{H_i}. If this, in turn, is successful, then we replace H_i by N_{H_i} in N.

Here is a summary of the algorithm described in [134]:

Algorithm 12.2.7 (Level-1 from dense rooted triples)   *Given a dense set of rooted triples R on $\mathcal{X}$, the aim is to construct a level-1 network N that contains R, if one exists.*

If $\mathcal{X}$ contains only one element, x, then return the tree consisting of a single node labeled x.

Otherwise, compute the Aho graph $AG(R)$.

If $AG(R)$ has only one connected component, then attempt to resolve the obstruction by collapsing collapsible sets of taxa and then postulating an appropriate reticulation, as described above. If this is not possible, the algorithm reports fail. Otherwise, recursively run the algorithm on each collapsed set of taxa.

Else, the graph $AG(R)$ must have two connected components. Recursively compute a network for each of them.

Create a root node ρ and connect it to the roots of the recursively computed networks.

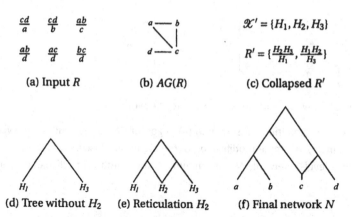

$$\frac{cd}{a} \quad \frac{cd}{b} \quad \frac{ab}{c}$$

$$\frac{ab}{d} \quad \frac{ac}{d} \quad \frac{bc}{d}$$

(a) Input R (b) $AG(R)$ (c) Collapsed R'

$\mathcal{X}' = \{H_1, H_2, H_3\}$

$R' = \{\frac{H_2 H_3}{H_1}, \frac{H_1 H_2}{H_3}\}$

(d) Tree without H_2 (e) Reticulation H_2 (f) Final network N

Figure 12.5   An example of how the level-1 algorithm works. (a) A dense set of rooted triples R on $\mathcal{X} = \{a, b, c, d\}$. (b) The Aho graph $AG(R)$. (c) There are three collapsible, maximal proper subsets in $\mathcal{X}$, $H_1 = \{a, b\}$, $H_2 = \{c\}$, and $H_3 = \{d\}$. Collapsing them produces the taxon set $\mathcal{X}'$ and dense set of rooted triples R'. (d) If we remove H_2 from $\mathcal{X}'$ then the remaining two taxa H_1 and H_3 can be fitted onto a tree T. (e) Attaching H_2, as shown here, produces a network that contains R'. (f) After uncollapsing $\mathcal{X}'$ we obtain the final phylogenetic network N.

We illustrate the algorithm in Figure 12.5. We start with the dense set of rooted triples R on $\mathcal{X} = \{a, b, c, d\}$ shown in Figure 12.5a. The Aho graph $AG(R)$ is connected, so R cannot be represented by a tree and we must attempt to resolve the obstruction by postulating a reticulation. To this end, we identify the set $\mathcal{H}$ of all collapsible, maximal proper subsets of $\mathcal{X}$. There is only one non-trivial collapsible subset, namely $\{a, b\}$, and two trivial ones, $\{c\}$ and $\{d\}$. We construct a taxon set $\mathcal{X}'$ in which each taxon is one of the three sets in $\mathcal{H}$, which we denote by $H_1 = \{a, b\}$, $H_2 = \{c\}$ and $H_3 = \{d\}$. The corresponding set of rooted triples is given by $R' = \{\frac{H_2 H_3}{H_1}, \frac{H_1 H_2}{H_3}\}$. This set is obtained by replacing all occurrences of a or b in R by H_1, all occurrences of c by H_2 and all occurrences of d by H_3, in R.

Removal of any of the three taxa in $\mathcal{X}'$ leads to a set of two taxa, which can always be represented by a tree. If we remove H_2 and then attach it by two edges, as shown in Figure 12.5e, then we obtain a network that contains all rooted triples in R'. Then, expanding the collapsed set $H_1 = \{a, b\}$ gives rise to the final output of the algorithm, a bifurcating level-1 network N that contains R, shown in Figure 12.5f.

The above algorithm will find a level-1 network representing R, if one exists. However, as formulated, it is not guaranteed to find a level-1 network that contains only a minimum number of reticulations. Superfluous reticulations occur when the postulated reticulate taxon is a collapsed set of taxa that contains some original

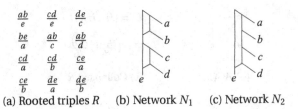

(a) Rooted triples R (b) Network N_1 (c) Network N_2

Figure 12.6 For the set of rooted triples R shown in (a), Algorithm 12.2.7 produces the level-1 network N_1 shown in (b). A simple modification of the algorithm described in the text is necessary to obtain a level-1 network N_2 that minimizes the number of reticulations, as displayed in (c).

taxa that do not necessarily have to be placed below the reticulation. Consider, for example, the input set of rooted triples R listed in Figure 12.6a. Algorithm 12.2.7 collapses the collapsible subset $\{c, d, e\}$ of taxa and then constructs the network N_1 shown in Figure 12.6b, which has two reticulations. This network contains R, but is not minimum, because the network N_2 in Figure 12.6c has only one reticulation and also contains R.

Algorithm 12.2.7 can be modified so as to always construct a level-1 network containing a minimum number of reticulations (if such a network exists). The modified version loops through all collapsible subsets H_r' of the postulated reticulate taxon H_r and attempts to place H_r', instead of H_r, below the reticulation. In the example displayed in Figure 12.6, $H_r' = \{e\}$ is a collapsible subset of $H_r = \{c, d, e\}$ and putting e below a reticulation leads to the minimum network N' that is displayed in Figure 12.6c. A dynamic programming algorithm can be used to find such an optimal network in polynomial time [134].

The described algorithm for constructing a level-1 network for a dense set of rooted triples R, if one exists, can be generalized to level-2. In [236], the authors describe an algorithm to compute in polynomial time, for every fixed k, the level-k network with minimum number of reticulations (if such a network exists).

12.3 The quartet-net method

The *quartet-net method* (or QNet, for short) [91], takes as input a collection of weighted quartet topologies on $\mathcal{X}$ and produces as output a set of weighted splits on $\mathcal{X} = \{x_1, \ldots, x_n\}$ that is circular, using a modification of the neighbor-net method, which is described in Section 10.4.

More precisely, for every quadruple $Q = \{w, x, y, z\} \subseteq \mathcal{X}$ of four distinct taxa the input must provide a non-negative weight $\omega(\hat{Q})$ for each of the three possible non-trivial splits (called *quartet topologies*) $\hat{Q}$, namely $\frac{\{w,x\}}{\{y,z\}}$, $\frac{\{w,y\}}{\{x,z\}}$ and $\frac{\{w,z\}}{\{x,y\}}$. Just as the neighbor-net method, the quartet-net method proceeds in three steps:

(i) The main computation is to determine a circular ordering Z of $\mathcal{X}$.

(ii) Then a set of weighted splits S is computed that respect the circular ordering Z.

(iii) Finally, the splits are processed by the circular network algorithm so as to obtain a split network N.

We assume that the reader is already familiar with our description of the neighbor-net method (see Section 10.4) and so this exposition is quite brief. Just as the neighbor-net method, the algorithm uses an auxiliary graph G whose node set is given by $\mathcal{X}$. Initially, the graph consists of n isolated nodes $x_1, \ldots, x_n$.

Whereas the neighbor-net method maintains the graph G as a collection of isolated nodes ("singles") and pairs of nodes connected by an edge ("couples"), the quartet-net method never removes any nodes but rather maintains G as a collection of paths and isolated nodes.

While the auxiliary graph G still consists of four or more connected components, the algorithm repeatedly does the following:

Choose a pair of paths P_i and P_j in G that maximizes the following *paths score:*

$$ps(P_i, P_j) = \sum_{\substack{P_k \neq P_i, P_j \\ P_l \neq P_i, P_j}} \sum_{\substack{x_i \in P_i, x_j \in P_j, \\ x_k \in P_k, x_l \in P_l}} \frac{\omega\left(\frac{\{x_i, x_j\}}{\{x_k, x_l\}}\right)}{|P_i||P_j||P_k||P_l|}, \tag{12.2}$$

where the first sum is over all pairs of other paths P_k and P_l and the second sum is over all choices of taxa, one from each path. Here $|P_i|$ denotes the number of nodes of the path P_i.

The next step is to decide which ends of the two paths are to be joined together by a new edge to merge the two paths. To discuss this, let us assume that the nodes in the two paths P_i and P_j are labeled as follows:

$$u_s - u_{s-1} - \cdots - u_1 - u_0 \text{ and } v_0 - v_1 - \cdots - v_{t-1} - v_t,$$

and that our goal is to evaluate the choice of connecting nodes u_0 and v_0 to obtain a new chain

$$u_s - u_{s-1} - \cdots - u_1 - u_0 - v_0 - v_1 - \cdots - v_{t-1} - v_t.$$

The *ends score* for this choice is obtained as the sum of weights over all quadruples that have three or four taxa in the new path and are compatible with the path:

$$es(u_0, v_0) = \sum_{0 \leq k < l \leq s} \sum_{0 \leq p \leq t} \sum_{x \in \mathcal{X} - (P_i \cup P_j)} \omega\left(\frac{\{u_k, v_p\}}{\{u_l, x\}}\right)$$

$$+ \sum_{0 \leq p \leq s} \sum_{0 \leq k < l \leq t} \sum_{x \in \mathcal{X} - (P_i \cup P_j)} \omega\left(\frac{\{u_p, v_k\}}{\{v_l, x\}}\right)$$

$$+ \sum_{0 \leq k < l \leq s} \sum_{0 \leq p < q \leq t} \omega\left(\frac{\{u_k, v_p\}}{\{u_l, v_q\}}\right). \tag{12.3}$$

The two ends that give rise to the largest score are merged.

When only three connected components are left in the graph, then we do not use the paths score to determine which two paths are to be merged next, but rather for each of the (up to) eight possible pairings of the (up to) six different path ends, we use the ends scores to decide how to merge all three components into a single cycle Z.

Once the circular ordering $Z = (x_1, x_2, \ldots, x_n)$ has been computed in this way, we construct the set S of all $\binom{n}{2} - n$ non-trivial splits on $\mathcal{X}$ that respect the circular ordering Z, just as in the neighbor-net method. Note that the quartet-net method does not provide any information on the weights of the n trivial splits.

The weights of the splits in S are calculated using the least squares method. To this end, we represent the quartet weights as a $3\binom{n}{4}$-dimensional vector

$$
\mathbf{w} = \begin{pmatrix} \omega\left(\frac{\{x_1, x_2\}}{\{x_3, x_4\}}\right) \\ \omega\left(\frac{\{x_1, x_2\}}{\{x_3, x_5\}}\right) \\ \cdots \\ \omega\left(\frac{\{x_{n-3}, x_{n-2}\}}{\{x_{n-1} x_n\}}\right) \end{pmatrix}. \tag{12.4}
$$

The set of quartets is represented by an $3\binom{n}{4} \times \left(\binom{n}{2} - n\right)$-dimensional matrix A in which each row of A corresponds to a quartet and each column corresponds to a split, that is,

$$
A_{(i,j,k,l)t} = \begin{cases} 1 & \text{if } \frac{\{x_i, x_j\}}{\{x_k, x_l\}} = S_t|_{\{x_i, x_j, x_k, x_l\}}, \\ 0 & \text{else.} \end{cases} \tag{12.5}
$$

Let $\mathbf{b}$ be the $\binom{n}{2}$-dimensional vector representing the weights of the splits. The ordinary least square estimate of $\mathbf{b}$ is given by

$$
\mathbf{b} = (A^T A)^{-1} A^T \mathbf{w}. \tag{12.6}
$$

Again, as in the case of the neighbor-net method, we are more interested in solving the least-squares approximation under the constraint that all entries of the solution vector are non-negative.

Finally, the circular ordering Z and set of splits S are passed to the circular network algorithm (see Section 7.2) to compute an actual split network N. We have the following result:

Theorem 12.3.1 (Consistency of quartet-net) *Let Q be a collection of weighted quartet trees on $\mathcal{X}$ arising from a distance matrix D. The set of weighted splits S computed by quartet-net represents D if and only if D is circular.*

The proof of this result can be found in [93]. Because compatible splits are always circular, it follows that the quartet-net method (combined with the circular network algorithm) always computes the correct phylogenetic tree, when given an input set that corresponds to a tree.

Chapter notes

The two methods described in this chapter, which construct a phylogenetic network directly from triples or quartets, are both very recent and so far have not been used in any biological studies. Moreover, actual rooted triple sets are rarely dense so the algorithms presented in Section 12.2 may be of limited use in practice. An alternative approach is to construct a level-k network that is consistent with as many of the input triples as possible, relaxing the requirement that the input set be dense. This is a NP-hard problem and a heuristic for $k = 1$ can be found in [115]. If given a dense set of triples that is fully consistent with some level-1 network, this algorithm computes such a network. Otherwise, it produces a level-1 network that is guaranteed to be consistent with at least $\frac{5}{12}$ of the input triples.

13

Drawing phylogenetic networks

The algorithms presented in the previous chapters compute a phylogenetic tree or network as a labeled graph, possibly with edge lengths. In this chapter, we discuss the problem of assigning coordinates to the nodes (and sometimes also to points along the edges) of a given phylogenetic tree or network so that it can be drawn by a computer program. The results presented here are based on [62, 77, 124].

13.1 Overview

Figure 13.1 shows the relationships between some of the main concepts introduced in this chapter. The focus of this chapter is on how to draw *phylogenetic trees* and *networks*.

A *rooted* network or tree can be depicted either as a *phylogram*, in which the edges are drawn to scale and thus represent, for example, distances or the number of mutations along an edge, or as a *cladogram*, which aims at representing only the topology of the network. Both types of drawings come in a number of different variants, including *triangular*, *rectangular* and *circular* diagrams. We need to distinguish between a *combining view* and a *transfer view* of a rooted phylogenetic network. In the former, all in-edges of a reticulate node are drawn as reticulate edges, whereas as in the latter one in-edge is drawn as a tree edge and all other in-edges are drawn as directed edges that indicate *transfer events*, see Section 13.6.

An *unrooted* network or tree is usually drawn to scale as a *radial* diagram.

We first discuss algorithms for drawing trees and then modify these so as to obtain algorithms for drawing phylogenetic networks. We do this in detail for the combining view of rooted phylogenetic networks and then indicate the modifications necessary to accommodate the transfer view, too.

13.2 Cladograms for rooted phylogenetic trees

In the context of phylogenetic trees, a *cladogram* is a representation of a rooted phylogenetic tree in which the main goal is to show the topology of the tree as

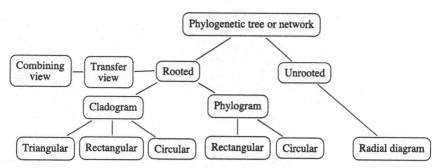

Figure 13.1   Overview of the main concepts introduced in this chapter.

clearly as possible, in particular, showing how clusters are nested within each other. Usually, the leaves of the tree are equally spaced along a base-line or around a circle (see Figure 13.2) and the internal nodes of the tree are placed at different levels that indicate how near (in terms of the number of intermediate edges) they are to the leaves of the tree. Edge lengths are chosen to facilitate this and are not interpreted, for example, in terms of distance or mutations.

Let T be a rooted phylogenetic tree on a taxon set $\mathcal{X}$, with root ρ. As usual, we assume that all edges of T are directed away from the root. The *level of a node* v is defined as follows: if v is a leaf, then we set $\text{level}_T(v) = 0$, and otherwise we set $\text{level}_T(v)$ to the maximum number of edges in any path from v to some leaf of T. We can compute the levels of all nodes in linear time: in a postorder traversal of T, set the level of any internal node v to the maximum level of any of its children, plus 1.

We define a *topological embedding* of T to be a mapping τ that assigns to each node $v \in V$ an ordering $\tau(v) = (v_1, \ldots, v_k)$ of the set of its children $\{v_1, \ldots, v_k\}$.

Any drawing of T defines such a topological embedding: For any node $v \neq \rho$, consider a small circle c around v. Starting at the point where c intersects the in-edge of v, list the target nodes of all out-edges in the order in which you encounter them when traveling once around c in anti-clockwise order. A similar construction produces an ordering for the root ρ, but, because the root does not have an in-edge, one may break the circular ordering around ρ at an arbitrary place to obtain a linear ordering of the children of ρ.

To construct a cladogram for T, one must first determine a topological embedding τ for T. Often, this is done only implicitly by using the order in which the nodes where attached to the tree during computation or input of the tree. All cladograms for trees can be constructed by some variant of the following basic approach.

Algorithm 13.2.1 (Triangular cladogram for trees)   *To compute a* triangular clado-gram, *as shown in Figure 13.2a, we define a map* $\kappa : V \to \mathbb{R}^2$, $\kappa(v) = (x(v), y(v))$

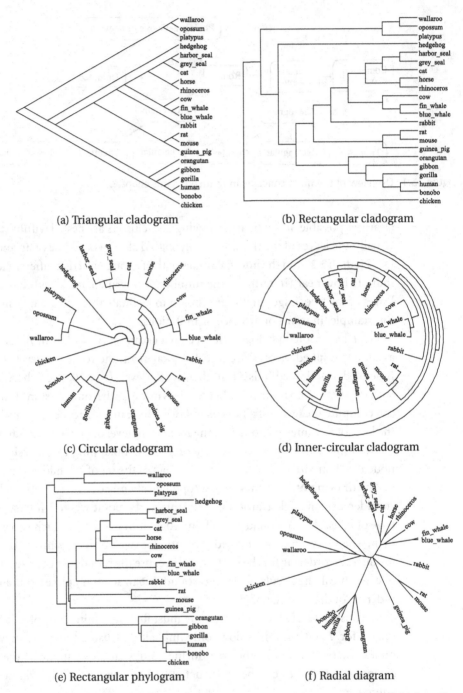

(a) Triangular cladogram

(b) Rectangular cladogram

(c) Circular cladogram

(d) Inner-circular cladogram

(e) Rectangular phylogram

(f) Radial diagram

Figure 13.2 Six different representations of a phylogenetic tree containing mammals and chicken. The first four diagrams are cladograms, the fifth is a phylogram and the sixth is an unrooted radial diagram.

that assigns Cartesian coordinates to each node v in a postorder traversal of the tree T. Starting at the root ρ, proceed as follows:

(i) Process all children of v in the order prescribed by $\tau(v) = (v_1, \ldots, v_k)$.

(ii) If v is a leaf, then set $\kappa(v) = (0, i)$, where i is the number of leaves visited so far.

(iii) Otherwise, let y_{min} and y_{max} be the minimum and maximum y-coordinates of all leaves below v. Set $\kappa(v) = (\frac{y_{min} - y_{max}}{2}, \frac{y_{min} + y_{max}}{2})$.

To draw the tree, for each edge (v, w), draw a line from v to w (that is, from $\kappa(v)$ to $\kappa(w)$, to be precise).

In a triangular cladogram it is desirable that all edges of a bifurcating tree have a slope of the same absolute value and that is why the x-coordinate of a node v is set to half the distance of the two most distant leaves below v. In other types of cladograms, the x-coordinate of a node v is determined by its level, that is, by that maximum number of edges on a path from v to some leaf below v.

Algorithm 13.2.2 (Rectangular cladogram for trees) To compute a rectangular cladogram, as shown in Figure 13.2b, replace step (iii) of Algorithm 13.2.1 with:

(iii') Otherwise, set $\kappa(v) = (-\text{level}_T(v), \bar{y})$, where $\bar{y}$ is the average y-coordinate of all children of v. v.

To draw the tree, for each edge $e = (v, w)$, first define a corner c with Cartesian coordinates $(x(v), y(w))$, then draw a line from v to c and from c to w.

A common variant of the rectangular cladogram is to use the average y-coordinate of the leaves under v, as in the computation of the triangular cladogram, rather than of the children of v.

To construct a circular cladogram, the leaves are assigned to points around a circle of radius $\text{level}_T(\rho)$ at equally spaced angles. Each internal node v is placed on a circle of radius $\text{level}_T(\rho) - \text{level}_T(v)$ at an angle that is computed as the average of all angles of the children of v.

Algorithm 13.2.3 (Circular cladogram for trees) To compute a circular cladogram, as shown in Figure 13.2c, we define a map $\phi : V \rightarrow \mathbb{R}^2$, $\phi(v) = (r(v), \theta(v))$ that assigns polar coordinates to each node v in a postorder traversal of the tree T. Starting at the root ρ, proceed as follows:

(i) Process all children of v in the order prescribed by $\tau(v) = (v_1, \ldots, v_k)$.

(ii) If v is a leaf, then set $\phi(v) = (\text{level}_T(\rho), 2\pi \frac{i}{n})$, where n is the total number of leaves and i is the number of leaves visited so far.

(iii) Otherwise, set $\phi(v) = (\text{level}_T(\rho) - \text{level}_T(v), \bar{\theta})$, where $\bar{\theta}$ is the average over all angles $\theta(v_i)$ of children v_i of v.

To draw the tree, for each edge $e = (v, w)$, first define a corner c with with polar coordinates $(r(v), \theta(w))$, then draw a circle section from v to c and a line from c to w.

An interesting visualization is obtained by "inverting" the circular cladogram. This is easily done by modifying the calculation of coordinates slightly. The new algorithm has one parameter, namely the radius R of the inner circle C on which the leaves are placed.

Algorithm 13.2.4 (Inner circular cladogram for trees)   *To compute an inner circular cladogram, as shown in Figure 13.2d, replace steps (ii) and (iii) of Algorithm 13.2.3 with:*

(ii') *If v is a leaf, then set $\phi(v) = (R, 2\pi \frac{i}{n})$, where n is the total number of leaves and i is the number of leaves visited so far, and R is the chosen radius of the inner circle.*

(iii') *Otherwise, set $\phi(v) = (R + \text{level}_T(v), \bar\theta)$, where $\bar\theta$ is the average over all angles $\theta(v_i)$ of children v_i of v.*

13.3 Cladograms for rooted phylogenetic networks

To obtain an algorithm for computing a cladogram for a rooted phylogenetic network, we first introduce the naive algorithm for drawing rooted phylogenetic networks and then discuss how to modify the naive algorithm appropriately.

13.3.1 The naive algorithm for rooted phylogenetic networks

In this section, we develop a naive algorithm for drawing rooted phylogenetic networks. We will see that this can be done by applying a tree drawing algorithm to an appropriate *guide tree* that is derived from the given rooted phylogenetic network.

Let N be a rooted phylogenetic network with root ρ. The first idea that comes to mind is to use as guide tree some tree T that is contained in N, that is, can be obtained from N by removing precisely all but one in-edge for each reticulate node r in N (see Lemma 1.4.3). Note that we do not suppress any resulting nodes of degree 2. All nodes of N are contained in T and so a set of coordinates for T computed using one of the above algorithms can be used to draw N. We refer to this as the *naive algorithm*:

Algorithm 13.3.1 (Naive algorithm for drawing rooted networks)   *Given a rooted phylogenetic network N, choose some tree T embedded in N (keeping internal nodes of degree two) and use a tree-drawing algorithm to compute coordinates for all nodes of N and draw the edges of T. Then draw all missing reticulated edges of N as straight lines, using these coordinates.*

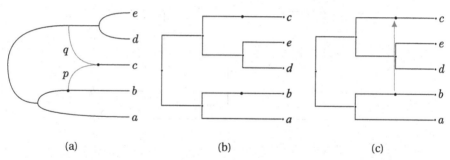

(a)   (b)   (c)

Figure 13.3 (a) The rooted phylogenetic network N to be drawn. Removal of the edge marked p produces a phylogenetic tree T, which is depicted as a rectangular cladogram in (b). Application
of the naive algorithm using T as a guide tree produces the rectangular cladogram shown
in (c).

The naive algorithm is a good starting point, but we cannot expect it to always
provide a useful result. We now discuss some of the problems that arise and how
they can be addressed. Consider the example in Figure 13.3. The network to be
drawn is depicted in (a). In the naive algorithm, one of the two reticulate edges
p and q leading to c is suppressed. Choosing p gives rise to the tree T, shown in
(b) as a cladogram computed using Algorithm 13.2.2. Note that T contains three
nodes of degree two, namely the root and the source and sink of the suppressed edge
(indicated by black disks). A drawing of N is then obtained by additionally drawing
a line to represent the suppressed edge p, as indicated in (c) by a vertical arrow.

The cladogram depicted in Figure 13.3c is not ideal for a number of reasons.
The first problem is the apparent visual asymmetry between the two reticulate
edges p and q. While p is drawn as a vertical line that cuts across the cladogram,
q is shown as an ordinary rectangular edge. In particular, it is undesirable that the
source node of one reticulation edge lies at the same level as the reticulate node,
whereas the source node of the other lies at a higher level (that is, has a smaller
x-coordinate). Note, however, that such an asymmetry is desirable when drawing
a rooted phylogenetic network in the transfer view, as discussed in Section 13.6.

The second problem is that the cladogram is not a *plane drawing* (in which no
two edges cross or intersect). While we do not expect that all rooted phylogenetic
networks can be represented by a plane drawing, for this particular example a plane
drawing is possible, as shown in Figure 13.3a.

13.3.2 Improving the naive algorithm

Now we discuss how the two problems exhibited by the naive algorithm can be
solved so as to obtain good algorithms for drawing cladograms of rooted phylogenetic networks.

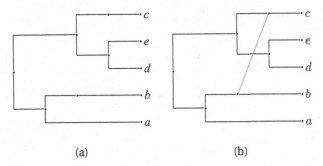

Figure 13.4 (a) Drawing of guide tree T obtained using the whole network N to determine node levels and (b) the improved network cladogram based on this.

The first problem, visual asymmetry, can be addressed by two simple modifications of the naive algorithm. First, in the computation of the level of a node, instead of using the guide tree T, one should rather use the complete network N. (The definition of the level of a node v in a rooted phylogenetic network N is just as in the case of a rooted tree, namely $\text{level}_N(v)$ is the number of edges in any longest path from v to any leaf of N.) With this modification, the source nodes of p and q will both be placed at a higher level than their common target node. Secondly, for visual clarity, reticulate edges should be rendered in a different style than tree edges. For example, in a rectangular cladogram, they can be shown as curves or diagonal lines, rather than rectangular lines.

The second problem, crossing edges as illustrated in Figure 13.3c, arises because the topological embedding of the guide tree is not well chosen. What is desirable, in general, is to produce a drawing of a rooted phylogenetic network N in which no tree edges intersect, and in which the number of crossings between reticulate edges and tree edges is minimized.

Because the naive algorithm uses all tree edges, it ensures that no two tree edges intersect. The problem of minimizing the number of crossings involving reticulate edges can be addressed by determining an appropriate topological embedding for the guide tree. To this end, we now introduce a different choice of guide tree T, which we call the *LSA* tree*. This tree is the LSA tree introduced in Section 6.6, but with all internal nodes of degree two and all unlabeled leaves kept. In more detail, the LSA* tree contains all tree edges and all nodes of N. Additionally, for each reticulate node r it contains an *auxiliary edge* $(lsa(r), r)$ that joins the LSA of r to r, as illustrated in Figure 13.5.

Now, let u be some internal node of the LSA* tree T and let $T_1, \ldots, T_{k-1}$ denote the set of all subtrees rooted at the children of u, as illustrated in Figure 13.6a. Moreover, let T_0 and T_k be the parts of T that are placed above and below u in the current layout of the tree, respectively, as indicated in Figure 13.6a. We want

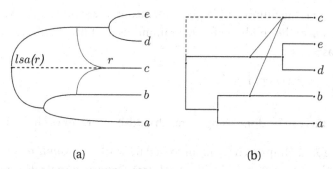

(a)   (b)

Figure 13.5 (a) The dotted line represents the auxiliary edge from $lsa(r)$ to r that is used in the LSA* tree T, giving rise to the cladogram shown in (b), for example.

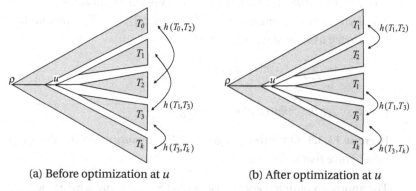

(a) Before optimization at u (b) After optimization at u

Figure 13.6 (a) For a given node u, we look at all subtrees $T_1, \ldots, T_{k-1}$ rooted at the children of u, and the two parts of the tree T_0 and T_k that lie above and below u in the topological embedding, respectively. Moreover, we consider the number $h(T_i, T_j)$ of reticulate edges that link any two such subtrees T_i and T_j, with $0 \leq i < j \leq k$. (b) The layout improvement algorithm changes the order of the children as illustrated here.

to determine a permutation π of the children of u that minimizes the number of times reticulate edges cross tree edges in the drawing of the network N. To this end, for each pair of tree parts T_i and T_j (with $0 \leq i < j \leq k$), let $h(T_i, T_j)$ denote the number of reticulate edges between some node in T_i and some node in T_j. We estimate the number of crossing caused by placing tree part T_i at position $\pi(i)$ and tree part T_j at position $\pi(j)$ as the number of reticulate edges between the two tree parts multiplied by the number of leaves that lie in the subtrees that are placed between the two, given by

$$H(i, j) = h(T_{\pi(i)}, T_{\pi(j)}) \times \sum_{i < t < j} \ell(T_{\pi(t)}), \tag{13.1}$$

where $\ell(T_{\pi(t)})$ is the number of leaves contained in the subtree $T_{\pi(t)}$. We choose a permutation π of the children of u that minimizes the total *stretch* associated with u, defined as

$$\sum_{0 \leq i < j \leq k} H(\pi(i), \pi(j)). \tag{13.2}$$

In practice, this can be done using a branch-and-bound approach.

Algorithm 13.3.2 (Improved layout of rooted networks) *Assume that we are given a rooted phylogenetic network N with a guide tree T and an initial topological embedding τ. The embedding τ can be improved by applying the following modification to each internal node u in a preorder traversal of T:*

  (i) *Determine the tree parts $T_0, \ldots, T_k$ associated with u.*
  (ii) *Determine a permutation π of the subtrees $T_1, \ldots, T_{k-1}$ that minimizes the total stretch associated with u, given by*

$$\sum_{0 \leq i < j \leq k} H(\pi(i), \pi(j)). \tag{13.3}$$

(iii) *Reorder the children of u according to π.*

Exercise 13.3.3 (Optimization goal) *What quantity does the above algorithm optimize? Prove that it does so.*

The approach outlined above, using the LSA* tree and a heuristic for determining a good topological embedding, already works very well in practice. However, for more complicated networks, an additional problem can occur that is not present in the above example. Assume that the network N contains a node x for which all out-edges are reticulate edges, as shown in Figure 13.7. Then, in the LSA* tree T associated with N the node x will be an unlabeled leaf, and application of Algorithm 13.2.2 will assign an integer y-coordinate to x. In the cladogram for N, this will result in an unequal spacing of leaves along the base line. To address this, after computing the initial coordinates for all leaves of T, one should reprocess the list of leaves such that "true leaves", namely ones that are also leaves in N, obtain adjacent integer y-coordinates, whereas the "false leaves" obtain intermediate, fractional y-coordinates. To see the resulting difference, compare Figures 13.7c and e.

In Figure 13.8 we show the result obtained by applying Algorithm 13.3.2 to the rooted phylogenetic network depicted in Figure 13.5b.

Based on the described improvements of the naive algorithm, we are now ready to give formals descriptions of the algorithms for computing rectangular and circular cladograms of rooted phylogenetic networks. Unfortunately there does not exist a

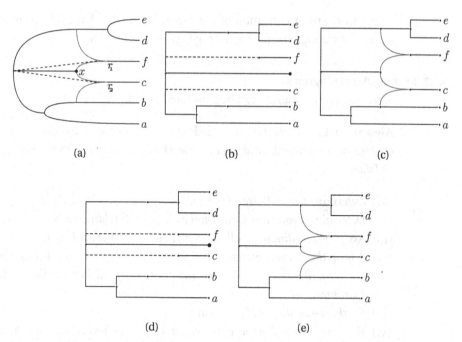

(a)   (b)   (c)

(d)   (e)

Figure 13.7 (a) A network N containing two reticulations r_1 and r_2, with auxiliary edges leading to r_1 and r_2 shown as dashed lines. (b) None of the reticulate edges are present in the LSA* tree T, thus T contains a new unlabeled leaf. A cladogram constructed without adjustment of the spacing between true leaves is shown in (c). Modification of the spacing of leaves in the LSA* tree (d) gives rise to a network cladogram with evenly spaced leaves (e).

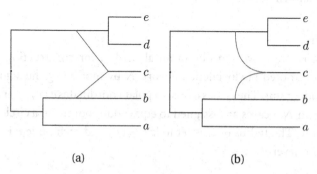

(a)   (b)

Figure 13.8 Cladogram obtained for rooted phylogenetic network using straight (a) or curved lines (b) to indicate reticulate edges.

straightforward modification of the Algorithm 13.2.1 that could be used to draw triangular cladograms for rooted phylogenetic networks.

13.3.3 Rectangular cladograms

A rectangular cladogram can be computed as follows.

Algorithm 13.3.4 (Rectangular cladogram for rooted networks) *A rectangular cladogram for a rooted phylogenetic network N, as shown in Figure 13.7e, is computed as follows:*

(i) *Determine the LSA* tree T for N.*
(ii) *Determine a good topological embedding for T relative to N.*
(iii) *Assign y-coordinates to all leaves of T, as in Algorithm 13.2.1.*
(iv) *Adjust the y-coordinates of the leaves of T so that true leaves obtain integer y-coordinates and all other leaves are assigned intermediate, fractional y-coordinates.*
(v) *Set the x-coordinate of each true leaf to 0.*
(vi) *For each internal node v of N, set $\kappa(v) = (-\text{level}_N(v), \bar{y})$, where $\bar{y}$ is the y-coordinate already assigned to v, if v is a false leaf, and $\bar{y}$ is the average y-coordinate of all children of v, otherwise.*

To draw the network, for each edge $e = (v, w)$, first define a corner c with Cartesian coordinates $(x(v), y(w))$. Using this, if e is a tree edge, then draw e as a rectangular edge, otherwise draw e as a quadratic curve (or, alternatively, as a straight line from v to w).

An example is shown in Figure 13.8.

13.3.4 Circular cladograms

As in the case of trees, we can obtain an algorithm for constructing a circular cladogram for a rooted phylogenetic network N by modifying the algorithm for rectangular cladograms. The goal is to assign polar coordinates $\phi(v) = (r(v), \theta(v))$ to each node v in N. Leaves are assigned to equidistant points on a circle C, rather than along a line. The radius of C is set to $\text{level}_N(\rho)$ and each node v is placed on a circle of radius $\text{level}_N(v)$.

Algorithm 13.3.5 (Circular cladogram for rooted networks) *A circular cladogram for a rooted phylogenetic network N, as shown in Figure 13.9b, is computed as follows:*

(i) *Determine the LSA* tree T for N.*
(ii) *Determine a good topological embedding for T relative to N.*
(iii) *Assign angle coordinates θ to all leaves of T, as in Algorithm 13.2.3.*

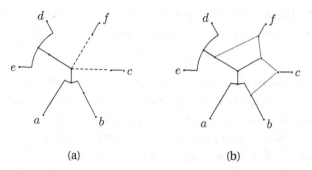

Figure 13.9 The LSA* tree T in (a) is used to generate the circular cladogram of the rooted phylogenetic network N shown in (b).

(iv) *Adjust the angles of the leaves of T so that true leaves obtain angles that are integer multiples $0, 1 \ldots, n-1$ of $\frac{2\pi}{n}$, where n is the number of leaves of N, and other leaves are assigned intermediate angles.*

(v) *Set the radius r of each true leaf to $\mathrm{level}_N(\rho)$.*

(vi) *For each internal node v of N, set $\phi(v) = (\mathrm{level}_N(\rho) - \mathrm{level}_N(v), \bar{\theta})$, where $\bar{\theta}$ is the angle already assigned to v, if v is a false leaf, and θ is the average over the angles assigned to the children of v in T, otherwise.*

To draw the network, for each tree edge $e = (v, w)$, first define a corner c with with polar coordinates $(r(v), \theta(w))$, then draw a circle section from v to c and a line from c to w. Draw all reticulate edges as straight lines.

An example is shown in Figure 13.9.

13.4 Phylograms for rooted phylogenetic trees

In the context of trees, a *phylogram* is a representation of a rooted phylogenetic tree in which the edges are drawn to scale to represent some form of evolutionary distance along edges, such as the number of mutations thought to have taken place, or the distances assigned by an algorithm such as neighbor-joining, see Section 3.14.

Let T be a rooted phylogenetic tree on a taxon set $\mathcal{X}$ and, as usual, we assume that all edges of T are directed away from the root ρ. We assume that each edge e has been given a *weight* or *length* $\omega(e) \geq 0$. Such weights are represented explicitly in the drawings: in a rectangular phylogram, the horizontal part of a rectangular edge is scaled to represent the corresponding weight, whereas in a circular diagram, the whole edge is scaled to represent the corresponding weight.

To compute a rectangular phylogram for T, as in Section 13.2 we make use of a topological embedding τ of T to assign y-coordinates to all nodes of T and then use a simple preorder traversal to assign x-coordinates to all nodes:

Algorithm 13.4.1 (Rectangular phylogram for trees)   *A rectangular phylogram for a phylogenetic tree T with edge weighting ω, as shown in Figure 13.2e, is computed as follows:*

(i) *In a postorder traversal, assign a y-coordinate to each node v as follows: if v is a leaf, then set $y(v) = i$, where i is the number of leaves visited so far, and otherwise set $y(v)$ equal to the average y-coordinate of all children of v.*

(ii) *Set $x(\rho) = 0$ and then, in a preorder traversal, for each edge $e = (v, v_i)$, set the x-coordinate of v_i to $x(v_i) = x(v) + \omega(e)$.*

Draw the tree as in Algorithm 13.2.2.

Polar coordinates for a circular phylogram can be computed by first assigning an angle to each node, as described in Algorithm 13.2.3, and then setting the radius of each node v to the path distance from the root ρ to v.

Algorithm 13.4.2 (Circular phylogram for trees)   *A circular phylogram for a phylogenetic tree T with edge weighting ω is computed as follows:*

(i) *In a postorder traversal, assign an angle coordinate $\theta(v)$ to each node v as follows: if v is the i-th leaf to be visited, then set $\theta(v) = 2\pi \frac{i}{n}$, where n is the total number of leaves, otherwise set $\theta(v)$ equal to the average angle assigned to all children of v.*

(ii) *Set $r(\rho) = 0$ and then, in a preorder traversal, for each edge $e = (v, v_i)$ set the radius coordinate of v_i to $r(v_i) = r(v) + \omega(e)$.*

Draw the edges as in Algorithm 13.2.3.

13.5 Phylograms for rooted phylogenetic networks

In this section, we discuss how to draw phylograms for rooted phylogenetic networks. Again, the starting point is the naive algorithm described in Section 13.3.1 and we obtain useful algorithms by solving the problems that the naive approach exhibits.

13.5.1 Rectangular phylograms

Let N be a rooted phylogenetic network with edge weights ω. We can obtain a rectangular phylogram for N by applying Algorithm 13.3.1 to the LSA* tree T of N. Just as in the case of cladograms, a better drawing can be obtained by modifying the vertical spacing of leaves of T so that true leaves of T obtain consecutive integer y-coordinates, whereas false leaves (ones that are leaves in T but not in N) obtain intermediate, fractional y-coordinates. Also, the topological embedding of

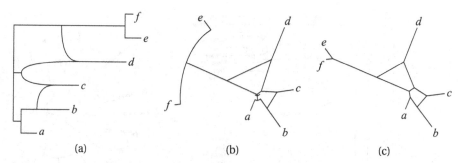

Figure 13.10 A rooted phylogenetic network shown as (a) a rectangular phylogram, (b) a circular phylogram and (c) a radial diagram. Whereas the root node in (b) is easily identified as the node
at which all tree edges would meet, if extended, in (c) the position of the root does not
distinguish it from other nodes.

T should be chosen so as to minimize the number of crossings involving reticulate
edges in N.

As in the case of cladograms, for visual clarity, we want the source node v of any
reticulate edge $e = (v, v_i)$ to have a strictly smaller x-coordinate than the target
node v_i. To ensure this, for any such v_i we set $x(v_i) = x_{max} + \Delta$, where x_{max} is
the largest x-coordinate of any parent node of v_i and $\Delta > 0$ is a user specified
offset parameter that controls how far reticulate nodes "stick out". We compute the
x-coordinates in a modified preorder traversal of T, in which we only process a
node v_i once all its parents in N have been processed.

Algorithm 13.5.1 (Rectangular phylogram for networks) *A rectangular phylogram for a rooted phylogenetic network N, as shown in Figure 13.10a is computed as
follows:*

(i) *Determine the LSA* tree T for N.*
(ii) *Determine a good topological embedding for T relative to N.*
(iii) *Assign and adjust the y-coordinates of all leaves of T, as in Algorithm 13.3.4.*
(iv) *Set $x(\rho) = 0$ and then proceed in a preorder traversal:*
  (a) *If the visited node v_i has precisely one in-edge $e = (v, v_i)$ in N, then set
  $x(v_i) = x(v) + \omega(e)$.*
  (b) *If v_i has more than one parent in N and they have all been assigned x-
  coordinates, set $x(v_i) = x_{max} + \Delta$, where x_{max} is the largest x-coordinate of
  any parent node of v_i and $\Delta > 0$ is the specified offset parameter.*
  (c) *If v_i has more than one parent in N and at least one of the parents has not
  yet been assigned coordinates, then place v_i back into the queue of nodes that
  are waiting to be processed.*

Draw the edges as in Algorithm 13.3.4.

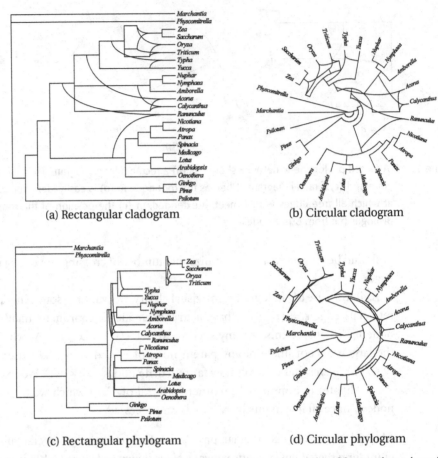

(a) Rectangular cladogram   (b) Circular cladogram

(c) Rectangular phylogram   (d) Circular phylogram

Figure 13.11 Four different representations of a phylogenetic network computed for 25 plants, based on [161]. The first two diagrams are cladograms and the other two are phylograms.

13.5.2 Circular phylograms

A simple modification of the previous algorithm can be used to compute a circular phylogram for a rooted phylogenetic network N. Again, a user-defined parameter Δ governs how much reticulate nodes are to rise above the source nodes of their incoming edges.

Algorithm 13.5.2 (Circular phylogram for networks) *A circular phylogram for a rooted phylogenetic network N, as shown in Figure 13.10b, is computed as follows:*

(i) *Determine the LSA* tree T for N.*

(ii) *Determine a good topological embedding for T relative to N.*

(iii) *Assign and adjust the angle coordinates of all leaves of T, as in Algorithm 13.3.5.*

(iv) *Set $r(\rho) = 0$ and then proceed in a preorder traversal: If v_i has precisely one in-edge $e = (v, v_i)$ in N, then set $r(v_i) = r(v) + \omega(e)$. Otherwise, when all*

parents of v_i in N have been assigned x-coordinates, set $r(v_i) = r_{max} + \Delta$,
where r_{max} is the largest radius coordinate of any parent node of v_i and $\Delta > 0$ is
the specified offset parameter.

Draw the edges as in Algorithm 13.3.5.

13.5.3 Examples

The four different drawings of rooted phylogenetic networks – rectangular and circular cladogram, and rectangular and circular phylogram – are illustrated in Figure 13.11, which shows a cluster network based on all clusters that appear in more than 20% of 59 gene trees computed for 25 plants [161].

13.6 Drawing rooted phylogenetic networks with transfer edges

The drawing algorithms for rooted phylogenetic networks described so far in this chapter depict *all* edges entering a particular reticulate node explicitly as reticulate edges. As discussed in Section 4.4.6, we call this type of illustration a *combining view* of a rooted phylogenetic network.

Now, suppose that a given rooted phylogenetic network N is to be interpreted as a species tree that has been augmented by some additional *transfer edges* representing putative horizontal gene transfer events, for example. Or, suppose we have a rooted phylogenetic tree describing the evolution of spoken languages, together with a set of transfer edges (also called *contact edges* in this context) that represent significant periods of contact between different languages. Then a combining view is not appropriate and what is required is a *transfer view*. We show examples of both kinds of views in Figure 13.12.

To compute a transfer view for a rooted phylogenetic network N, we need some additional input information. For each reticulate node v we require that exactly one in-edge has been chosen as the tree edge leading to v. Under these circumstances, the network N can be decomposed into two parts:

(i) A rooted phylogenetic tree $T(N)$ that contains all nodes of N, and whose edge set consists of all tree edges of N and exactly one chosen in-edge for each reticulate node.

(ii) The set of all *transfer* edges of N consisting of all reticulate edges of N that were not chosen as tree edges.

We can modify the algorithms described in this chapter to produce transfer views. Instead of using the LSA* tree as a guide tree, we use the tree $T(N)$. When drawing cladograms, there is another minor modification that must be made to the algorithms: the level of a node v that is the source of a transfer edge (v, v_i)

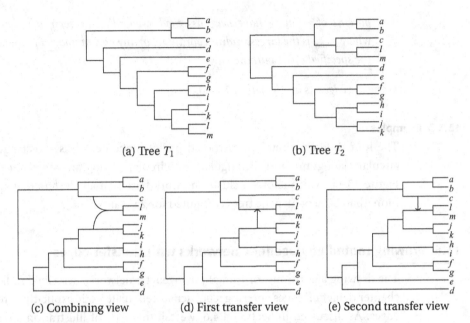

(a) Tree T_1　　　　　　　　　(b) Tree T_2

(c) Combining view　　　(d) First transfer view　　　(e) Second transfer view

Figure 13.12 In (a) and (b) we show two slightly different rooted phylogenetic trees on $\mathcal{X} = \{a, \ldots, m\}$. In (c) we display a rooted phylogenetic network that contains both trees and is drawn in a combining view. In (d) we show a transfer view of the same network, in which the reticulate edge coming from the lineage to c has been chosen as a tree edge, whereas in (e) we show the result of choosing the other reticulate edge as a tree edge.

should be set to the level to the target node of the edge, v_i, if possible. Optionally, the transfer edges can be drawn as arrows to indicate the direction of the transfer.

Exercise 13.6.1 (Combining view and transfer view in Dendroscope)　*Write down the Newick strings for the three networks shown in Figure 13.12. Enter them into the Dendroscope program (see Section 14.4) and compare the resulting visualizations.*

13.7 Radial diagrams for unrooted trees

Another popular tree-drawing method is the radial diagram, which is often employed when the root of a phylogenetic tree is unknown, unclear or of little interest. It is computed by first assigning angles to nodes, as in Algorithm 13.4.2, and then translating nodes in the direction of their angles in a preorder traversal. Although the algorithm that we describe processes the tree from the root down, the resulting drawing is independent of the choice of root (up to congruence) and thus the result of this algorithm is considered a diagram, rather than a phylogram. This algorithm is known as the *equal angle* algorithm [174].

Algorithm 13.7.1 (Radial diagram for trees)   *A radial diagram for a phylogenetic tree T, as shown in Figure 13.2f, is computed as follows:*

(i) *In a postorder traversal, assign an angle $\theta(v)$ to each node v as follows: if v is the i-th leaf to be visited, then set $\theta(v) = 2\pi \frac{i}{n}$, where n is the total number of leaves, otherwise set $\theta(v)$ equal to the average angle assigned to all children of v.*

(ii) *Set $(x(\rho), y(\rho)) = (0, 0)$ and then, in a preorder traversal, for each edge $e = (v, v_i)$, compute the coordinates of v_i by translating the point $(x(v), y(v))$ through a distance of $\omega(e)$ in the direction of $\theta(v_i)$.*

(iii) *Draw every edge (v, w) as a straight line from v to w.*

It is possible to adapt this algorithm to the task of drawing radial diagrams for rooted phylogenetic networks, see Figure 13.10c. However, the position of the root is particularly important for a rooted phylogenetic network and so this "unrooted" visualization is of limited use in practice. In the following section, we discuss the problem of drawing split networks, which are unrooted and are usually drawn in a radial fashion.

13.8 Radial diagrams for split networks

In Chapter 5, we discuss how to construct a split network from a set of splits. In this section, we look at the problem of assigning coordinates to the nodes of a split network.

Let S be a set of splits on a taxon set $\mathcal{X}$, and let $\omega : S \to \mathbb{R}$ be a weighting of the splits. Recall from Chapter 5 that a split network $N = (V, E, \sigma, \lambda)$ representing S is given by a split graph $(G = (V, E), \sigma : E \to S)$ and a node labeling $\lambda : \mathcal{X} \to V$, with the property that for any split $S = A \mid B \in S$ the deletion of every edge e with $\sigma(e) = S$ produces a graph consisting of precisely two connected components, one containing $\lambda(A)$ and the other containing $\lambda(B)$.

Of course, if S is compatible, then the canonical split network N associated with S is a phylogenetic tree and we can obtain a radial phylogram using Algorithm 13.7.1. Otherwise, if S contains a split S that is incompatible with $k > 0$ other splits, then S will be represented by at least k edges. For visual clarity, we want these edges to be represented by parallel lines of length $\omega(S)$, as specified in the following definition.

Definition 13.8.1 (Embedding of a split network)   *Let S be a set of splits on $\mathcal{X}$ with weights ω. An embedding of a split network $N = (V, E, \sigma, \lambda)$ for S is given by a map $\kappa : V \to \mathbb{R}^2$, $\kappa(v) = (x(v), y(v))$ that has the following property:*

■ *For any two edges $e = (v, v_i) \in E$ and $e' = (v', v_i') \in E$ representing the same split $S \in S$, that is with $\sigma(e) = \sigma(e') = S$, we have that the line segments $(\kappa(v), \kappa(v_i))$ and $(\kappa(v'), \kappa(v_i'))$ are parallel and have the same length $\omega(S)$.*

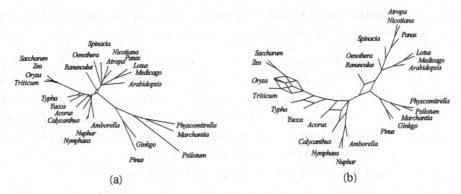

(a) (b)

Figure 13.13 (a) A radial diagram representation of a split network computed for 25 plants, based on [161]. Replacing all split weights by uniform weights leads to the diagram shown in (b).

To ensure that all edges representing a given split are drawn in parallel, we assign an angle $\theta(S)$ to every split $S \in \mathcal{S}$ and use this to determine the direction of all edges representing S. One way to assign such angles is to do so randomly, but this, of course, may lead to an overly complicated drawing with many unnecessary edge crossings. In practice, a good way to proceed is to first greedily select a large subset of splits $S' \subseteq \mathcal{S}$ that is circular (as defined in Chapter 5) and then to set $\theta(S) = \frac{2\pi}{n} \frac{(p+q)}{2} = \frac{\pi(p+q)}{n}$ for all $S = \frac{\{x_p,...,x_q\}}{\mathcal{X} - \{x_p,...,x_q\}} \in S'$. This is done under the assumption, without loss of generality, that the circularity of S' is with respect to the ordering $(x_1, \ldots, x_n)$. All other splits are assigned random angles.

Now, given a split network N and an assignment of weights ω, an embedding of a split network N can be computed by the following algorithm:

Algorithm 13.8.2 (Radial diagram for split networks) *Let S be a set of splits on $\mathcal{X}$ with weighting $\omega : S \to \mathbb{R}^{\geq 0}$. A radial diagram for a split network $N = (V, E, \sigma, \lambda)$ for S is computed as follows:*

(i) *Assign an angle θ to each split $S \in S$.*
(ii) *Choose an arbitrary starting node ρ and set $(x(\rho), y(\rho)) = (0, 0)$.*
(iii) *Starting at ρ, perform a preorder traversal of N as follows: On entering a node v_i via an edge $e = (v, v_i)$, compute the coordinates of v_i by translating the point $(x(v), y(v))$ through a distance of $\omega(e)$ in the direction of $\theta(\sigma(e))$. Then, process all nodes u adjacent to v_i that have not yet been processed and for whom the split $\sigma(f)$, with $f = (v_i, u)$, does not occur as the split of any edge on the current traversal path from ρ to v_i.*
(iv) *Draw every edge (v, w) as a straight line from v to w.*

An example is shown in Figure 13.13. Note that if the input set of splits S is circular and the split network was computed using the circular network algorithm, then the assignment of coordinates computed by this algorithm will produce a plane network.

Chapter notes

A summary of different strategies for drawing phylogenetic tree can be found in [77]. The algorithms described there are extended to unrooted phylogenetic networks in [62] and to rooted phylogenetic networks in [124]. Some strategies for improving the drawing of a split network can be found in [82]. The presented algorithms for drawing rooted and unrooted phylogenetic networks are implemented in the programs Dendroscope [120] and SplitsTree [125], respectively.

Software

In this chapter, we give an overview of the software available for computing phylogenetic networks. We first discuss some of the currently most widely used programs in slightly more detail and then briefly summarize many of the others.

14.1 SplitsTree

SplitsTree (or, more precisely, SplitsTree4) is an interactive tool for computing unrooted phylogenetic networks from sequences, distances, trees or splits [122, 125]. The program provides a wide range of methods for computing distances from a multiple sequence alignment of nucleotide sequences, amino acid sequences and some other types of sequences. Methods for computing splits or a network from different types of data include: split decomposition, neighbor-net, median network, median-joining, minimum spanning network, consensus split network, Z-closure super network, the convex hull algorithm and the circular network algorithm.

The program is designed around a *processing pipeline* that determines the order in which data is processed, namely: *sequences* → *distances* → *trees* → *splits* → *network*. Processing can start anywhere in this chain and can also skip some of the intermediate steps. The processing pipeline is illustrated in Figure 14.1. The program uses the *Nexus* file format [170] to represent the different blocks data, but also supports import and export in several other file formats.

The program is designed as an interative tool for exploring a dataset, providing different filters for including or excluding different taxa, trees or splits. Using a plugin concept that makes it easy to add new algorithms to the program, it contains many other methods in addition to those mentioned in Figure 14.1.

The aim of the first published version of SplitsTree [122] was to provide an implementation of the split decomposition and supporting methods. In 2006, a completely new version of the program was published [125], which is written in Java and provides a much wider range of methods, as listed above. Installers for all three major operating systems (Microsoft® Windows®, MacOS® and Unix®) can

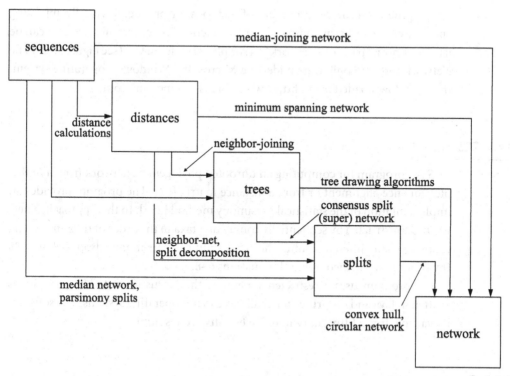

Figure 14.1 Some of the main features of the SplitsTree processing pipeline. The input data can be
aligned sequences, distances, a set of trees, a set of splits or a network. Data is processed
from top-left to bottom-right. Arrows between different types of data are labeled by the
types of methods that can be applied. For example, the neighbor-net method can be used
to compute a set of splits from distances. SplitsTree currently does not contain any method
for computing one or more trees directly from sequences, but it is possible to call an
external program to do this, for example, the program DNAPars [75] to compute a set of
most-parsimonious trees.

be downloaded from: http://www.splitstree.org. All split networks illustrated in
this book were computed using SplitsTree, see, in particular, Figures 4.3 and 11.3.

14.2 Network

Network is an interactive program for computing unrooted phylogenetic networks
from sequence data. It provides implementations of the median network, the
reduced median network, and the median-joining algorithm. For an example of a
reduced median network constructed by this program, see Figure 4.4. The program
supports the following types of input: DNA sequences, amino-acid sequences, STR
(short tandem repeat) or microsatellite data, and language/linguistic or manuscript
data. Network uses its own file format.

The program provides a number of methods for producing visually appealing and clearer networks, including a "star contraction" preprocessing step that can be used to contract parts of a network to a simplified skeleton tree (see Figure 9.12). The Network software only runs under the Microsoft® Windows® operating system. It can be downloaded from: http://www.fluxus-engineering.com.

14.3 TCS

TCS is a program for computing an unrooted phylogenetic network from a multiple sequence alignment or from a distance matrix [52]. The program provides an implementation of the statistical parsimony method [233]. In this approach, a network is constructed by sequentially connecting taxa in the order of their increasing character-state differences until the "probability of parsimony" drops below 95% (by default), as defined in [233], Equations 6–8.

The program uses a restricted version of the Nexus format. The program is written in Java and thus runs under all three major operating systems. The software is available from: http://darwin.uvigo.es/software/tcs.html.

14.4 Dendroscope

The program *Dendroscope* was originally developed as an interactive tool for drawing phylogenetic trees [130], in particular, large trees containing up to 1 million nodes. In version 2, the program has been extended so as to allow the computation and visualization of rooted phylogenetic *networks*, as well [120].

The input to Dendroscope is a file containing one or more phylogenetic trees or rooted phylogenetic networks, each described in the Newick or extended-Newick format, as discussed in Sections 3.18 and 4.5, respectively. The program allows the user to scroll through the file and to look at each of the trees or networks, using one of the eight different visualizations offered by the program. The visualizations are based on the algorithms described in Chapter 13. Each tree or network can be rerooted, pruned, colored or edited in a number of different ways. Most of the drawings of rooted phylogenetic networks shown in this book were produced using Dendroscope.

In addition to the ability to visualize rooted phylogenetic networks, Dendroscope also provides a number options for computing a rooted phylogenetic network from a collection of input trees, these include the computation of a cluster network, a galled network or a level-k network, as described in Sections 8.1, 8.4 and 8.5, respectively.

To handle the case when the input trees do not all contain exactly the same set of taxa $\mathcal{X}$, the program contains an implementation of the Z-closure algorithm (as described in Section 11.2). This algorithm determines all splits represented by the input trees and repeatedly applies the Z-closure rule to grow all partial splits to full splits on $\mathcal{X}$.

Using this, Dendroscope allows one to compute a rooted phylogenetic network from a set of trees, even when the trees contain different sets of taxa. Moreover, the program provides a computation of (rooted versions of) the strict consensus and majority consensus trees in this case.

Additionally, Dendroscope provides implementations of two different methods of computing a rooted phylogenetic network to obtain a single-labeled representation of a multi-labeled tree, as described in Section 11.7. An upcoming version of Dendroscope will contain an implementation of an algorithm for computing all hybridization networks for two trees (see Section 11.5).

Dendroscope is implemented in Java and installers for all three major operating systems are available from: http://www.dendroscope.org.

14.5 Other programs

Above, we describe some of the currently most widely used tools for computing phylogenetic networks. There are more than 40 software packages available for performing calculations in the context of phylogenetic networks. Many of these can be characterized as "proof-of-concept" implementations that have not (yet) been engineered so as to provide a robust and user-friendly tool that can be used by biologists in the lab. Here we give a short overview over some of the other programs and packages, based on a database maintained by Philippe Gambette at: http://www.atgc-montpellier.fr/phylnet/programs.

Arlequin is a widely used program for population genetics data analysis [71]. It also provides an implementation of the minimum spanning network, as discussed in Section 9.6.1.

Beagle is a small collection of programs for analyzing the minimum number of recombinations required for a multiple alignment of binary sequences [169].

Bio-PhyloNetwork is a BioPerl package that implements a number of algorithms on rooted phylogenetic networks, in particular different methods for comparing them [42].

CombineTrees combines a set of maximum parsimony trees into a single rooted phylogenetic network using a method called *UMP* [45].

EEEP ("efficient evaluation of edit paths") aims at reconciling a rooted reference tree and an unrooted test tree by performing a minimum number of SPR operations on the reference tree [18]. As discussed in Section 11.5, SPR moves are closely related to horizontal transfer events and so the reconciliation path output by the program corresponds to a set of horizontal gene transfers between organisms in the reference tree.

GalledTree is a commandline program that determines whether a given multiple alignment of binary sequences can be represented by a galled tree [98].

HybridInterleave is a Java program that calculates the minimum number of reticulations required by a hybridization network to represent two rooted phylogenetic trees [24], as discussed in Section 11.5.

LEVEL2 is a Java program for constructing a level-2 phylogenetic network for a set of rooted triples [133].

MARLON ("minimum amount of reticulation level one network") attempts to compute a level-1 phylogenetic network, with the minimum number of reticulate nodes, from a set of rooted triples [134], see Section 12.2.

MY-CLOSURE computes a supernetwork from a set of partial phylogenetic trees using so-called closure operations [92].

Notung provides a framework for incorporating duplication/loss parsimony into phylogenetic tasks, such as reconciling a gene tree with a species tree or estimating upper and lower bounds on the time of duplication [49].

PADRE is a Java package for the computation and graphical representation of rooted phylogenetic networks obtained from one or more multi-labeled trees [167], see Section 11.7.

PEGAS ("population and evolutionary genetics analysis system") is an R package for the analysis of population genetics data [193]. It provides functions for standard population genetic methods and for plotting haplotype networks.

PhyloNet is a suite of tools for computing and analyzing reticulate evolutionary relationships [234]. It includes tools for inferring horizontal gene transfers, for detecting recombination breakpoints in a multiple sequence alignment and for comparing rooted phylogenetic networks.

QNet is a Java program that implements the quartet-net method [91], see Section 12.3.

Recodon is a population genetics simulator that generates samples of nucleotide and codon sequences from populations with recombination, migration and growth, after generating their genealogy (a rooted phylogenetic network) under the coalescent framework [4].

RecPars performs a parsimony analysis of a set of DNA sequences [106]. It tries to find the best phylogenies for different segments of the sequences, thereby postulating a recombination event between these segments.

Serial NetEvolve is a simulator that evolves serially sampled sequences along a randomly generated rooted phylogenetic tree or network [35].

SHRUB constructs a minimum recombination network that represents a multiple alignment of binary sequences under the infinite sites assumption [225].

Sliding MinPD reconstructs phylogenetic networks of serially sampled sequences by combining minimum pairwise distance measures with automated recombination detection based on a sliding window approach [36].

Spectronet provides implementations of split-based methods such as the (reduced) median network and the so-called pruned-median network [116].

T-Rex allows one to compute a tree with short-cut edges for a given distance matrix [171], see Section 10.5.

Treemapper is a program for reconciling host and parasite trees (or also gene trees and species trees) using the concept of a *jungle* [48].

Glossary

Abstract phylogenetic network A phylogenetic network used to visualize incompatiblities in a dataset, such as a split network or a haploytpe network, 71

Adams consensus tree A rooted consensus tree in which the maximal clusters are intersections of maximal clusters of the input trees, 65

Aho algorithm Computes the rooted phylogenetic tree encoded by a set of rooted triples, if it exists, 301

Ancestral recombination graph Graph describing the history of a set of sequences under the coalescent-with-recombination model, 236

Bayesian tree inference Computes a posterior probability distribution of phylogenetic trees for a multiple sequence alignment, under a given model of evolution and prior distribution of trees, 45

Bicombination A node in a rooted phylogenetic network of indegree two, 140

Biconnected component A maximal subgraph that is induced by a set of edges and cannot be disconnected by deleting a single node, 6

Bifurcation A node in a rooted phylogenetic tree or network of outdegree two, 140

BME tree An unrooted phylogenetic tree with balanced edge lengths and minimum tree length, 56

Buneman graph A canonically defined split network, 89

Circular splits A set of splits that can be represented by an outer-labeled planar split network, 102

Clade A monophyletic group, 11

Cladogram A drawing of a rooted phylogenetic tree or network aimed at clearly representing its topology, that is, showing how clusters are nested within each other, and in which edge lengths do not have an interpretation, 312

Cluster A proper subset of a set of taxa, 128

Cluster network A rooted phylogenetic network that provides a hardwired representation of a set of (not necessarily compatible) clusters, 134

Combining view Drawing of a rooted phylogenetic network in which reticulate edges are shown as combining edges, 81

Compatibility Two splits, characters or clusters, are compatible, if they both can occur together in some phylogenetic tree, 90

Consensus split network A split network used to represent all splits that occur in a given proportion of a set of input trees, 266

DAG A directed graph that is free of directed cycles, 8

Decomposition property A phylogenetic network is decomposable if there is a bijection between its biconnected components and the connected components of the incompatibility graph of the represented data, 151

Dendroscope A computer program for rooted phylogenetic trees and networks, 334

Directed graph A set of nodes and a set of directed edges between nodes, 4

Displayed trees distance Distance based on the number of represented trees by which two rooted phylogenetic networks differ, 158

Distance matrix A symmetric non-negative function that satisfies the triangle inequality, 50

Distance-based method Computes a phylogenetic tree or network from a distance matrix, 50

DLT event A gene duplication, loss or horizontal transfer event, 289

DLT scenario A mapping of a gene tree on to a species tree using gene duplications, losses and transfers, 294

Edit distance The minimum number of edit operations (such as insertions, deletions and replacements) required to transform one sequence into another, 14

Explicit phylogenetic network A phylogenetic network used to represent a putative evolutionary history involving reticulate events, such as a hybridization network, 71

FastME method A heuristic for computing a BME tree from distances, 59

Galled network A rooted phylogenetic network in which each reticulation has a tree cycle, 162

Galled tree A rooted phylogenetic network in which each non-trivial biconnected component properly contains exactly one reticulate node, 156

Gene conversion The conversion of one allele to another due to base mismatch repair during recombination, x

Haplotype network An unrooted phylogenetic network used to show relationships between closely related sequences, 75

Hardwired cluster distance Distance based on the number of hardwired clusters by which two rooted phylogenetic networks differ, 173

Hardwired network A rooted phylogenetic network interpreted in the hardwired sense in which all reticulate edges are considered to be *on*, 148

Hasse diagram A drawing of a partially ordered set, organized in layers from top to bottom, 133

Homologous sequences A set of biomolecular sequences that have evolved from a common ancestral sequence, 13

Horiztonal gene transfer The transfer of genes from one organism to another, as opposed to the more usual vertical transfer in which an organism obtains its genes from its parents or parent, ix

Hybridization The production of offspring from parents of different lineages, x

Hybridization network A rooted phylogenetic network used to explain the incongruence of two gene trees in terms of hybridization events, 275

Incompatibility graph A graph whose nodes represent splits, or clusters, and whose edges represent pairs of incompatible splits, or clusters, respectively, 91

Level-*k* network A bicombining, rooted phylogenetic network in which each biconnected component contains at most *k* reticulations, 160

Lowest stable ancestor The last node on all paths from the root to a given node, 142

LSA consensus tree A consensus tree that is obtained as the LSA tree of the cluster network of a collection of rooted phylogenetic trees, 143

Majority consensus tree A phylogenetic tree that represents the set of all splits present in more than half of all members of a collection of unrooted phylogenetic trees, 57

Maximum likelihood estimation Computes the phylogenetic tree that maximizes the likelihood of generating a given multiple sequence alignment under a given model of evolution, 64

Maximum parsimony method Computes a phylogenetic tree that provides the most parsimonious explanation of the differences in a multiple sequence alignment, 33

Median network A canonical split network that represents a multiple alignment of binary sequences, 220

Median-joining network An unrooted phylogenetic network that represents a multiple sequence alignment, 230

Molecular clock The molecular clock hypothesis states that the mutation rate is constant over all sites of sequences and over all branches of a phylogenetic tree, 52

Monophyletic group A group that contains all (and only) the descendants of a common ancestor and the ancestor itself, 11

Multi-labeled tree A phylogenetic tree in which taxa appear more than once, 287

Multicombination A node in a rooted phylogenetic network of indegree three or more, 139

Multifurcation A node in a rooted phylogenetic tree or network of outdegree three or more, 140

Multiple sequence alignment A display of sequences, one per row, in which gaps have been inserted so that putatively homologous positions in the sequences are placed in the same column, 20

Neighbor-joining method An agglomerative algorithm for computing unrooted phylogenetic trees from distances, 54

Neighbor-net method An algorithm that constructs a set of circular splits for a distance matrix, 254

Nested-labels distance Distance based on the number of nested labels by which two rooted phylogenetic networks differ, 180

Network software A computer program for calculating median networks and median-joining networks, 333

Parsimony splits An algorithm for computing a set of weakly compatible splits for a given multiple sequence alignment, 218

Path-multiplicity distance Distance for rooted phylogenetic networks that is based on counting the number of different paths from internal nodes to leaves, 179

Phylogenetic network Any graph that is used to represent evolutionary relationships (either abstractly or explicitly) between taxa, 69

Phylogenetic tree A tree aimed at representing evolutionary relationships between the taxa that label its leaves, 25

Phylogram A drawing of a rooted phylogenetic tree or network in which edges are drawn to scale to represent some form of evolutionary distance, 323

Reassortment The swapping of genetic material between viruses that have co-infected a host cell, x

Reassortment network A rooted phylogenetic network that uses reassortment events in the description of the evolutionary history of viruses, 80

Recombination The exchange of genes between chromosomes, usually of homologous genes between homologous chromosomes, x

Recombination network A rooted phylogenetic network in which reticulate nodes represent putative recombination events, 233

Reticulate evolution Evolution involving events such as hybridization, horizontal gene transfer, reassortment or recombination, ix

Reticulation A node in a rooted phylogenetic network that has more than one in-edge, 139

Rooted network alignment A mapping between similar nodes of two rooted phylogenetic networks, 181

Rooted phylogenetic network Any rooted graph that represents evolutionary relationships (either abstractly or explicitly) between taxa, 76

Rooted phylogenetic tree A phylogenetic tree for which a node has been specified as root, thus giving a direction of time to the tree, 26

Rooted triple A bifurcating, rooted phylogenetic tree on three taxa, 300

Sequence-based method Any method that computes a phylogenetic tree (or network) from a multiple sequence alignment, 33

Softwired cluster distance Distance based on the number of softwired clusters by which two rooted phylogenetic networks differ, 174

Softwired network A rooted phylogenetic network interpreted in the softwired sense in which reticulate edges can be switched *on* or *off*, 148

Split A bipartition of a set of taxa, for example, induced by an edge of an unrooted phylogenetic tree, 87

Split decomposition An algorithm for computing a set of weakly compatible splits for a given distance matrix, 111

Split encoding The set of all splits associated with an unrooted phylogenetic tree, 89

Split graph The underlying graph of a split network, 94

Split network A network representing a set of splits in which taxa appear as node labels and splits are represented by arrays of parallel edges, 96

SplitsTree A computer program for computing split networks, 332

Strict consensus tree A consensus tree that is obtained from the set of splits that are present in all input trees, 63

Subnetwork distance Distance based on the number of isomorphic subnetworks by which two rooted phylogenetic networks differ, 178

Substitution matrix Assigns a score to the substitution of one character state by another one in an alignment, 15

Super network A phylogenetic network constructed from phylogenetic trees on overlapping taxon sets, 271

Taxon set A set of taxonomic units or *taxa* that represent groups of organisms, 11

TCS A computer program for calculating an unrooted phylogenetic network for sequences, 334

Transfer view Drawing of a rooted phylogenetic network that uses transfer edges, 81

Tree A connected graph with no (undirected) cycles, 7

Tree-child property Every internal node in a rooted phylogenetic network has a child that is a tree node, 164

Tree-sibling property Every reticulate node in a rooted phylogenetic network has a sibling that is a tree node, 166

Tripartition distance Distance based on the number of tripartions by which two rooted phylogenetic networks differ, 175

Undirected graph A set of nodes and a set of undirected edges between nodes, 3

Unrooted phylogenetic network Any unrooted graph that is used to (abstractly) represent evolutionary relationships between taxa, 71

Unrooted phylogenetic tree A phylogenetic tree for which a root has not been specified and thus the direction of time is undefined, 25

UPGMA A distance-based method that computes a rooted phylogenetic tree in which all leaves have the same distance from the root, 52

Weak compatibility Three splits are weakly compatible, if they fulfill a certain relaxation of the compatibility requirement for two splits, 106

References

[1] E. N. Adams III. Consensus techniques and the comparison of phylogenetic trees. *Systematic Zoology*, 21(4):390–397, 1972.

[2] E. N. Adams III. N-trees as nesttings: complexity, similarity, and consensus. *Journal of Classification*, 3:299–317, 1986.

[3] A. V. Aho, Y. Sagiv, T. G. Szymanski, and J. D. Ullman. Inferring a tree from lowest common ancestors with an application to the optimization of relational expressions. *SIAM Journal on Computing*, 10(3):405–421, 1981.

[4] M. Arenas and D. Posada. Recodon: coalescent simulation of coding DNA sequences with recombination, migration and demography. *BMC Bioinformatics*, 8:458, 2007.

[5] S. C. Ayling and T. A. Brown. Novel methodology for construction and pruning of quasi-median networks. *BMC Bioinformatics*, 9:115, 2008.

[6] H. -J. Bandelt. Generating median graphs from boolean matrices. In Y. Dodge, ed., *L1-Statistical Analysis*, pp. 305–309. North-Holland, 1992.

[7] H. -J. Bandelt. Phylogenetic networks. *Verhandl. Naturwiss. Vereins Hamburg (NF)*, 34:51–71, 1994.

[8] H. -J. Bandelt and A. Dress. A relational approach to split decomposition. In O. Opitz, B. Lausen, and R. Klar, eds., *Information and Classification*, pp. 123–131. Springer, 1993.

[9] H. -J. Bandelt and A. W. M. Dress. A canonical decomposition theory for metrics on a finite set. *Advances in Mathematics*, 92:47–105, 1992.

[10] H. -J. Bandelt, P. Forster, and A. Röhl. Median-joining networks for inferring intraspecific phylogenies. *Molecular Biology and Evolution*, 16:37–48, 1999.

[11] H. J. Bandelt, P. Forster, B. C. Sykes, and M. B. Richards. Mitochondrial portraits of human population using median networks. *Genetics*, 141(2):743–753, 1995.

[12] H. -J. Bandelt, V. Macaulay, and M. Richards. Median networks: speedy construction and greedy reduction, one simulation, and two case studies from human mtDNA. *Molecular Phylogenetics and Evolution*, 16(1):8–28, 2000.

[13] E. Bapteste, M. O'Malley, R. Beiko, *et al.* Prokaryotic evolution and the tree of life are two different things. *Biology Direct*, 4(1):34, 2009.

[14] A. Barbrook, C. Howe, and N. Blake. The phylogeny of the Canterbury tales. *Nature*, 394:839, 1998.

[15] M. Baroni, S. Grüenewald, V. Moulton, and C. Semple. Bounding the number of hybridization events for a consistent evolutionary history. *Mathematical Biology*, 51:171–182, 2005.

[16] M. Baroni, C. Semple, and M. Steel. Hybrids in real time. *Systematic Biology*, 55(1):46–56, 2006.

[17] J. -P. Barthélemy and A. Guénoche. *Trees and Proximity Representations*. Wiley, 1991.

[18] R. Beiko and N. Hamilton. Phylogenetic identification of lateral genetic transfer events. *BMC Evolutionary Biology*, 6(1):15, 2006.

[19] C. Berthouly, G. Leroy, T. N. Van, *et al.* Genetic analysis of local Vietnamese chickens provides evidence of gene flow from wild to domestic populations. *BMC Genetics*, 10(1):1, 2009.

[20] J. Bohne, A. W. M. Dress, and S. Fischer. A simple proof for de Bruijn's dualization principle. *Sankhya. Series A*, 54(Special Issue):77–84, 1992.

[21] S. Bokhari and D. Janies. Reassortment networks for investigating the evolution of segmented viruses. *IEEE/ACM Transactions on Computational Biology and Bioinformatics*, 99(RapidPosts), 2008.

[22] T. Bonfert, R. Rupp, and D. H. Huson. Phylogenetic networks from multi-labeled trees. Poster presented at German Conference on Bioinformatics, Halle (Saale), 2009.

[23] K. S. Booth and G. S. Luecker. Testing for the consecutive ones property, interval graphs and graph planarity using pq-tree algorithms. *Journal of Computer and System Sciences*, 13:335–379, 1976.

[24] M. Bordewich, S. Linz, K. St. John, and C. Semple. A reduction algorithm for computing the hybridization number of two trees. *Evolutionary Bioinformatics*, 3:86–98, 2007.

[25] M. Bordewich and C. Semple. On the computational complexity of the rooted subtree prune and regraft distance. *Annals of Combinatorics*, 8(4):409–423, 2005.

[26] M. Bordewich and C. Semple. Computing the hybridization number of two phylogenetic trees is fixed-parameter tractable. *IEEE/ACM Transactions on Computational Biology and Bioinformatics*, 4(3):458–466, 2007.

[27] M. Bordewich and C. Semple. Computing the minimum number of hybridisation events for a consistent evolutionary history. *Discrete Applied Mathematics*, 155(8):914–928, 2007.

[28] L. Boto. Horizontal gene transfer in evolution: facts and challenges. *Proceedings of the Royal Society B: Biological Sciences*, 277(1683):819–827, 2010.

[29] N. G. de Bruijn. Dualization of multigrids. *Journal de Physique*, 47(7 Suppl. Colloq. C3):C3–9–C3–18, 1986.

[30] D. Bryant. *Hunting for trees, building trees and comparing trees: theory and method in phylogenetic analysis*. PhD thesis, Department of Mathematics, University of Canterbury, 1997.

[31] D. Bryant and V. Moulton. NeighborNet: an agglomerative method for the construction of planar phylogenetic networks. In R. Guig and D. Gusfield, eds., *Algorithms in Bioinformatics: Proceedings of the Second International Workshop on Algorithms in Bioinformatics (WABI)*, vol. 2452 of *LNCS*, pp. 375–391, 2002.

[32] D. Bryant and V. Moulton. Neighbor-Net: an agglomerative method for the construction of phylogenetic networks. *Molecular Biology and Evolution*, 21(2):255–265, 2004.

[33] D. Bryant, V. Moulton, and A. Spillner. Consistency of the Neighbor-Net algorithm. *Algorithms for Molecular Biology*, 2(1):8, 2007.

[34] A. Brysting, B. Oxelman, K. Huber, V. Moulton, and C. Brochmann. Untangling complex histories of genome mergings in high polyploids. *Systematic Biology*, 56(3):467–476, 2007.

[35] P. Buendia and G. Narasimhan. Serial NetEvolve: a flexible utility for generating serially-sampled sequences along a tree or recombinant network. *Bioinformatics*, 22(18):2313–2314, 2006.

[36] P. Buendia and G. Narasimhan. Sliding MinPD: building evolutionary networks of serial samples via an automated recombination detection approach. *Bioinformatics*, 23(22):2993–3000, 2007.

[37] P. Buneman. The recovery of trees from measures of dissimilarity. In F. R. Hodson, D. G. Kendall, and P. Tautu, eds., *Mathematics in the Archaeological and Historical Sciences*, pp. 387–395. Edinburgh University Press, 1971.

[38] G. Cardona, M. Llabrés, F. Rosselló, and G. Valiente. A distance metric for a class of tree-sibling phylogenetic networks. *Bioinformatics*, 24(13):1481–1488, 2008.

[39] G. Cardona, M. Llabrés, F. Rosselló, and G. Valiente. Metrics for phylogenetic networks I: generalizations of the Robinson-Foulds metric. *IEEE/ACM Transactions on Computational Biology and Bioinformatics*, 6(1):46–61, 2009.

[40] G. Cardona, M. Llabrés, F. Rosselló, and G. Valiente. On Nakhleh's metric for reduced phylogenetic networks. *IEEE/ACM Transactions on Computational Biology and Bioinformatics*, 6(4):629–638, 2009.

[41] G. Cardona, F. Rosselló, and G. Valiente. Comparison of tree-child phylogenetic networks. *IEEE/ACM Transactions on Computational Biology and Bioinformatics*, 6:552–569, 2007.

[42] G. Cardona, F. Rosselló, and G. Valiente. A Perl package and an alignment tool for phylogenetic networks. *BMC Bioinformatics*, 9:175, 2008.

[43] G. Cardona, F. Rosselló, and G. Valiente. Extended Newick: it is time for a standard representation of phylogenetic networks. *BMC Bioinformatics*, 9:532, 2008.

[44] F. Casado, C. R. Bonvicino, and H. N. Seuanez. Phylogeographic analyses of *Callicebus lugens* (Platyrrhini, Primates). *Journal of Heredity*, 98(1):88–92, 2007.

[45] I. Cassens, P. Mardulyn, and M. C. Milinkovitch. Evaluating intraspecific "network" construction methods using simulated sequence data: do existing algorithms outperform the global maximum parsimony approach? *Systematic Biology*, 54(3):363–372, 2005.

[46] J. Castresana. Selection of conserved blocks from multiple alignments for their use in phylogenetic analysis. *Molecular Biology and Evolution*, 17(4):540–552, 2000.

[47] M. Castrucci, I. Donatelli, L. Sidoli, *et al.* Genetic reassortment between avian and human influenza A viruses in Italian pigs. *Virology*, 193(1):503–506, 1993.

[48] M. A. Charleston. Jungles: a new solution to the host/parasite phylogeny reconciliation problem. *Mathematical Biosciences*, 149:191–223, 1998.

[49] K. Chen, D. Durand, and M. Farach-Colton. Notung: dating gene duplications using gene family trees. *Journal of Computational Biology*, 7(3):429–447, 2000.

[50] B. Chor and T. Tuller. Finding a maximum likelihood tree is hard. *Journal of the ACM*, 53(5):722–744, 2006.

[51] C. Choy, J. Jansson, K. Sadakane, and W. -K. Sung. Computing the maximum agreement of phylogenetic networks. *Theoretical Computer Science*, 335(1):93–107, 2005.

[52] M. Clement, D. Posada, and K. A. Crandall. TCS: a computer program to estimate gene genealogies. *Molecular Ecology*, 9:1657–1659, 2000.

[53] J. Cohen. Mathematics is biology's next microscope, only better; biology is mathematics' next physics, only better. *PLoS Biology*, 2(12):e439, 2004.

[54] L. A. David and E. J. Alm. Insights into the origins of microbial gene families. In press. Software: AnGST, 2010.

[55] C. Delwiche. Tracing the thread of plastid diversity through the tapestry of life. *American Naturalist*, 154:S164–S177, 1999.

[56] C. F. Delwiche and J. D. Palmer. Rampant horizontal transfer and duplication of rubisco genes in eubacteria and plastids. *Molecular Biology and Evolution*, 13:873–882, 1996.

[57] R. Desper and O. Gascuel. Fast and accurate phylogeny reconstruction algorithms based on the minimum-evolution principle. *Journal of Computational Biology*, 9(5):687–705, 2002.

[58] R. Desper and O. Gascuel. Theoretical foundation of the balanced minimum evolution method of phylogenetic inference and its relationship to weighted least-squares tree fitting. *Molecular Biology and Evolution*, 21(3):587–598, 2004.

[59] C. Dessimoz, D. Margadant, and G. H. Gonnet. DLIGHT: lateral gene transfer detection using pairwise evolutionary distances in a statistical framework. In *Research in Computational Molecular Biology: Proceedings of the 12th International Conference on Research in Computational Molecular Biology (RECOMB)*, vol. 4955, *LNCS*, pp. 315–330. Springer, 2008.

[60] T. Disotell. Discovering human history from stomach bacteria. *Genome Biology*, 4(5):213, 2003.

[61] W. F. Doolittle. Phylogenetic classification and the Universal Tree. *Science*, 284:2124–2128, 1999.

[62] A. W. M. Dress and D. H. Huson. Constructing splits graphs. *IEEE/ACM Transactions in Computational Biology and Bioinformatics*, 1(3):109–115, 2004.

[63] A. W. M. Dress, V. Moulton, and W. Terhalle. T-theory: an overview. *European Journal of Combinatorics*, 17(2-3):161–175, 1996.

[64] R. Durbin, S. R. Eddy, A. Krogh, and G. Mitchison. *Biological Sequence Analysis*. Cambridge University Press, 1998.

[65] I. Ebersberger, P. Galgoczy, S. Taudien, *et al.* Mapping human genetic ancestry. *Molecular Biology and Evolution*, 24(10):2266–2276, 2007.

[66] R. C. Edgar. MUSCLE: multiple sequence alignment with high accuracy and high throughput. *Nucleic Acids Research*, 32(5):1792–97, 2004.

[67] A. W. F. Edwards and L. L. Cavalli-Sforza. The reconstruction of evolution. *Annals of Human Genetics*, 27:105–106, 1963.

[68] A. W. F. Edwards and L. L. Cavalli-Sforza. Reconstruction of evolutionary trees. In V. H. Heywood and J. McNeill, eds., *Phenetic and Phylogenetic Classification*, vol. 6, pp. 67–76. Systematics Association, London, 1964.

[69] J. Escobar, A. Cenci, C. Scornavacca, *et al.* Combining supermatrix and supertree in *Triticeae* (Poaceae). Unpublished manuscript, 2010.

[70] G. F. Estabrook, F. R. McMorris, and C. A. Meacham. Comparison of undirected phylogenetic trees based on subtrees of four evolutionary units. *Systematic Zoology*, 34(2):193–200, 1985.

[71] L. Excoffier, G. Laval, and S. Schneider. Arlequin ver. 3.0: an integrated software package for population genetics data analysis. *Evolutionary Bioinformatics Online*, 1:47–50, 2005.

[72] L. Excoffier and P. Smouse. Using allele frequencies and geographic subdivision to reconstruct gene trees within a species: molecular variance parsimony. *Genetics*, 136:343–359, Jan. 1994.

[73] J. Felsenstein. Evolutionary trees from DNA sequences: a maximum likelihood approach. *Journal of Molecular Evolution*, 17(6):368–376, 1981.

[74] J. Felsenstein. Statistical inference of phylogenies (with discussion). *Journal of the Royal Statistical Society: Series A (Statistics in Society)*, 146:246–272, 1983.

[75] J. Felsenstein. PHYLIP: phylogeny inference package (version 3.2). *Cladistics*, 5:164–166, 1989.

[76] J. Felsenstein. An alternating least squares approach to inferring phylogenies from pairwise distances. *Systematic Biology*, 46(1):101–111, March 1997.

[77] J. Felsenstein. *Inferring Phylogenies*. Sinauer Associates, Inc., 2004.

[78] J. Felsenstein and G. A. Churchill. A hidden Markov model approach to variation among sites in rate of evolution. *Molecular Biology and Evolution*, 13(1):93–104, 1996.

[79] W. M. Fitch. Toward defining the course of evolution: minimum change for a specified tree topology. *Systematic Zoology*, 20:406–416, 1971.

[80] W. M. Fitch. Networks and viral evolution. *Journal of Molecular Evolution*, 44:65–75, 1997.

[81] W. M. Fitch and E. Margoliash. Construction of phylogenetic trees. *Science*, 155(3760):279–284, 1967.

[82] P. Gambette and D. H. Huson. Improved layout of phylogenetic networks. *IEEE/ACM Transactions on Computational Biology and Bioinformatics*, 5(3):472–479, 2008.

[83] M. R. Garey and D. S. Johnson. *Computers and intractability, a guide to the theory of NP-completeness*. Bell Telephone Laboratories, Inc., 1979.

[84] O. Gascuel and M. Steel. Neighbor-joining revealed. *Molecular Biology and Evolution*, 23(11):1997–2000, 2006.

[85] D. Gömöry and L. Paule. Reticulate evolution patterns in western-Eurasian beeches. *Botanica Helvetica*, 2010.

[86] P. Górecki. Reconciliation problems for duplication, loss and horizontal gene transfer. In *Proceedings of the Eighth Annual Internation Conference on Research in Computational Molecular Biology (RECOMB)*, pp. 316–325. ACM Press, 2004.

[87] Grass Phylogeny Working Group. Phylogeny and subfamilial classification of the grasses (Poaceae). *Annals of the Missouri Botanical Garden*, 88(3):373–457, 2001.

[88] P. J. Green. Reversible jump Markov Chain Monte Carlo computation and Bayesian model determination. *Biometrika*, 82(4):711–732, 1995.

[89] R. C. Griffiths and P. Marjoram. Ancestral inference from samples of DNA sequences with recombination. *Journal of Computational Biology*, 3:479–502, 1996.

[90] R. C. Griffiths and P. Marjoram. An ancestral recombination graph. In P. Donnelly and S. Tavaré, eds., *Progress in Population Genetics and Human Evolution*, vol. 87, *IMA Volumes of Mathematics and its Applications*, pp. 257–270. Springer, 1997.

[91] S. Grünewald, K. F. amd A. W. M. Dress, and V. Moulton. QNet: an agglomerative method for the construction of phylogenetic networks from weighted quartets. *Molecular Biology and Evolution*, 24(2):532–538, 2007.

[92] S. Grünewald, K. T. Huber, and Q. Wu. Novel closure rules for constructing phylogenetic super-networks. *Bulletin of Mathematical Biology*, 70:1906–1924, 2008.

[93] S. Grünewald, V. Moulton, and A. Spillner. Consistency of the QNet algorithm for generating planar split networks from weighted quartets. *Discrete Applied Mathematics*, 157(10):2325–2334, 2009.

[94] S. Grünewald, M. A. Steel, and M. S. Swenson. Closure operations in phylogenetics. *Mathematical Biosciences*, 208(2):521–537, 2007.

[95] D. Gusfield. Optimal, efficient reconstruction of root-unknown phylogenetic networks with constrained and structured recombination. *Journal of Computer and System Sciences*, 70:381–398, 2005.

[96] D. Gusfield and V. Bansal. A fundamental decomposition theory for phylogenetic networks and incompatible characters. In *Research in Computational Molecular Biology: Proceedings of the Ninth International Conference on Research in Computational Molecular Biology (RECOMB)*, vol. 3500 of *LNCS*, pp. 217–232. Springer, 2005.

[97] D. Gusfield, V. Bansal, V. Bafna, and Y. S. Song. A decomposition theory for phylogenetic networks and incompatible characters. *Journal of Computational Biology*, 14(10):1247–1272, 2007.

[98] D. Gusfield, S. Eddhu, and C. Langley. Efficient reconstruction of phylogenetic networks with constrained recombinations. In *Proceedings of the IEEE Computer Society Conference on Bioinformatics (CSB)*, p. 363. IEEE Computer Society, 2003.

[99] D. Gusfield, S. Eddhu, and C. Langley. Optimal, efficient reconstruction of phylogenetic networks with constrained recombination. *Journal of Bioinformatics and Computational Biology*, 2(1):173–213, 2004.

[100] M. Hallett, J. Largergren, and A. Tofigh. Simultaneous identification of duplications and lateral transfers. In *Proceedings of the Eight International Conference on Research in Computational Molecular Biology (RECOMB)*, pp. 347–356. ACM Press, 2004.

[101] J. A. Hartigan. Minimum mutation fits to a given tree. *Biometrics*, 29(1):53–65, 1973.

[102] D. L. Hartl and A. G. Clark. *Principles of Population Genetics, Fourth Edition*. Sinauer Associates, Inc., 2006.

[103] M. Hasegawa, H. Kishino, and T. Yano. Dating of the human-ape splitting by a molecular clock of mitochondrial DNA. *Journal of Molecular Evolution*, 22(2):160–174, 1985.

[104] W. K. Hastings. Monte Carlo sampling methods using Markov chains and their applications. *Biometrika*, 57:97–109, 1970.

[105] J. Hein. Reconstructing evolution of sequences subject to recombination using parsimony. *Mathematical Biosciences*, 98(2):185–200, 1990.

[106] J. Hein. A heuristic method to reconstruct the history of sequences subject to recombination. *Journal of Molecular Evolution*, 36:396–405, 1993.

[107] J. Hein, T. Jiang, L. Wang, and K. Zhang. On the complexity of comparing evolutionary trees. *Discrete Applied Mathematics*, 71(1–3):153–169, 1996.

[108] J. Hein, M. H. Schierup, and C. Wiuf. *Gene Genealogies, Variation and Evolution: A Primer in Coalescent Theory*. Oxford University Press, February 2005.

[109] M. D. Hendy and D. Penny. Branch and bound algorithms to determine minimal evolutionary trees. *Mathematical Biosciences*, 59(2):277–290, 1982.

[110] J. G. Henikoff and S. Henikoff. Blocks database and its applications. *Methods in Enzymology*, 266:88–105, 1996.

[111] E. Hilario and J. P. Gogarten. Horizontal transfer of ATPase genes: the tree of life becomes a net of life. *Biosystems*, 31:111–119, 1993.

[112] B. Holland, K. Huber, V. Moulton, and P. J. Lockhart. Using consensus networks to visualize contradictory evidence for species phylogeny. *Molecular Biology and Evolution*, 21:1459–1461, 2004.

[113] B. Holland and V. Moulton. Consensus networks: a method for visualizing incompatibilities in collections of trees. In G. Benson and R. Page, eds., *Workshop on Algorithms in Bioinformatics (WABI)*, vol. 2812, pp. 165–176. Springer, 2003.

[114] E. C. Holmes, M. Worobey, and A. Rambaut. Phylogenetic evidence for recombination in dengue virus. *Molecular Biology and Evolution*, 16:405–409, 1999.

[115] K. Huber, L. van Iersel, S. Kelk, and R. Suchecki. A practical algorithm for reconstructing level-1 phylogenetic networks. In *IEEE/ACM Transactions on Computational Biology and Bioinformatics*, 2010.

[116] K. T. Huber, M. Langton, D. Penny, V. Moulton, and M. Hendy. Spectronet: a package for computing spectra and median networks. *Applied Bioinformatics*, 1:159–161, 2002.

[117] K. T. Huber, B. Oxelman, M. Lott, and V. Moulton. Reconstructing the evolutionary history of polyploids from multilabeled trees. *Molecular Biology and Evolution*, 23(9):1784–1791, 2006.

[118] R. R. Hudson and N. L. Kaplan. Statistical properties of the number of recombination events in the history of a sample of DNA sequences. *Genetics*, 111:147–164, 1985.

[119] J. P. Huelsenbeck, F. Ronquist, R. Nielsen, and J. P. Bollback. Bayesian inference of phylogeny and its impact on evolutionary biology. *Science*, 294:2310–2314, 2001.

[120] D. Huson and C. Scornavacca. Dendroscope 2: a program for computing and drawing rooted phylogenetic trees and networks. In preparation, software available from: www.dendroscope.org, 2011.

[121] D. Huson and R. Rupp. Summarizing multiple gene trees using cluster networks. In *Algorithms in Bioinformatics: Proceedings of the Eighth International Workshop on Algorithms in Bioinformaticsy (WABI)*, vol. 5251, *LNBI*, pp. 211–225, 2008.

[122] D. H. Huson. SplitsTree: analyzing and visualizing evolutionary data. *Bioinformatics*, 14(10):68–73, 1998.

[123] D. H. Huson. Split networks and reticulate networks. In O. Gascuel and M. A. Steel, eds., *Reconstructing Evolution: New Mathematical and Computational Advances*, pp. 247–276. Oxford University Press, 2007.

[124] D. H. Huson. Drawing rooted phylogenetic networks. *IEEE/ACM Transactions on Computational Biology and Bioinformatics*, 6(1):103–109, 2009.

[125] D. H. Huson and D. Bryant. Application of phylogenetic networks in evolutionary studies. *Molecular Biology and Evolution*, 23:254–267, 2006.

[126] D. H. Huson, T. Dezulian, T. Kloepper, and M. A. Steel. Phylogenetic super-networks from partial trees. *IEEE/ACM Transactions in Computational Biology and Bioinformatics*, 1(4):151–158, 2004.

[127] D. H. Huson, T. Kloepper, P. J. Lockhart, and M. A. Steel. Reconstruction of reticulate networks from gene trees. In *Research in Computational Molecular Biology: Proceedings of the Ninth International Conference on Research in Computational Molecular Biology (RECOMB)*, vol. 3500, *LNCS*, pp. 233–249. Springer, 2005.

[128] D. H. Huson and T. H. Klöpper. Computing recombination networks from binary sequences. *Bioinformatics*, 21(suppl. 2):ii159–ii165, 2005.

[129] D. H. Huson and T. H. Klöpper. Beyond galled trees: decomposition and computation of galled networks. In *Research in Computational Molecular Biology: Proceedings of the 11th International Conference on Research in Computational Molecular Biology (RECOMB)*, vol. 4453 of *LNCS*, pp. 211–225. Springer, 2007.

[130] D. H. Huson, D. C. Richter, C. Rausch, *et al.* Dendroscope: an interactive viewer for large phylogenetic trees. *BMC Bioinformatics*, 8(1):460, 2007.

[131] D. H. Huson, R. Rupp, V. Berry, P. Gambette, and C. Paul. Computing galled networks from real data. *Bioinformatics*, 25(12):i85–i93, 2009.

[132] D. H. Huson, M. A. Steel, and J. Whitfield. Reducing distortion in phylogenetic networks. In P. Bücher and B. M. E. Moret, editors, *Algorithms in Bioinformatics. Proceedings of the 6th International Workshop on Algorithms in Bioinformatics (WABI)*, vol. 4175, *LNBI*, pp. 150–161. Springer, 2006.

[133] L. van Iersel, J. Keijsper, S. Kelk, L. Stougie, F. Hagen, and T. Boekhout. Constructing level-2 phylogenetic networks from triplets. In *Research in Computational Molecular Biology: Proceedings of the 12th International Conference on Research in Computational Molecular Biology (RECOMB)*, vol. 4955, *LNCS*, pp. 450–462. Springer, 2008.

[134] L. van Iersel and S. Kelk. Constructing the simplest possible phylogenetic network from triplets. *Algorithmica*, 2009. DOI 10.1007/s00453-009-9333-0.

[135] L. van Iersel and S. Kelk. When two trees go to war. In preparation, 2011.

[136] L. van Iersel, S. Kelk, R. Rupp, and D. Huson. Phylogenetic networks do not need to be complex: using fewer reticulations to represent conflicting clusters. *Bioinformatics*, 26(12):i124–131, 2010.

[137] J. Isbell. Six theorems about injective metric spaces. *Commentarii Mathematici Helvetici*, 39(1):65–76, 1964.

[138] J. Jansson. On the complexity of inferring rooted evolutionary trees. *Electronic Notes in Discrete Mathematics*, 7:50–53, 2001.

[139] J. Jansson, N. B. Nguyen, and W. -K. Sung. Algorithms for combining rooted triplets into a galled phylogenetic network. *SIAM Journal on Computing*, 35(5):1098–1121, 2006.

[140] J. Jansson and W. -K. Sung. Inferring a level-1 phylogenetic network from a dense set of rooted triplets. In *Computing and Combinatorics: Proceedings of the 10th Annual International Conference on Computing and Combinatorics (COCOON)*, vol. 3106, *LNCS*, pp. 462–471. Springer, 2004.

[141] G. Jin, L. Nakhleh, S. Snir, and T. Tuller. Efficient parsimony-based methods for phyloge-netic network reconstruction. In *Proceedings of the European Conference on Computational Biology (ECCB)*, vol. 23, *Bioinformatics*, pp. e123–e128, 2006.

[142] G. Jin, L. Nakhleh, S. Snir, and T. Tuller. Maximum likelihood of phylogenetic networks. *Bioinformatics*, 22(21):2604–2611, 2006.

[143] G. Jin, L. Nakhleh, S. Snir, and T. Tuller. Inferring phylogenetic networks by the maxi-mum parsimony criterion: a case study. *Molecular Biology and Evolution*, 24(1):324–337, 2007.

[144] T. H. Jukes and C. R. Cantor. Evolution of protein molecules. In H. N. Munro, ed., *Mammalian Protein Metabolism*, pages 21–132. Academic Press, 1969.

[145] I. A. Kanj, L. Nakhleh, C. Than, and G. Xia. Seeing the trees and their branches in the network is hard. *Theoretical Computer Science*, 401(1–3):153–164, 2008.

[146] R. M. Karp. Reducibility among combinatorial problems. In R. E. Miller and J. W. Thatcher, eds., *Complexity of Computer Computations*, pp. 85–103. Plenum Press, 1972.

[147] L. Kils-Hutten, R. Cheynier, S. Wain-Hobson, and A. Meyerhans. Phylogenetic reconstruc-tion of intrapatient evolution of human immunodeficiency virus type 1: predominance of drift and purifying selection. *Journal of General Virology*, 82(7):1621–1627, 2001.

[148] M. Kimura. A simple method for estimating evolutionary rates of base substitutions through comparative studies of nucleotide sequences. *Journal of Molecular Evolution*, 16(2):111–120, 1980.

[149] M. Kimura. Estimation of evolutionary distances between homologous nucleotide sequences. *Proceedings of the National Academy of Sciences USA*, 78(1):454–458, 1981.

[150] M. Kimura and J. Crow. The number of alleles that can be maintained in a finite popula-tion. *Genetics*, 49:725–738, 1964.

[151] T. Kivisild, M. Bamshad, K. Kaldma, *et al.* Deep common ancestry of Indian and western Eurasian mtDNA lineages. *Current Biology*, 9:1331–1334, 1999.

[152] T. Klöpper. *Algorithms for the calculation and visualization of phylogenetic networks*. PhD thesis, Center for Bioinformatics, University of Tübingen, 2008.

[153] S. Koblmuller, N. Duftner, K. Sefc, *et al.* Reticulate phylogeny of gastropod-shell-breeding cichlids from Lake Tanganyika: the result of repeated introgressive hybridization. *BMC Evolutionary Biology*, 7(1):7, 2007.

[154] M. A. Koch, C. Dobes, C. Kiefer, *et al.* Supernetwork identifies multiple events of plastid *trn*F(GAA) pseudogene evolution in the Brassicaceae. *Molecular Biology and Evolution*, 24(1):63–73, 2007.

[155] J. Kruskal. On the shortest spanning tree of a graph and the traveling salesman problem. *Proceedings of the American Mathematical Society*, 7:48–50, 1956.

[156] H. W. Kuhn. The Hungarian method for the assignment problem. *Naval Research Logistics Quarterly*, 2:83–97, 1955.

[157] V. Kunin, L. Goldovsky, N. Darzentas, and C. A. Ouzounis. The net of life: reconstructing the microbial phylogenetic network. *Genome Research*, 15(7):954–959, 2005.

[158] C. Lanave, G. Preparata, C. Saccone, and G. Serio. A new method for calculating evolu-tionary substitution rates. *Journal of Molecular Evolution*, 20(1):86–93, 1984.

[159] R. Lapointe, K. Tanaka, W. E. Barney, *et al.* Genomic and morphological features of a banchine polydnavirus: comparison with bracoviruses and ichnoviruses. *Journal of Virology*, 81(12):6491–6501, 2007.

[160] B. Larget and D. L. Simon. Markov Chain Monte Carlo algorithms for the Bayesian analysis of phylogenetic trees. *Molecular Biology and Evolution*, 16(6):750–759, Jul. 1999.

[161] J. Leebens-Mack, L. A. Raubeson, L. Cui, *et al.* Identifying the basal angiosperm node in chloroplast genome phylogenies: sampling one's way out of the Felsenstein Zone. *Molecular Biology and Evolution*, 22(10):1948–1963, 2005.

[162] P. Legendre and V. Makarenkov. Reconstruction of biogeographic and evolutionary networks using reticulograms. *Systematic Biology*, 51(2):199–216, 2002.

[163] P. Lemey, M. Salemi, and A. -M. Vandamme, eds. *The Phylogenetic Handbook*, 2nd ed. Cambridge University Press, 2009.

[164] T. Lengauer and R. E. Tarjan. A fast algorithm for finding dominators in a flowgraph. *ACM Transactions on Programming Languages and Systems*, 1(1):121–141, 1979.

[165] C. R. Linder and L. H. Rieseberg. Reconstructing patterns of reticulate evolution in plants. *American Journal of Botany*, 91:1700–1708, 2004.

[166] S. Linz and C. Semple. A cluster reduction for computing the subtree distance between phylogenies. *Annals of Combinatorics*, in press, 2010.

[167] M. Lott, A. Spillner, K. T. Huber, and V. Moulton. Padre: a package for analyzing and displaying reticulate evolution. *Bioinformatics*, 25(9):1199–1200, 2009.

[168] E. Luks. Isomorphism of graphs of bounded valence can be tested in polynomial time. *Journal of Computer and System Sciences*, 25:42–65, 1982.

[169] R. B. Lyngso and Y. S. Song. Minimum recombination histories by branch and bound. In *Algorithms in Bioinformatics: Proceedings of the Fifth International Workshop on Algorithms in Bioinformaticsy (WABI)*, vol. 3692, *LNBI*, pp. 239–250. Springer, 2005.

[170] D. R. Maddison, D. L. Swofford, and W. P. Maddison. NEXUS: an extendible file format for systematic information. *Systematic Biology*, 46(4):590–621, 1997.

[171] V. Makarenkov. T-REX: reconstructing and visualizing phylogenetic trees and reticulation networks. *Bioinformatics*, 17(7):664–668, 2001.

[172] V. Makarenkov and P. Legendre. From a phylogenetic tree to a reticulated network. *Journal of Computational Biology*, 11(1):195–212, 2004.

[173] W. Martin. Mosaic bacterial chromosomes: a challenge en route to a tree of genomes. *Bioessays*, 21(2):99–104, 1999.

[174] C. A. Meacham. Theoretical and computational considerations of the compatibility of qualitative taxonomic characters. In J. Felsenstein, ed., *Numerical Taxonomy*, vol. G1. Springer, 1983.

[175] N. Metropolis, A. W. Rosenbluth, M. N. Rosenbluth, A. H. Teller, and E. Teller. Equation of state calculations by fast computing machines. *Journal of Chemical Physics*, 21(6):1087–1092, 1953.

[176] C. M. Miller-Butterworth, D. S. Jacobs, and E. H. Harley. Strong population substructure is correlated with morphology and ecology in a migratory bat. *Nature*, 424(6945):187–191, July 2003.

[177] S. Mitra, J. A. Gilbert, D. Field, and D. H. Huson. Comparison of multiple metagenomes using phylogenetic networks based on ecological indices. *ISME Journal*, doi:10.1038/ismej.2010.51, 2010 (see http://www.nature.com/ismej/journal/vaop/ncurrent/abs/ismej201051a.html).

[178] G. Moore, M. Goodman, and J. Barnabas. An iterative approach from the standpoint of the additive hypothesis to the dendrogram problem posed by molecular data sets. *Journal of Theoretical Biology*, 38:423–457, 1973.

[179] B. M. E. Moret, L. Nakhleh, T. Warnow, *et al*. Phylogenetic networks: modeling, reconstructibility, and accuracy. *IEEE/ACM Transactions on Computational Biology and Bioinformatics*, 1(1):13–23, 2004.

[180] D. Morrison. Networks in phylogenetic analysis: new tools for population biology. *International Journal for Parasitology*, 35(5):567–582, 2005.

[181] D. Morrison. Phylogenetic networks in systematic biology (and elsewhere). In R. Mohan, ed., *Research Advances in Systematic Biology*, pp. 1–48. Global Research Network, Trivandrum, India, 2010.

[182] V. Moulton and K. T. Huber. Split networks. a tool for exploring complex evolutionary relationships in molecular data. In P. Lemey, M. Salemi, and A. -M. Vandamme, eds., *The Phylogenetic Handbook*, 2nd edn. pp. 631–653. Cambridge University Press, 2009.

[183] L. Nakhleh. A metric on the space of reduced phylogenetic networks. *IEEE/ACM Transactions on Computational Biology and Bioinformatics*, 99(RapidPosts), 2009.

[184] L. Nakhleh. Evolutionary phylogenetic networks: models and issues. In L. L. S. Heath and N. Ramakrishnan, eds., *The Problem Solving Handbook for Computational Biology and Bioinformatics*. Springer, 2010.

[185] L. Nakhleh, T. Warnow, and C. R. Linder. Reconstructing reticulate evolution in species: theory and practice. In *Proceedings of the Eighth International Conference on Research in Computational Molecular Biology (RECOMB)*, pp. 337–346. ACM Press, 2004.

[186] S. B. Needleman and C. D. Wunsch. A general method applicable to the search for similarities in the amino acid sequences of two proteins. *Journal of Molecular Biology*, 48:443–453, 1970.

[187] M. Nei. Estimation of average heterozygosity and genetic distance from a small number of individuals. *Genetics*, 89:583–590, 1978.

[188] C. Notredame, D. G. Higgins, and J. Heringa. T-Coffee: a novel method for fast and accurate multiple sequence alignment. *Journal of Molecular Biology*, 302(1):205–217, 2000.

[189] H. Ochman, J. G. Lawrence, and E. A. Groisman. Lateral gene transfer and the nature of bacterial innovation. *Nature*, 405(6784):299–304, 2000.

[190] K. O'Donnell, H. Kistler, B. Tacke, and H. Casper. Gene genealogies reveal global phylogeographic structure and reproductive isolation among lineages of *Fusarium graminearum*, the fungus causing wheat scab. *PNAS*, 97(14):7905–7910, 2000.

[191] R. D. M. Page. On islands of trees and the efficacy of different methods of branch swapping in finding most-parsimonious trees. *Systematic Biology*, 42(2):200–210, 1993.

[192] R. D. M. Page. Modified mincut supertrees. In R. Guigó and D. Gusfield, eds., *Algorithms in Bioinformatics: Proceedings of the 2nd International Workshop on Algorithms in Bioinformatics (WABI)*, vol. 2452 of *LNCS*, pp. 537–552, 2002.

[193] E. Paradis. pegas: an R package for population genetics with an integrated-modular approach. *Bioinformatics*, 26(3):419–420, 2010.

[194] L. Parida, M. Melé, F. Calafell, and J. Bertranpetit. Estimating the ancestral recombinations graph (ARG) as compatible networks of SNP patterns. *Journal of Computational Biology*, 15(9):1133–1153, 2008.

[195] N. Patterson, D. J. Richter, S. Gnerre, E. S. Lander, and D. Reich. Genetic evidence for complex speciation of humans and chimpanzees. *Nature*, 441:1103–1108, 2006.

[196] Y. Pauplin. Direct calculation of a tree length using a distance matrix. *Journal of Molecular Biology*, 51(1):41–47, 2000.

[197] P. J. Planet, S. C. Kachlany, D. H. Fine, R. DeSalle, and D. H. Figurski. The widespread colonization island of *Actinobacillus actinomycetemcomitans*. *Nature Genetics*, 34:193–198, 2003.

[198] D. Posda and K. A. Crandall. The effect of recombination on the accuracy of phylogeny estimation. *Journal of Molecular Evolution*, 54:396–402, 2002.

[199] R. C. Prim. Shortest connection networks and some generalizations. *Bell System Technology Journal*, 36:1389–1401, 1957.

[200] S. J. Prohaska, C. Fried, C. T. Amemiya, F. H. Ruddle, G. P. Wagner, and P. F. Stadler. The shark HoxN cluster is homologous to the human HoxD cluster. *Journal of Molecular Evolution*, 58:212–217, 2004.

[201] L. H. Rieseberg. Hybrid origins of plant species. *Annual Review of Ecology Evolution and Systematics*, 28:359–389, 1997.

[202] L. H. Rieseberg, S. J. E. Baird, and K. A. Gardner. Hybridization, introgression, and linkage evolution. *Plant Molecular Biology*, 42:205–224, 2000.

[203] D. F. Robinson. Comparison of labeled trees with valency three. *Journal of Combinatorial Theory, Series B*, 11(2):105–119, 1971.

[204] D. F. Robinson and L. R. Foulds. Comparison of phylogenetic trees. *Mathematical Biosciences*, 53:131–147, 1981.

[205] F. Rodríguez, J. L. Oliver, A. Marín, and J. R. Medina. The general stochastic model of nucleotide substitution. *Journal of Theoretical Biology*, 142(4):485–501, 1990.

[206] A. Rokas, B. L. Williams, N. King, and S. B. Carroll. Genome-scale approaches to resolving incongruence in molecular phylogenies. *Nature*, 425:798–804, 2003.

[207] D. Ruths and L. Nakhleh. Recombination and phylogeny: effects and detection. *International Journal of Bioinformatics Research and Applications*, 1(2):202–212, 2005.

[208] N. Saitou and M. Nei. The Neighbor-Joining method: a new method for reconstructing phylogenetic trees. *Molecular Biology and Evolution*, 4:406–425, 1987.

[209] M. Salemi, T. De Oliveira, V. Courgnaud, *et al.* Mosaic genomes of the six major primate lentivirus lineages revealed by phylogenetic analyses. *Journal of Virology*, 77(13):7202–7213, 2003.

[210] D. Sankoff. Minimal mutation trees of sequences. *SIAM Journal of Applied Mathematics*, 28:35–42, 1975.

[211] M. H. Schierup and J. Hein. Consequences of recombination on traditional phylogenetic analysis. *Genetics*, 156:879–891, 2000.

[212] C. Scornavacca, V. Berry, V. Lefort, E. J. P. Douzery, and V. Ranwez. PhySIC_IST: cleaning source trees to infer more informative supertrees. *BMC Bioinformatics*, 9(8):413, 2008.

[213] C. Semple. Reconstructing minimal rooted trees. *Discrete Applied Mathematics*, 127(3):489–503, 2003.

[214] C. Semple. Hybridization networks. In O. Gascuel and M. Steel, eds., *Reconstructing Evolution: New Mathematical and Computational Advances*, pp. 277–314. Oxford University Press, 2007.

[215] C. Semple and M. Steel. A supertree method for rooted trees. *Discrete Applied Mathematics*, 105(1–3):147–158, 2000.

[216] C. Semple and M. Steel. Cyclic permutations and evolutionary trees. *Advances in Applied Mathematics*, 32(4):669–680, 2004.

[217] C. Semple and M. A. Steel. *Phylogenetics*. Oxford University Press, 2003.

[218] T. F. Smith and M. S. Waterman. Identification of common molecular subsequences. *Journal of Molecular Biology*, 147:195–197, 1981.

[219] P. Sneath. Cladistic representation of reticulate evolution. *Systematic Zoology*, 24(3):360–368, 1975.

[220] R. R. Sokal and C. D. Michener. A statistical method for evaluating systematic relationships. *University of Kansas Scientific Bulletin*, 28:1409–1438, 1958.

[221] Y. S. Song. A concise necessary and sufficient condition for the existence of a galled-tree. *IEEE/ACM Transactions in Computational Biology and Bioinformatics*, 3(2):186–191, 2006.

[222] Y. S. Song and J. Hein. Parsimonious reconstruction of sequence evolution and haplotype blocks: Finding the minimum number of recombination events. In *Algorithms in Bioinformatics: Proceedings of the Third International Workshop on Algorithms in Bioinformatics (WABI)*, vol. 2812, *LNBI*, pp. 287–302, 2003.

[223] Y. S. Song and J. Hein. On the minimum number of recombination events in the evolutionary history of DNA sequences. *Journal of Mathematical Biology*, 48:160–186, 2004.

[224] Y. S. Song and J. Hein. Constructing minimal ancestral recombination graphs. *Journal of Computational Biology*, 12(2):147–69, 2005.

[225] Y. S. Song, Y. Wu, and D. Gusfield. Efficient computation of close lower and upper bounds on the minimum number of recombinations in biological sequence evolution. *Bioinformatics*, 21:413–422, 2005.

[226] M. Spencer, E. A. Davidson, A. C. Barbrook, and C. J. Howe. Phylogenetics of artificial manuscripts. *Journal of Theoretical Biology*, 227(4):503 – 511, 2004.

[227] M. A. Steel and A. M. Hamel. Finding a maximum compatible tree is NP-hard for sequences and trees. *Applied Mathematics Letters*, 9(2):55–59, 1996.

[228] K. Strimmer and V. Moulton. Likelihood analysis of phylogenetic networks using directed graphical models. *Molecular Biology and Evolution*, 17(6):875–881, 2000.

[229] J. A. Studier and K. J. Keppler. A note on the neighbor-joining method of Saitou and Nei. *Molecular Biology and Evolution*, 5(6):729–731, 1988.

[230] D. L. Swofford, G. J. Olsen, P. J. Waddell, and D. M. Hillis. Chapter 11: Phylogenetic inference. In D. M. Hillis, C. Moritz, and B. K. Mable, eds., *Molecular Systematics*, 2nd edn., pp. 407–514. Sinauer Associates, Inc., 1996.

[231] M. Syvanen. Cross-species gene transfer; implications for a new theory of evolution. *Journal of Theoretical Biology*, 112(2):333–43, 1985.

[232] K. Tamura and M. Nei. Estimation of the number of nucleotide substitutions in the control region of mitochondrial DNA in humans and chimpanzees. *Molecular Biology and Evolution*, 10(3):512–526, 1993.

[233] A. R. Templeton, K. A. Crandall, and C. F. Sing. A cladistic analysis of phenotypic associations with haplotypes inferred from restriction endonuclease mapping and DNA sequence data III cladogram estimation. *Genetics*, 132:619–633, 1992.

[234] C. Than, D. Ruths, and L. Nakhleh. PhyloNet: a software package for analyzing and reconstructing reticulate evolutionary relationships. *BMC Bioinformatics*, 9:322, 2008.

[235] J. D. Thompson, D. G. Higgins, and T. J. Gibson. CLUSTAL W: improving the sensitivity of progressive multiple sequence alignment through sequence weighting, position-specific gap penalties and weight matrix choice. *Nucleic Acids Research*, 22:4673–4680, 1994.

[236] T. -H. To and M. Habib. Level-k phylogenetic networks are constructable from a dense triplet set in polynomial time. In *Combinatorial Pattern Matching: Proceeding of the 20th Annual Symposium Combinatorial Pattern Matching (CPM)*, vol. 5577, LNCS, pp. 275–288, 2009.

[237] A. Tofigh, M. Hallet, and J. Lagergren. Simultaneous identification of duplications and lateral gene transfers. *IEEE/ACM Transactions on Computational Biology and Bioinformatics*, 99(PrePrints), 2010.

[238] A. Tofigh and J. Lagergren. Inferring duplications and lateral gene transfers- an algorithm for parametric tree reconciliation. Unpublished manuscript, 2009.

[239] J. Wagele and C. Mayer. Visualizing differences in phylogenetic information content of alignments and distinction of three classes of long-branch effects. *BMC Evolutionary Biology*, 7(1):147, 2007.

[240] L. Wang and T. Jiang. On the complexity of multiple sequence alignment. *Journal of Computational Biology*, 1(4):337–348, 1994.

[241] L. Wang, K. Zhang, and L. Zhang. Perfect phylogenetic networks with recombination. *Journal of Computational Biology*, 8(1):69–78, 2001.

[242] D. Weigel and R. Mott. The 1001 genomes project for *Arabidopsis thaliana*. *Genome Biology*, 10(5):107, 2009.

[243] J. Whitfield, S. Cameron, D. Huson, and M. Steel. Filtered Z-closure supernetworks for extracting and visualizing recurrent signal from incongruent gene trees. *Systematic Biology*, 57(6):939–947, 2008.

[244] B. Y. Wu. Constructing the maximum consensus tree from rooted triples. *Journal of Combinatorial Optimization*, 8(1):29–39, 2004.

[245] G. -S. Wu, Y. -G. Yao, K. -X. Qu, *et al.* Population phylogenomic analysis of mitochondrial DNA in wild boars and domestic pigs revealed multiple domestication events in East Asia. *Genome Biology*, 8(11):R245, 2007.

[246] Z. Yang. Among-site rate variation and its impact on phylogenetic analyses. *Trends in Ecology and Evolution*, 11:367–372, 1996.

[247] A. Zharkikh. Estimation of evolutionary distances between nucleotide sequences. *Journal of Molecular Evolution*, 39(3):315–329, 1994.

Index